Numerical Methods and Modelling for Engineering

Richard Khoury • Douglas Wilhelm Harder

Numerical Methods and Modelling for Engineering

 Springer

Richard Khoury
Lakehead University
Thunder Bay, ON, Canada

Douglas Wilhelm Harder
University of Waterloo
Waterloo, ON, Canada

ISBN 978-3-319-79331-3   ISBN 978-3-319-21176-3   (eBook)
DOI 10.1007/978-3-319-21176-3

Printed on acid-free paper

This Springer imprint is published by Springer Nature
The registered company is Springer International Publishing AG Switzerland

Conventions

Throughout this textbook, the following conventions are used for functions and variables:

Object	Format	Example
Scalar number (real or complex)	Lowercase italics	$x = 5$
Vector	Lowercase bold	$\mathbf{v} = [x_0, x_1, \ldots, x_i, \ldots, x_{N-1}]$
Matrix	Uppercase bold	$$\mathbf{M} = \begin{bmatrix} x_{0,0} & x_{0,1} & \cdots & x_{0,j} & \cdots & x_{0,N-1} \\ x_{1,0} & & & x_{1,j} & & x_{1,N-1} \\ \cdots & & & \cdots & & \cdots \\ x_{i,0} & x_{i,1} & \cdots & x_{i,j} & \cdots & x_{i,N-1} \\ \cdots & & & \cdots & & \cdots \\ x_{M-1,0} & x_{M-1,1} & \cdots & x_{M-1,j} & \cdots & x_{M-1,N-1} \end{bmatrix} = \begin{bmatrix} \mathbf{v}_0 \\ \mathbf{v}_1 \\ \cdots \\ \mathbf{v}_i \\ \cdots \\ \mathbf{v}_{M-1} \end{bmatrix}$$
Scalar-valued function of scalar	Lowercase italics with lowercase italics	$f(x) = 5x + 2$
Scalar-valued function of vector	Lowercase italics with lowercase bold	$f(\mathbf{v}) = f(x_0, x_1, \ldots, x_i, \ldots, x_{N-1})$ $= 5x_0 + 2x_1 + \cdots + 7x_i + \cdots + 4x_{N-1} + 3$
Vector-valued function of scalar	Lowercase bold with lowercase italic	$\mathbf{f}(x) = [f_0(x), f_1(x), \ldots, f_i(x), \ldots, f_{N-1}(x)]$
Vector-valued function of vector	Lowercase bold with lowercase bold	$\mathbf{f}(\mathbf{v}) = [f_0(\mathbf{v}), f_1(\mathbf{v}), \ldots, f_i(\mathbf{v}), \ldots, f_{N-1}(\mathbf{v})]$

(continued)

Object	Format	Example
Matrix-valued function of scalar	Uppercase bold with lowercase italics	$\mathbf{M}(x) = \begin{bmatrix} f_{0,0}(x) & f_{0,1}(x) & \cdots & f_{0,j}(x) & \cdots & f_{0,N-1}(x) \\ f_{1,0}(x) & & & f_{1,j}(x) & & f_{1,N-1}(x) \\ \cdots & & & & & \cdots \\ f_{i,0}(x) & f_{i,1}(x) & \cdots & f_{i,j}(x) & \cdots & f_{i,N-1}(x) \\ \cdots & & & & & \cdots \\ f_{M-1,0}(x) & f_{M-1,1}(x) & \cdots & f_{M-1,j}(x) & \cdots & f_{M-1,N-1}(x) \end{bmatrix}$
Matrix-valued function of vector	Uppercase bold with lowercase bold	$\mathbf{M}(\mathbf{v}) = \begin{bmatrix} f_{0,0}(\mathbf{v}) & f_{0,1}(\mathbf{v}) & \cdots & f_{0,j}(\mathbf{v}) & \cdots & f_{0,N-1}(\mathbf{v}) \\ f_{1,0}(\mathbf{v}) & & & f_{1,j}(\mathbf{v}) & & f_{1,N-1}(\mathbf{v}) \\ \cdots & & & & & \cdots \\ f_{i,0}(\mathbf{v}) & f_{i,1}(\mathbf{v}) & \cdots & f_{i,j}(\mathbf{v}) & \cdots & f_{i,N-1}(\mathbf{v}) \\ \cdots & & & & & \cdots \\ f_{M-1,0}(\mathbf{v}) & f_{M-1,1}(\mathbf{v}) & \cdots & f_{M-1,j}(\mathbf{v}) & \cdots & f_{M-1,N-1}(\mathbf{v}) \end{bmatrix}$

Acknowledgements

Thanks to the following for pointing out mistakes, providing suggestions, or helping to improve the quality of this text:

- Khadijeh Bayat
- Dan Busuioc
- Tim Kuo
- Abbas Attarwala
- Prashant Khanduri
- Matthew Chan
- Christopher Olekas
- Jaroslaw Kuszczak
- Chen He
- Hans Johannes Petrus Vanleeuwen
- David Smith
- Jeff Teng
- Roman Kogan
- Mohamed Oussama Damen
- Rudko Volodymyr
- Vladimir Rutko
- George Rizkalla
- Alexandre James
- Scott Klassen
- Brad Murray
- Brendan Boese
- Aaron MacLennan

Contents

List of Figures

List of Tables

Chapter 1
Modelling and Errors

1.1 Introduction

As an engineer, be it in the public or private sector, working for an employer or as an independent contractor, your job will basically boil down to this: you need to solve the problem you are given in the most efficient manner possible. If your solution is less efficient than another, you will ultimately have to pay a price for this inefficiency. This price may take many forms. It could be financial, in the form of unnecessary expenses. It could take the form of less competitive products and a reduced market share. It could be the added workload due to problems stemming from the inefficiencies of your work. It could be an intangible but very real injury to your professional reputation, which will be tarnished by being associated to inefficient work. Or it could take many other forms, all negative to you.

How can you make sure that your proposed solutions are as efficient as possible? By representing the problem you are working on with an appropriate mathematical model and then solving this model to find the optimal solution while being mindful of the numerical errors that will necessarily crop in. Numerical methods, the algorithms presented in this textbook, are the tools you can use to this end.

Numerical methods are a set of mathematical modelling tools. Each method allows you to solve a specific type of problem: a root-finding problem, an optimization problem, an integral or derivative problem, an initial value problem, or a boundary value problem. Once you have developed a proper model and understanding of the problem you are working on, you can break it down into a set of these problems and apply the appropriate numerical method. Each numerical method encompasses a set of algorithms to solve the mathematical problem it models given some information and to a known error bound. This will be an important point throughout the book: none of the algorithms that will be shown can allow you to find the exact perfect solution to the problems, only approximate solutions with known error ranges. Completely eliminating the errors is impossible;

© Springer International Publishing Switzerland 2016
R. Khoury, D.W. Harder, *Numerical Methods and Modelling for Engineering*,
DOI 10.1007/978-3-319-21176-3_1

rather, as an engineer, it is your responsibility to design systems that are tolerant of errors. In that mindset, being able to correctly measure the errors is an important advantage.

1.2   Simulation and Approximation

Before we can meaningfully talk about solving a model and measuring errors, we must understand the modelling process and the sources of error.

Engineers and scientists study and work in the physical world. However, exactly measuring and tracking every value of every variable in the natural world, and its complete effect on nature, is a completely impossible task. Consequently, all engineers and scientists work on different *models* of the physical world, which track every variable and natural phenomenon we need to be aware of for our given tasks to a level of accuracy we require for our work. This implies that different professionals will work using different models; for instance, while an astrophysicist studying the movement of galaxies and a quantum physicist studying the collision of subatomic particles are studying the same physical world, they use completely different models of it in their work.

Selecting the proper model for a project is the first step of the *modelling cycle* shown in Fig. 1.1. All models stem in some way from the real physical world we live in and are meant to represent some aspect of it. Once a proper model has been selected, an implementation of it has to be made. Today, this is synonymous with writing a software version of the model to run on a computer, but in the past a simpler implementation approach was used, which consisted in writing down all necessary equations on paper or on a blackboard. Whichever implementation method is used, it will include variables that are placeholders for real-world values. Consequently, the next step is to look to the physical world again and take measurements of these values to fill in the variables. Finally, we are ready for the final step, simulation. At this step, we execute the implementation of the model with the measured values to get an output result. Whether this execution is done by a

Fig. 1.1 The modelling loop of reality to engineering approximation

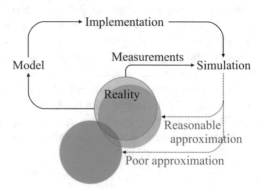

computer running software or an engineer diligently computing values on their slide rule is dependent on the implementation method chosen, but that is of little consequence. In either case, the final result will be an approximation of part of reality that the model was meant to represent.

One thing that is evident from Fig. 1.1, and should also be understood from our previous explanation, is that the approximation obtained at the end of the modelling cycle is not a 100 % perfect reflection of reality. This is unavoidable: every step of the modelling cycle includes errors, and so the final result cannot possibly be perfect. But more fundamentally, engineers do not seek to have a perfect model of the world; they seek to have a model that is good enough to complete a project correctly and efficiently. Correctness is not synonymous with perfection. For example, Newtonian physics is not a perfect model of the universe, but it is correct enough to have been the foundation of countless successful systems over the centuries and to still be commonly used today despite its limitations being well known now. Likewise, efficiency can stem from a simpler (and less accurate) model. Relativistic physics is unarguably a more accurate model of the universe than Newtonian physics, but trying to apply it when designing a factory or a road will only lead to an inefficient waste of effort and time (and thus money) to model undetectable relativistic effects.

Nonetheless, all approximations are not equal. Again referring to Fig. 1.1, it can be seen that some approximations will greatly overlap with reality, while others will have little if anything in common with reality. While an engineer does not seek perfect and complete overlap between the approximation and reality, there must be overlap with the parts of reality that matter, the ones that affect the project. To build a factory, the approximation of the weight that can be supported by load-bearing structures must be an accurate representation of reality, and the approximation of the weight expected to be put on that structure must be an accurate prediction of the weight of upper floors, of machinery and people on these floors, and of snow accumulations on the roof in winter. The difference between a good and a bad approximation is the difference between a successful project and a disastrous one. It is critically important that an engineer be always aware of the errors that creep into the modelling cycle and of the resulting inaccuracies in the approximation, to insure that the approximation is still valid and the project is successful.

It was noted earlier that one reason why the approximation will always differ from reality is because of errors introduced in each step of the modelling cycle. There are four steps in this cycle: model selection, implementation, measurement, and simulation. There are likewise four possible types of errors: model errors, implementation errors, measurement errors, and simulation errors.

Model errors come from selecting an inappropriate model for a given application. An inappropriate model is one that does not include all the aspects of the physical world that will influence the project being done. Designing docks using a model of the shoreline that does not include tides and a skyscraper using a model of the city that does not include winds are examples of model errors. One of the most famous examples of model errors occurred in 1964, when Arno Penzias and Robert Wilson were experimenting with a supersensitive antenna. After filtering out all sources of interference accounted for in their model, they found that there was still a low steady noise being detected by their receiver, and that it was 100 times more intense than what the model had predicted they would find. This noise, it was later found, was the cosmic microwave background radiation of the universe left over from the Big Bang, which their model did not include and which, as a result, threw off their entire predictions. But to be fair, the Big Bang was still a recently proposed hypothesis at that time, and one that a large portion of the scientific community did not yet accept, and so we cannot fault Penzias and Wilson for selecting a model that did not include it. In fact, their accidental discovery earned them a Nobel Prize in 1978. Most engineer errors do not have such positive outcomes however.

Implementation errors occur when the software code representing the model in a computer is poorly built. This can be the result of algorithmic errors, of passing values in the wrong order through a software interface, of legacy code being used in a way it was never meant for, and of many more potential issues. These errors are usually detected and corrected through proper software quality assurance (SQA) methods within the project, and consequently SQA is an important component of software engineering good practice. Implementation errors have notably plagued space exploration agencies worldwide and are blamed for some of the most famous space disasters. The explosion of the European Ariane 5 rocket shortly after takeoff in 1996 was due to a 64-bit value in a new inertial reference system being passed into a 16-bit value in a legacy control system. When the value exceeded the 16-bit limit, the control system failed and the rocket shot off course and destroyed itself and its cargo, a loss of some $500 million. Likewise, NASA lost its $328 million Mars Climate Orbiter in 1999 because of poorly documented software components. The Orbiter's instruments took measurements in imperial units and relayed them to the control software, which was designed to handle metric units. Proper documentation of these two modules to explicitly name the units used by each, along with proper code review, would have caught this problem easily; instead, it went unnoticed until the Orbiter crashed into Mars.

Once an appropriate model has been selected and correctly implemented, it is necessary to fill in unknown variables with values measured from the real world to represent the current problem. Measurement errors occur at this stage, when the measurements are inaccurate. In a sense, measurement errors will always occur; while one can avoid model errors through proper research and implementation errors through a proper SQA process, measuring tools will always have limited precision and consequently the measurements themselves will always have some errors. However, care can still be taken in a number of ways: by including error bounds on the measures rather than treating them as exact values, by running the computations on worst-case scenarios in addition to more likely average scenarios, by designing in safety buffers, and of course by making sure the measurements are taken properly in the first place. This was not the case in 1979, by a Finnish team tasked to build a lighthouse on Märket Island. This island is on the border between Sweden and Finland and had been neatly divided by a treaty between the two nations 170 years before. The team was tasked with building the new lighthouse on the Finnish side of the island, but because of improper geographical measurements, they built it on the Swede side accidentally. Rectifying the situation after construction required reopening the century-old treaty between the two nations to negotiate new borders that remained fair for territory size, coast lines, fishery claims, and more. And while the two nations resolved the issue peacefully, to their credit, accidentally causing an international incident is not a line any team leader would want to include on their resume.

Even once a proper model has been selected, correctly implemented and populated with accurate measurements, errors can still occur. These final errors are simulation errors that are due to the accumulation of inaccuracies over the execution of the simulation. To understand the origin of these errors, one must remember that a simulation on a computer tries to represent reality, a continuous-valued and infinite world, with a discrete set of finite values, and then predicts what will happen next in this world using approximation algorithms. Errors are inherent and unavoidable in this process. Moreover, while the error on an individual value or algorithm may seem so small as to be negligible, these errors accumulate with each other. An individual value may have a small error, but then is used in an algorithm with its own small error and the result has the error of both. When a proper simulation uses dozens of values and runs algorithms hundreds of times, the errors can accumulate to very significant values. For example, in 1991, the Sleipner A oil platform under construction in Norway collapsed because of simulation errors. The problem could be traced back to the approximation in a finite element function in the model; while small, this error then accumulated throughout the simulation so that by the end the stress predicted on the structure by the model was 47 % less than reality. Consequently, the concrete frame of the oil platform was designed much too weak, sprung a leak after it was submersed under water, and caused the entire platform to sink to the bottom of a fjord. The shock of the platform hitting the bottom of the fjord caused a seismic event of 3.0 on the Richter scale about $700 million in damages.

This book focuses on simulation errors. Throughout the work, it will present not only algorithms to build simulations and model reality but their error values in order to account for simulation errors in engineering work.

1.3   Error Analysis

1.3.1   Precision and Accuracy

Before talking about errors, it is necessary to lay down some formal vocabulary. The first are the notions of *precision* and *accuracy*, two words that are often used interchangeably by laypeople. In engineering, these words have different, if related, meanings. Precision refers to the number of digits an approximation uses to represent a real value, while accuracy refers to how close to the real value the approximation is.

An example can help clarify these notions. Imagine a car with two speedometers, an analogue one and a digital one. The digital one indicates the car's speed at every 0.1 km/h, while the analogue one only indicates it at every 1 km/h. When running an experiment and driving the car at a constant 100 km/h, it is observed that the digital speedometer fluctuates from 96.5 to 104.4 km/h, while the analogue one only fluctuates from 99 to 101 km/h. In this example, the digital speedometer is more precise, as it indicates the speed with one more digit than the analogue one, but the analogue speedometer is more accurate, as it is closer to the real value of 100 km/h than the digital one.

While precision and accuracy measure two different and independent aspects of our values, in practice it makes sense to use precision to reflect accuracy. Adding additional digits of precision that cannot be accurately measured consists simply in adding noise in our values. This was the case in the previous example, with the digital speedometer showing a precision of 0.1 km/h when it couldn't accurately measure the speed to more than 3 or 4 km/h. On the other hand, if a value can be accurately measured to a great precision, then these digits should be included. If the car's speed is accurately measured to 102.44 km/h, then reporting it to a lesser precision at 102 km/h not only discards useful information, it actually reduces accuracy by rounding known figures.

Consequently, the accuracy of a measure is usually a function of the last digit of precision. When a speedometer indicates the car's speed to 0.1 km/h, it implies that it can accurately measure its speed to that precision. In fact, given no other information except a value, it is implied that the accuracy is half the last digit of precision. For example, a car measured as going to 102.3 km/h is implied to have been accurately measured to ±0.05 km/h to get that precision. This accuracy is called the *implied precision* of the measure. In our example, this means that the real speed of the car is somewhere in the range from 102.25 to 103.35 km/h and cannot be obtained any more accurately than that.

1.3.2   Absolute and Relative Error

The next important term to introduce is that of *error*. The error is the value of the inaccuracy on a measure. If it is given with the same units as the measure itself, then it is an *absolute error*. More formally, given a real measure and an approximation, the absolute error is the difference between the approximation and the real value:

$$E_{\text{abs}} = |\text{approximation} - \text{value}| \tag{1.1}$$

It can be realized at this point that the implied precision introduced in the previous subsection is also a measure of absolute error. Absolute error has the benefit of being immediately clear and related to the measure being evaluated. However, it is also inherently vague when it comes to determining if that measure is accurate or not. Given a distance with an absolute error of 3 m, one can get an immediate sense of the precision that was used to measure it and of how far apart the two objects might be, but is this accurate enough? The answer is that it depends on the magnitude of the distance being measured. An absolute error of 3 m is incredibly accurate when measuring the thousands of metres of distance between two cities, but incredibly inaccurate when measuring the fraction of a metre distance between your thumb and index finger. The notion of *relative error*, or absolute error as a ratio of the value being measured, introduces this difference:

$$E_{rel} = \left| \frac{\text{approximation} - \text{value}}{\text{value}} \right| \tag{1.2}$$

Unlike absolute error, which is given in the same units as the value being measured, relative error is given as a percentage of the measured value.

Example 1.1
What is the maximum and minimum resistance of a resistor labelled "brown, grey, brown, red"?

Solution
Given the colour code, the resistor is 180 Ω with a tolerance of ±2 %. In order words, the resistance value is approximated as 180 Ω and the relative error on this approximation is 2 %. Putting these values in the relative error formula (1.2) to solve for the real value:

$$E_{rel} = \frac{|180\Omega - r|}{|r|} = 0.02 \Rightarrow r = \begin{cases} 176.5\,\Omega \\ 183.7\,\Omega \end{cases}$$

The resistor's minimum and maximum resistance values are 176.5 and 183.7 Ω, respectively, and the real resistance value is somewhere in that range. It can be noted that the absolute error on the resistance value is 3.6 Ω, which is indeed 2 % of 180 Ω.

Example 1.2
A car's speedometer indicates a current speed of 102 km/h. What is the relative error on that measure?

Solution
The implied precision on the measure is half the last decimal, or ±0.5 km/h. The real speed is in the interval from 101.5 to 102.5 km/h. The relative error is computed from these two bounds:

(continued)

Example 1.2 (continued)

$$\frac{|102 - 101.5|}{|101.5|} = 0.004926$$

$$\frac{|102 - 102.5|}{|102.5|} = 0.004878$$

Thus, the relative error is 0.4926 %.

1.3.3 Significant Digits

When trying to determine which of two approximations of a value is more accurate, it is intuitive to compare each to the correct value digit by digit and pick the approximation with the greatest number of digits in common with the value. Given two approximations of the constant π, one at 3.142 and one at 3.1416, the second would be intuitively preferred because it has four digits in common with the real value of π, as opposed to three digits in the first one. The set of correct digits in the approximation are called *significant digits*, and the intuitive idea of preferring an approximation with more significant digits is entirely valid.

However, a simple count of significant digits is not always enough; operations such as rounding decimals can cause approximations with fewer significant digits to be better. For example, for the value 2.0000, the approximation 2.9999 is much worse than the approximation 1.9999, despite the fact the former has one significant digit and the latter has none.

A better alternative to counting decimals is to look at the order of magnitude of the relative error of the approximations. This order of magnitude is proportional to the number of significant digits, without being misled by the rollover of values due to rounding. More formally, one would look for the integer value of n that satisfies the following inequality on the order of magnitude of the relative error:

$$E_{\text{rel}} \leq 0.5 \times 10^{-n} \tag{1.3}$$

This integer n is the actual number of significant digits that we are looking for. Given multiple approximations for a value, the most accurate one is the one with the highest value of n. Moreover, very bad approximations that yield a positive power of 10 in Eq. (1.3) and therefore negative values of n are said to have no significant digits.

Example 1.3
Given two approximations 2.9999 and 1.9999 for the real value 2.0000, which has the greatest number of significant digits?

Solution
First, compute the relative error of each approximation:

$$\frac{|2.9999 - 2.0000|}{|2.0000|} = 0.49995$$

$$\frac{|1.9999 - 2.0000|}{|2.0000|} = 0.00005$$

Next, find the maximum exponent n for the inequalities on the order of magnitude in Eq. (1.3):

$$0.49995 \leq 0.5 \times 10^0$$
$$0.00005 \leq 0.5 \times 10^{-4}$$

This tells us that the approximation of 1.9999 has four significant digits, while the approximation of 2.9999 has none. This is despite the fact that the value of 2.9999 has one digit in common with the real value of 2.0000 while 1.9999 has none. However, this result is in line with mathematical sense: 1.9999 is only 0.0001 off from the correct value, while 2.9999 is off by 0.9999.

1.3.4 Big O Notation

When it comes to measuring the error caused by mathematical algorithms, trying to compute an exact value often proves impractical. The mathematical formula that is implemented in the algorithm may be a summation of many (or an infinite number of) terms, making exact computation difficult (or impossible). Moreover, constant coefficients multiplying some terms of the mathematical formula may make a less-efficient algorithm appear more efficient in a special range of values, making the results only valid in that special case rather than a general conclusion. In this case, it is better to represent the error in general terms rather than try to compute an exact value. The general form we will use in this book is the *big O notation*.

Big O is a function of a variable of the equation being studied and is a measure of worst-case growth rate as the variable tends towards infinity or of decline rate as the variable tends towards zero. It is very commonly used in software engineering to measure the growth of time (computation) and space (memory) cost of software algorithms as the input size increases towards infinity. In that context, an algorithm with a smaller big O value is one whose time and space cost will increase more slowly with input size and thus should be preferred to another algorithm with a

Table 1.1 Sample functions and big O values

Function	Big O growth rate	Big O decline rate
$f(x) = 6x^4 + 3x^3 - 17x^2 + 4x$	$O(x^4)$	$O(x)$
$f(x) = -42x^4 + 17x^2$	$O(x^4)$	$O(x^2)$
$f(x) = 1050x^3$	$O(x^3)$	$O(x^3)$
$f(x) = -8$	$O(1)$	$O(1)$

greater big O value. It is important to note again that this is a general rule and does not account for special cases, such as specific input values for which an algorithm with a greater big O value might outperform one with a smaller big O value.

The generalization power of big O in that case comes from the fact that, given a mathematical sequence, it only keeps the term with the greatest growth rate, discarding all other terms and the coefficient multiplying that term. Table 1.1 gives some example of functions with their big O growth rates in the second column. In all these functions, the term with the greatest growth rate is the one with the greatest exponent. The first and second functions have the same big O value despite the fact they would give very different results mathematically, because they both have the same highest exponent x^4, and both the constant multiplying that term and all other terms are abstracted away. The third function will clearly give a greater result than either of the first two for a large range of lower values of x, but that is merely a special case due to the coefficient multiplying the x^3 of that equation term. Beyond that range in the general case, values of x^4 will be greater than x^3 multiplying a constant, and so the third function's $O(x^3)$ is considered lesser than $O(x^4)$. The fourth function is a constant; it returns the same value regardless of the input value of x. Likewise, its big O value is a constant $O(1)$. When the goal is to select the function with the least growth rate, the one with the lowest big O value is preferred.

The mathematical formula and algorithms used for modelling are also evaluated against a variable to obtain their big O values. However, unlike their software engineering counterparts, they are not measured against variable input sizes; their inputs will always be the measured values of the model. Rather, the variable will be the size of the simulation step meant to approximate the continuous nature of the natural world. Whether the model simulates discrete steps in time, in space, in frequency, in pressure, or in some other attribute of the physical world, the smaller the step, the more natural the simulation will be. In this context, big O notation is thus measuring a decline rate instead of a growth rate, and the value of x becomes smaller and tends towards zero. In that case, the term of the equation with greater exponents will decline more quickly than those with lesser exponents. Big O notation will thus estimate the worst-case decline value by keeping the lowest exponent term, discarding all other terms and constants multiplying that term. This yields the third column of Table 1.1, and the equation with the greatest big O exponent, rather than the lowest one, will be preferred. That equation is the one that will allow the error of the formula to decrease the fastest as the step size is reduced.

Big O notation will be used in this book to measure both the convergence rate of algorithms and their error rate. In fact, these two notions are interchangeable in this context: an algorithm converges on a solution by reducing the error on its approximation of this solution, and the rate at which it converges is the same as the rate at which it reduces the approximation error.

1.4   Summary

The main focus of this chapter has been to introduce and formally define several notions related to error measurement. The chapter began by introducing the four steps of the modelling cycle, namely, model selection to implementation, measurements, and simulation, along with the errors that can be introduced at each step. It then defined the vocabulary of error measurement, precision, accuracy, and implied precision. And finally it presented formal measures of error, namely, relative and absolute error, significant digits, and big O notation.

1.5   Exercises

1. Your partner uses a ruler to measure the length of a pencil and states that the length is 20.35232403 cm. What is your response to the given precision?
2. Given two approximations of the constant π, as 3.1417 and 3.1392838, which has the greatest precision? Which has the greatest accuracy?
3. Which number has more precision and which has more accuracy as an approximation of e, 2.7182820135423 or 2.718281828?
4. The distance between two cities is given as approximately 332 mi. As 1 mi $= 1.609344$ km exactly, it follows that the distance is approximately 534.302208 km. Discuss this conversion with respect to precision and accuracy.
5. What is approximately the absolute and relative error of 3.14 as an approximation of the constant π?
6. What are the absolute and relative errors of the approximation 22/7 of π? How many significant digits does it have?
7. What are the absolute and relative errors of the approximation 355/113 of π? How many significant digits does it have?
8. A resistor labelled as 240 Ω is actually measured at 243.32753 Ω. What are the absolute and relative errors of the labelled value?
9. The voltage in a high-voltage transmission line is stated to be 2.4 MV while the actual voltage may range from 2.1 to 2.7 MV. What is the maximum absolute and relative error of voltage?

10. A capacitor is labelled as 100 mF, whereas it is actually measured to be 108.2532 mF. What are the absolute and relative errors of the label? To how many significant digits does the label approximate the actual capacitance?

11. Of 3.1415 and 3.1416, which has more significant digits as an approximation of the constant π?

12. What is the number of significant digits of the label 240 Ω when the correct value is 243.32753 Ω?

13. To how many significant digits is the approximation 1.998532 when the actual value is 2.001959?

Chapter 2
Numerical Representation

2.1 Introduction

The numerical system used in the Western World today is a place-value base-10 system inherited from India through the intermediary of Arabic trade; this is why the numbers are often called Arabic numerals or more correctly Indo-Arabic numerals. However, this is not the only numerical system possible. For centuries, the Western World used the Roman system instead, which is a base-10 additive-value system (digits of a number are summed and subtracted from each other to get the value represented), and that system is still in use today, notably in names and titles. Other civilizations experimented with other bases: some precolonial Australian cultures used a base-5 system, while base-20 systems arose independently in Africa and in pre-Columbian America, and the ancient Babylonians used a base-60 counting system. Even today, despite the prevalence of the base-10 system, systems in other bases continue to be used every day: degrees, minutes, and seconds are counted in the base-60 system inherited from Babylonian astrologers, and base-12 is used to count hours in the day and months (or zodiacs) in the year.

When it comes to working with computers, it is easiest to handle a base-2 system with only two digits, 0 and 1. The main advantage is that this two-value system can be efficiently represented by an open or closed electrical circuit that measures 0 or 5 V, or in computer memory by an empty or charged capacitor, or in secondary storage by an unmagnetized or magnetized area of a metal disc or an absorptive or refractive portion of an optical disc. This base-2 system is called *binary*, and a single *bi*nary dig*it* is called a *bit*.

It should come as no surprise, however, that trying to model our infinite and continuous real world using a computer which has finite digital storage, memory, and processing capabilities will lead to the introduction of errors in our modelling. These errors will be part of all computer results; no amount of technological advancement or upgrading to the latest hardware will allow us to overcome them. Nonetheless, no one would advocate for engineers to give up computers altogether!

© Springer International Publishing Switzerland 2016
R. Khoury, D.W. Harder, *Numerical Methods and Modelling for Engineering*,
DOI 10.1007/978-3-319-21176-3_2

It is only necessary to be aware of the errors that arise from using computers and to account for them.

This chapter will look in details at how binary mathematics work and how computers represent and handle numbers. It will then present the weaknesses that result from this representation and that, if ignored, can compromise the quality of engineering work.

2.2   Decimal and Binary Numbers

This section introduces binary numbers and arithmetic. Since we assume the reader to be intimately familiar with decimal (base-10) numbers and arithmetic, we will use that system as a bridge to binary.

2.2.1   Decimal Numbers

Our decimal system uses ten ordered digits 0, 1, 2, 3, 4, 5, 6, 7, 8, and 9 to represent any number as a sequence:

$$d_n d_{n-1} d_{n-2} \ldots d_1 d_0 . d_{-1} d_{-2} \ldots \tag{2.1}$$

where every d_i is a digit, n is an integer, and $d_n \neq 0$. The sequence is place valued, meaning that the value of d_i in the complete number is the value of its digit multiplied by 10 to the power i. This means the value of the complete number can be computed as

$$\sum_{i=-\infty}^{n} d_i \times 10^i \tag{2.2}$$

The digits d_0 and d_{-1}, the first digit multiplied by 10 to a negative power, are separated by a point called the decimal point. The digit d_n, which has the greatest value in the total number, is called the *most significant digit*, while the digit d_i with the lowest value of i and therefore the lowest contribution in the total number is called the *least significant digit*.

It is often inconvenient to write numbers in the form of Eq. (2.1), especially when modelling very large or very small quantities. For example, the distance from the Earth to the Sun is 150,000,000,000 m, and the radius of an electron is 0.0000000000000028 m. For this reason, numbers are often represented in *scientific notation*, where the non-zero part is kept, normally with one digit left of the decimal point and a maximum of m digits on the right, and the long string of zeroes is simplified using a multiplication by a power of 10. The number can then be written as

$$d_0.d_{-1}d_{-2}\ldots d_{-m} \times 10^n \tag{2.3}$$

or equivalently but more commonly as

$$d_0.d_{-1}d_{-2}\ldots d_{-m}en \tag{2.4}$$

The $m+1$ digits that are kept are called the *mantissa*, the value n is called the *exponent*, and the letter "e" in Eq. (2.4) stands for the word "exponent", and normally $m < n$. Using scientific notation, the distance from the Earth to the Sun is 1.5×10^{11} m and the radius of the electron is 2.8×10^{-15}.

We can now define our two basic arithmetic operations. The rule to perform the addition of two decimal numbers written in the form of Eq. (2.1) is to line up the decimal points and add the digits at corresponding positions. If a digit is missing from a position in one of the numbers, it is assumed to be zero. If two digits sum to more than 9, the least significant digit is kept in that position and the most significant digit carries and is added to the digits on the left. An addition of two numbers in scientific notations is done first by writing the two numbers at the same exponent value, then adding the two mantissas in the same way as before. To multiply two decimal numbers written in the form of Eq. (2.1), multiply the first number by each digit d_i of the second number and multiply that partial result by 10^i, then sum the partial results together to get the total. Given two numbers in scientific notation, multiply the two mantissas together using the same method, and add the two exponents together.

2.2.2 Binary Numbers

A binary system uses only two ordered digits (or bits, short for "binary digits"), 0 and 1, to represent any number as a sequence:

$$b_n b_{n-1} b_{n-2} \ldots b_1 b_0 . b_{-1} b_{-2} \ldots \tag{2.5}$$

where every b_i is a bit, n is an integer, and $b_n \neq 0$. The sequence is place valued, meaning that the value of b_i in the complete number is the value of its digit multiplied by 2 to the power i. This means the value of the complete number can be computed as

$$\sum_{i=-\infty}^{n} b_i \times 2^i \tag{2.6}$$

The digits b_0 and b_{-1}, the first digit multiplied by 2 to a negative power, are separated by a point; however, it would be wrong to call it a decimal point now since this is not a decimal system. In binary it is called the *binary point*, and a more general term for it independent of base is the *radix point*. We can define a binary scientific notation as well, as

$$b_0.b_{-1}b_{-2}\ldots b_{-m} \times 2^n \tag{2.7}$$

$$b_0.b_{-1}b_{-2}\ldots b_{-m}en \tag{2.8}$$

The readers can thus see clear parallels with Eqs. (2.1)–(2.4) which define our decimal system. Likewise, the rules for addition and multiplication in binary are the same as in decimal, except that digits carry whenever two 1s are summed.

Since binary and decimal use the same digits 1 and 0, it can lead to ambiguity as to whether a given number is written in base 2 or base 10. When this distinction is not clear from the context, it is habitual to suffix the numbers with a subscript of their base. For example, the number 110 is ambiguous, but 110_{10} is one hundred and ten in decimal, while 110_2 is a binary number representing the number 6 in decimal. It is not necessary to write that last value as 6_{10} since 6_2 is nonsensical and no ambiguity can exist.

Example 2.1
Compute the addition and multiplication of 3.25 and 18.7 in decimal and of 1011.1 and 1.1101 in binary.

Solution
Following the first addition rule, line up the two numbers and sum the digits as follows:

$$\begin{array}{r} 3.25 \\ +18.7 \\ \hline 21.95 \end{array}$$

The second addition rule requires writing the numbers in scientific notation with the same exponent. These two numbers in scientific notations are 3.25×10^0 and 1.87×10^1, respectively. Writing them in the same exponent would change the first one to 0.325×10^1. Then the sum of the mantissa gives

$$\begin{array}{r} 0.325 \\ +1.87 \\ \hline 2.195 \end{array}$$

for a final total of 2.195×10^1, the same result as before.

(continued)

Example 2.1 (continued)

The multiplication of two numbers in decimal using the first rule is done by summing the result of partial multiplications as follows:

$$
\begin{array}{r}
3.25 \\
\times 18.7 \\
\hline
2.275 \\
26 \\
+32.5 \\
\hline
60.775
\end{array}
$$

To perform the operation using the second rule, write the numbers in scientific notation as 3.25×10^0 and 1.87×10^1, respectively, then multiply the mantissas as before:

$$
\begin{array}{r}
3.25 \\
\times 1.87 \\
\hline
0.2275 \\
2.6 \\
+3.25 \\
\hline
6.0775
\end{array}
$$

and sum the exponents $(0+1)$ to get a final result of 6.0775×10^1 as before.

Working in binary, the rules apply in the same way. The binary sum, using the first rule, gives

$$
\begin{array}{r}
1011.1 \\
+\,1.1101 \\
\hline
1101.0101
\end{array}
$$

Writing the numbers in scientific notation gives 1.0111×2^3 and 1.1101×2^0, respectively. Using the second rule, the multiplication of the mantissas gives

$$
\begin{array}{r}
1.0111 \\
\times\,1.1101 \\
\hline
0.00010111 \\
0. \\
0.010111 \\
0.10111 \\
+\,1.0111 \\
\hline
10.10011011
\end{array}
$$

and the sum of the exponents gives 3, for a total of 10.10011011×2^3, better written as 1.010011011×2^4.

2.2.3   Base Conversions

The conversion from binary to decimal can be done simply by computing the summation from Eq. (2.6).

The conversion from decimal to binary is much more tedious. For a decimal number N, it is necessary to find the largest power k of 2 such that $2^k \leq N$. Add that power of 2 to the binary number and subtract it from the decimal number, and continue the process until the decimal number has been reduced to 0. This will yield the binary number as a summation of the form of Eq. (2.6).

Example 2.2
Convert the binary number 101.101 to decimal, and then convert the result back to binary.

Solution
Convert the number to decimal by writing it in the summation form of Eq. (2.6) and computing the total:

$$101.101 = 1 \times 2^2 + 0 \times 2^1 + 1 \times 2^0 + 1 \times 2^{-1} + 0 \times 2^{-2} + 1 \times 2^{-3}$$
$$= 4 + 0 + 1 + 0.5 + 0 + 0.125$$
$$= 5.625$$

Converting 5.625 back to binary requires going through the algorithm steps:

$$\text{step } 1: \quad N = 5.625 \quad k = 2 \quad 2^k = 4 \leq N \quad N - 2^k = 1.625$$
$$\text{step } 2: \quad N = 1.625 \quad k = 0 \quad 2^k = 1 \leq N \quad N - 2^k = 0.625$$
$$\text{step } 3: \quad N = 0.625 \quad k = -1 \quad 2^k = 0.5 \leq N \quad N - 2^k = 0.125$$
$$\text{step } 4: \quad N = 0.125 \quad k = -3 \quad 2^k = 0.125 \leq N \quad N - 2^k = 0$$
$$5.625 = 4 + 1 + 0.5 + 0.125$$
$$= 1 \times 2^2 + 1 \times 2^0 + 1 \times 2^{-1} + 1 \times 2^{-3}$$
$$= 101.101$$

2.3   Number Representation

It is unreasonable and often impossible to store numbers in computer memory to a maximum level of precision by keeping all their digits. One major issue is that the increasing number of digits that results from performing complex calculations will quickly consume the computer's memory. For example, the product of the two 11-digit numbers $1.2345678901 \times 2.3456789012$ equals the 21-digit number 2.89589985190657035812, and the sum of two 9-digit numbers

123456789.0 + 0.123456789 equals the 18-digit 123456789.123456789 where the two summands each have 9 significant digits but the sum requires 18 significant digits. Moreover, certain numbers, such as π, have an infinite number of digits that would need to be stored in memory for a maximum level of precision, which is quite simply impossible to do. Numbers in the natural world may be infinite, but computer memory is not.

It is therefore necessary for modern computers to truncate the numbers stored in their memory. This truncation will naturally cause a loss in precision in the values stored and will be a source of errors for any computer model of the real world. And this problem will be unavoidable on any computer, no matter how advanced and powerful (at least, until someone invents a computer with infinite memory and instantaneous processing). Nonetheless, the way numbers are represented and stored in the computer will lead to different levels of seriousness of these errors, and one would be wise to pick a representation scheme that minimizes problems.

Since all number representation schemes are not equal, it is necessary to begin by defining four important requirements to compare the different schemes by:

1. A scheme must represent numbers using a fixed amount of memory. This is an unavoidable requirement of modern computers.
2. A scheme must allow the user to represent a range of values both very large and very small, in order to accommodate models of any aspect of the world. The real world has an infinite range of values, which is something that is impossible to represent given the requirement of using a fixed amount of memory. However, the greater the range of values that a scheme can be represented, the better.
3. A scheme must be able to represent numbers, within the range it can handle, with a small relative error. Truncation to accommodate a fixed amount of memory will necessarily lead to errors, but keeping these errors to a minimum is always preferable. This requirement does not take into account the error on numbers outside the range of values the scheme can represent, as this error can be infinite.
4. A scheme must allow software to efficiently test for equality and relative magnitude between two numbers. This is a computing requirement rather than a memory requirement. Number comparisons are the fundamental building block of software, and given two otherwise equivalent representation scheme, one that allows more efficient comparisons will lead to more efficient software and runtime performances.

2.3.1 Fixed-Point Representation

Perhaps the easiest method of storing a real number is by storing a fixed number of digits before and after the radix point, along with its sign (0 or 1 to represent a positive or negative number respectively). For the sake of example, we will assume three digits before the point and three after, thus storing a decimal number $\pm d_2d_1d_0.d_{-1}d_{-2}d_{-3}$. For example, the constant π would be stored as 0003142. This representation is called *fixed-point representation*. It clearly satisfies the first requirement of using a fixed amount of memory. Moreover, the fourth requirement of efficiently comparing two numbers for relative magnitude can be achieved by simply comparing the two numbers digit by digit from left to right and stopping as soon as one is greater than the other.

Unfortunately, fixed-point representation does not perform well on the other two requirements. For the second requirement, the range of values that this notation can represent is very limited. The largest value that can be stored is ± 999.999 and the smallest one is ± 000.001, which are neither very large nor very small. And for the third requirement, the relative error on some values within the range can be very large. For instance, the value 0.0015 is stored as 0.002 with a staggering relative error of 0.33.

2.3.2   Floating-Point Representation

Using the same amount of memory as fixed-point representation, it would be a lot more efficient to store numbers in scientific notation and use some of the stored digits to represent the mantissa and some to represent the exponent. Using the same example as before of storing six digits and a sign, this could give the decimal number $\pm d_0.d_{-1}d_{-2}d_{-3} \times 10^{E_0E_1-49}$. For reasons that will become clear shortly, the exponent digits are stored before the mantissa digits, so in this representation, the constant π would be stored as 0493142. This representation is called *floating-point representation*. The value 49 subtracted in the exponent is called the *bias* and is half the maximum value the exponent can take.

Floating-point representation satisfies all four requirements better than fixed-point representation. The first requirement is clearly satisfied by using exactly as much memory as fixed-point representation. The range of values covered to satisfy the second requirement is much larger than before: a two-digit exponent can cover 100 orders of magnitude, and thanks to the introduction of the bias, numbers can have exponents ranging from the minuscule 10^{-49} to the massive 10^{50}. As for the third requirement, the maximum relative error of any real number in the interval $[1.000 \times 10^{-49},\ 9.999 \times 10^{50}]$ is $1/2001 \approx 0.0005$. Finally, by adding the requirement that the first digit of the mantissa d_0 must be different from zero and thanks to the fact the exponent digits are stored first, it is still possible to efficiently compare the relative magnitude of two numbers by comparing them digit by digit from left to right and stopping as soon as one is greater than the other. The added requirement that the first mantissa bit be different from zero is required; otherwise it would always be necessary to read all digits of the numbers and to perform some computations to standardize them, as the same real number could be stored in several different ways. For instance, the number 3 could be stored as 0493000,

0500300, 0510030, or 0520003, and only by subtracting the bias and shifting the mantissa appropriately does it become evident that all four values are the same. The requirement that the first bit of the mantissa be non-zero insures that only the first of these four representations is legal.

The requirement that the first bit of the mantissa must be non-zero introduces a surprising new problem: representing the real value 0 in floating-point representation is a rule-breaking special case. Moreover, given that each floating-point value has a sign, there are two such special values at 0000000 and 1000000. Floating-point representation uses this to its advantage by actually defining two values of zero, a positive and a negative one. A positive zero represents a positive number smaller than the smallest positive number in the range, and a negative zero represents a negative number greater than the greatest negative number in the range.

It is also possible to include an additional exception to the rule that the first digit of the mantissa must be non-zero, in the special case where a number is so small that it cannot be represented while respecting the rule. In the six-digit example, this would be the case, for example, for the number 1.23×10^{-50}, which could be represented as 0000123 but only with a zero as the first digit of the mantissa. Allowing this type of exception is very attractive; it would increase the range of values that can be represented by several orders of magnitude at no cost in memory space and without making relative comparisons more expensive. But this is no free lunch: the cost is that the mantissa will have fewer digits, and thus the relative error on the values in this range will be increased. Nonetheless, this trade-off can sometimes be worthwhile. A floating-point representation that allows this exception is called *denormalized*.

Example 2.3
Represent 10! in the six-digit floating-point format.

Solution
First, compute that $10! = 3628800$, or 3.6288×10^6 in scientific notation. The exponent is thus 55 to take into account the bias of -49, the mantissa rounded to four digits is 3.629, and the positive sign is a 0, giving the representation 0553629.

Example 2.4
What number is represented, using the six-digit floating-point format, by 1234567?

Solution
The leading 1 indicates that it is a negative number, the exponent is 23 and the mantissa is 4.567. This represents the number $-4.567 \times 10^{23-49} = -4.567 \times 10^{-26}$.

2.3.3   Double-Precision Floating-Point Representation

The representation most commonly used in computers today is *double*, short for "double-precision floating-point format," and formally defined in the IEEE 754 standard. Numbers are stored in binary (as one would expect in a computer) over a fixed amount of memory of 64 bits (8 bytes). The name comes from the fact this format uses double the amount of memory that was allocated to the original floating-point format (float) numbers, a decision that was made when it was found that 4 bytes was not enough to allow for the precision needed for most scientific and engineering calculations.

The 64 bits of a double number comprise, in order, 1 bit for the sign (0 for positive numbers, 1 for negative numbers), 11 bits for the exponent, and 52 bits for the mantissa. The maximum exponent value that can be represented with 11 bits is 2047, so the bias is 1023 (01111111111_2), allowing the representation of numbers in the range from 2^{-1022} to 2^{1023}. And the requirement defined previously, that the first digit of the mantissa cannot be 0, still holds. However, since the digits are now binary, this means that the first digit of the mantissa must always be 1; it is consequently not stored at all, and all 52 bits of the mantissa represent digits after the radix point following an implied leading 1. This means also that double cannot be a denormalized number representation.

For humans reading and writing 64-bit-long binary numbers can be tedious and very error prone. Consequently, for convenience, the 64 bits are usually grouped into 16 sets of 4 bits, and the value of each set of 4 bits (which will be between 0 and 15) is written using a single hexadecimal (base-16) digit. The following Table 2.1 gives the equivalences between binary, hexadecimal, and decimal.

Table 2.1 Binary, hexadecimal, and decimal number conversions

Binary	Hexadecimal	Decimal
0000	0	0
0001	1	1
0010	2	2
0011	3	3
0100	4	4
0101	5	5
0110	6	6
0111	7	7
1000	8	8
1001	9	9
1010	a	10
1011	b	11
1100	c	12
1101	d	13
1110	e	14
1111	f	15

Converting a decimal real number to double format is done by converting it to scientific notation binary, rounding the mantissa to 52 bits after the radix point, and adding the bias of 1023 to the exponent. The double number is the assembled, starting with the sign bit of 0 or 1 if the number is positive or negative, respectively, then the exponent, and finally the 52 bits of the mantissa after the radix point, discarding the initial 1 before the radix point. On the other hand, converting a double number into a decimal real is done first by splitting it into three parts, the sign, exponent, and mantissa. The bias of 1023 is subtracted from the exponent, while a leading one is appended at the beginning of the mantissa. The real number is then computed by converting the mantissa to decimal, multiplying it by 2 to the correct exponent, and setting the appropriate positive or negative sign based on the value of the sign bit.

Example 2.5

Convert the double c066f40000000000 into decimal.

Solution

First, express the number into binary, by replacing each hexadecimal digit with its binary equivalent:

c	0	6	6	f	4	0	0	0	0	0	0	0	0	0	0
1100	0000	0110	0110	1111	0100	0000	0000	0000	0000	0000	0000	0000	0000	0000	0000

Next, consider the three components of a double number:

1. The first bit is the sign. It is 1, meaning the number is negative.
2. The next 11 bits are the exponent. They are $10000000110_2 = 1030_{10}$. Recall also that the bias in double is 1023_{10}, which means the number is to the power $2^{1030-1023} = 2^7$.
3. The remaining 52 bits are for the mantissa. They are 0110111101 0000000000000000000000000000000000000000000. Adding in the implied leading 1 before the radix point and discarding the unnecessary trailing zeros, this represents the mantissa $1.0110111101_2 = 1.4345703125_{10}$.

The number in decimal is thus $-1.4345703125 \times 2^7 = -183.625$.

Example 2.6

Find the double representation of the integer 289.

Solution

First, note that the number is positive, so the sign bit is 0.

Next, convert the number to binary: $289 = 256 + 32 + 1 = 2^8 + 2^5 + 2^0 = 100100001_2$. In scientific notation, this becomes $1.00100001_2 \times 2^8$ (the radix point must move eight places to the left). The exponent for the number

(continued)

Example 2.6 (continued)

is thus 1031 (10000000111_2) since $1031 - 1023 = 8$. And the mantissa, taking out the leading 1 that will be implied by double notation, is 00100001_2 followed by 44 more zeros to get the total 52 bits.

Putting it all together, the double representation is 0 10000000111 0010000 10000000000000000000000000000000000000000000000.

Example 2.7

Find the double-precision floating-point format of $-324/33$ given that its binary representation is:

$-1001.1101000101110100010111010001011101000101110100010111 0$
$100010111010001\ldots$

Solution

The number is negative, so the sign bit is 1.

The radix point must be moved three spots to the left to produce a scientific-format number, so the exponent is $3_{10} = 11_2$. Adding the bias gives $01111111111 + 11 = 10000000010$.

Finally, rounding the infinite number to 53 bits and removing the leading 1 yield the 52 bits of the mantissa, 0011101000101110100010111010001 01110100010111010100011.

Putting it all together, the double representation is:

1 10000000010 0011101000101110100010111010001011101000101110100011

2.4   Limitations of Modern Computers

Since computers try to represent the infinite range of real numbers using the finite set of floating-point numbers, it is unavoidable that some problems will arise. Three of the most common ones are explored in this section.

2.4.1   Underflow and Overflow

Any number representation format that is restricted to a limited amount of memory will necessarily only be able to represent a finite range of numbers. In the case of double, that range goes from negative to positive 1.8×10^{308} at the largest magnitudes and at negative to positive 5×10^{-324} at the smallest magnitudes around zero. Any number greater or smaller than these extremes falls outside of the number scheme and cannot be represented. This problem is called *overflow* when the number is greater than the maximum value and *underflow* when it is lesser than the smallest value.

Fig. 2.1 C code generating
the double special values

```
double MaxVal = 1.8E307;

double MinVal = 5E-324;



double PlusInf = MaxVal * 10;

double MinusInf = MaxVal * -1 * 10;

double PlusZero = MinVal / 10;

double MinusZero = MinVal * -1 / 10;
```

The name "overflow" comes from a figurative imagery of the problem. Picture the largest number that can be represented in 8-bit binary, 11111111_2. Adding 1 to that value causes a chain of carry-overs: the least significant bit flips to 0 and a 1 carries over to the next position and causes that 1 to flip and carry a 1 over to the next position, and so on. In the end the most significant bit flips to zero and a 1 is carried over, except it has no place to carry to since the value is bounded to 8 bits; the number is said to "over flow". As a result, instead of 100000000_2, the final result of the addition is only 00000000_2. This problem will be very familiar to the older generation of gamers: in many 8-bit RPG games, players who spent too much time levelling up their characters might see a level-255 (11111111_2) character gain a level and fall back to its starting level. This problem was also responsible for the famous kill screen in the original *Pac-Man* game, where after passing level 255, players found themselves in a half-formed level 00.

Overflow and underflow are well-known problems, and they have solutions defined in the IEEE 754 standard. That solution is to define four special values: a positive infinity as 7ff0000000000000, a negative infinity as fff0000000000000, a positive zero as 0000000000000000, and a negative zero as 8000000000000000 (both of which are different from each other and from an actual value of zero). Converting these to binary will show the positive and negative infinity values to be the appropriate sign bit with all-1 exponent bits and all-zero mantissa bits, while the positive and negative zero values are again the appropriate sign bit with all-zero exponent and mantissa bits. Whenever a computation gives a result that falls beyond one of the four edges of the double range, it is replaced by the appropriate special value. The sample code in the next Fig. 2.1 is an example that will generate all four special values.

2.4.2   Subtractive Cancellation

Consider the following difference: $3.523 - 3.537 = 0.014$. Using the six-digit floating-point system introduced previously, these numbers are represented by 0493523, 0493537, and 0471400, respectively. All three numbers appear to have the same precision, with four decimal digits in the mantissa. However, 3.523 is really a truncated representation of any number in the range [3.5225, 3.5235], as any number in that five-digit range will round to the four-digit representation 3.523. Likewise, the second number 3.537 represents the entire range [3.5365, 3.5375]. The maximum relative error on any of these approximations is 0.00014, so they are not a problem. However, when considering the ranges, the result of the subtraction is not 0.014 but actually could be any value in the range [0.013, 0.015]. The result 0.014 has no significant digits. Worse, as an approximation of the range of results, 0.014 has a relative error of 0.071, more than 500 times greater than the error of the initial values.

This phenomenon where the subtraction of similar numbers results in a significant reduction in precision is called *subtractive cancellation*. It will occur any time there is a subtraction of two numbers which are almost equal, and the result will always have no significant digits and much less precision than either initial numbers.

Unlike overflow and underflow, double format does not substitute the result of such operations with a special value. The result of 0.014 in the initial example will be stored and used in subsequent computations as if it were a precise value rather than a very inaccurate approximation. It is up to the engineers designing the mathematical software and models to check for such situations in the algorithms and take steps to avoid them.

Example 2.8

Consider two approximations of π using the six-digit floating-point representation: 3.142 and 3.14. Subtract the second from the first. Then, compute the relative error on both initial values and on the subtraction result.

Solution

$$3.142 - 3.14 = 0.002$$

However, in six-digit floating-point representation, 3.142 (0493142) represents any value in [3.1415, 3.1425] and 3.14 (0493140) represents any value in the range [3.1395, 3.1405]. Their difference is any number in the range [0.001, 0.003]. The result of 0.002 has no significant digits.

Compared to $\pi = 3.141592654\ldots$, the value 3.142 has a relative error of 0.00013 and the value 3.14 has a relative error of 0.0051. The correct result of the subtraction is $\pi - 3.14 - 0.001592654\ldots$, and compared to that result, 0.002 has a relative error of 0.2558, 50 times greater than the relative error of 3.14.

2.4.3   Non-associativity of Addition

In mathematics, *associativity* is a well-known fundamental property of additions which states that the order in which additions are done makes no difference on the final result. Formally

$$(a + b) + c = a + (b + c) \tag{2.9}$$

This property no longer holds in floating-point number representation. Because of the truncation required to enforce a fixed amount of memory for the representation, larger numbers will dominate smaller numbers in the summation, and the order in which the larger and smaller numbers are introduced into the summation will affect the final result.

To simplify, consider again the six-digit floating-point representation. For an example of a large number dominating a smaller one, consider the sum 5592 $+ 0.7846 = 5592.7846$. In six-digit floating-point representation, these numbers are written 0525592 and 0487846, respectively, and there is no way to store the result of their summation entirely as it would require ten digits instead of six. The result stored is actually 0525593, which corresponds to the value 5593, a rounded result that maintains the more significant digits of the larger number and discards the less significant digits of the smaller number. The larger number has eclipsed the smaller one in importance. The problem becomes even worse in the summation $5592 + 0.3923 = 5592.3923$. The result will again need to be cropped to be stored in six digits, but this time it will be rounded to 5592. This means that the summation has left the larger number completely unchanged!

These two rounding issues are at the core of the problem of *non-associativity* when dealing with sequence of sums as in Eq. (2.9). The problem is that not just the final result, but also every partial result summed in the sequence, must be encoded in the same representation and will therefore be subject to rounding and loss.

Consider the sum of three values, $5592 + 0.3923 + 0.3923 = 5592.7846$. As before, every value in the summation can be encoded in the six-digit floating-point format, but the final result cannot. However, the order in which the summation is computed will affect what the final result will be. If the summation is computed as

$$5592 + (0.3923 + 0.3923) = 5592 + 0.7846 = 5592.7846 \tag{2.10}$$

then there is no problem in storing the partial result 0.7846, and only the final result needs to be rounded to 5593. However, if the summation is computed as

$$(5592 + 0.3923) + 0.3923 = 5592.3923 + 0.3923 = 5592.7846 \tag{2.11}$$

then there is a problem, as the partial result 5592.3923 gets rounded to 5592, and the second part of the summation then becomes $5592 + 0.3923$ again, the result of which again gets rounded to 5592. The final result of the summation has changed

Fig. 2.2 C++ code
suffering from
non-associativity

```
double sum = 0.0;
for ( int i = 0; i < 100000; i++ ) {
    sum += 0.1;
}
```

because of the order in which the partial summations were computed, in clear violation of the associativity property.

Example 2.9

Using three decimal digits of precision, add the powers of 2 from 0 to 17 in the order from 0 to 17 and then in reverse order from 17 to 0. Compute the relative error of the final result of each of the two summations.

Solution

Recall that with any such system, numbers must be rounded before and after any operation is performed. For example, $2^{10} = 1024 = 1020$ after rounding to three significant digits. Thus, the partial sums in increasing order are

1 3 7 15 31 63 127 255 511 1020 2040 4090 8190 16400 32800 65600 131000 262000

while in decreasing order, they are

131000 196000 229000 245000 253000 257000 259000 260000 261000 261000 261000 261000 261000 261000 261000 261000 261000 261000

The correct value of this sum is $2^{18} - 1 = 262{,}143$. The relative error of the first sum is 0.00055, while the relative error of the second sum is 0.0044. So not only are the results different given the order of the sum, but the sum in increasing order gives a result an order of magnitude more accurate than the second one.

Like with subtractive cancellation, there is no special value in the double format to substitute in for non-associative results, and it is the responsibility of software engineers to detect and correct such cases when they happen in an algorithm. In that respect, it is important to note that non-associativity can be subtly disguised in the code. It can occur, for example, in a loop that sums a small value into a total at each increment of a long process, such as in the case illustrate in Fig. 2.2. As the partial total grows, the small incremental addition will become rounded off in later iterations, and inaccuracies will not only occur, they will become worse and worse as the loop goes on.

The solution to this problem is illustrated in Example 2.9. It consists in sorting the terms of summations in order of increasing value. Two values of the same magnitude summed together will not lose precision, as the digits of neither number will be rounded off. By summing together smaller values first, the precision of the partial total is maintained, and moreover the partial total grows in magnitude and can then be safely added to larger values. This ordering ensures that the cumulative sum of many small terms is still present in the final total.

2.5   Summary

Modern engineering modelling is done on computers, and this will continue to be the case long into the foreseeable future. This means that the first source of errors in any model is not in the algorithms used to compute and approximate it, but in the format used by the computer to store the very values that are modelled. It is simply impossible to store the infinite continuous range of real numbers using the finite set of discrete values available to computers, and this fundamental level of approximation is also the first source of errors that must be considered in any modelling process.

In engineering, a less accurate result with a predictable error is better than a more accurate result with an unpredictable error. This was one of the main reasons behind standardizing the format of floating-point representations on computers. Without standardization, the same code run on many machines could produce different answers. IEEE 754 standardized the representation and behaviour of floating-point numbers and therefore allowed better prediction of the error, and thus, an algorithm designed to run within certain tolerances will perform similarly on all platforms. Without standardization, a particular computation could have potentially very different results when run on different machines. Standardization allows the algorithm designer to focus on a single standard, as opposed to wasting time fine-tuning each algorithm for each different machine.

The properties of the double are specified by the IEEE 754 technical standard. For additional information, there are a few excellent documents which should be read, especially "Lecture Notes on the Status of IEEE Standard 754 for Binary Floating-Point Arithmetic" by Prof W. Kahan and "What Every Computer Scientist Should Know about Floating-Point Arithmetic" by David Goldberg.

2.6   Exercises

1. Represent the decimal number 523.2345 in scientific notation.
2. What is the problem if we don't use scientific notation to represent 1.23e10?
3. Add the two binary integers 100111_2 and 1000110_2.
4. Add the two binary integers 1100011_2 and 10100101101_2.
5. Add the two binary numbers 100.111_2 and 10.00110_2.
6. Add the two binary numbers 110.0011_2 and 10100.101101_2.
7. Multiply the two binary integers 100111_2 and 1010_2.
8. Multiply the two binary integers 1100011_2 and 10011_2.
9. Multiply the two binary numbers 10.1101_2 by 1.011_2.
10. Multiply the two binary numbers 100.111_2 and 10.10_2.
11. Multiply the two binary numbers 1100.011_2 and 10.011_2.
12. Convert the binary number 10010.0011_2 to decimal.
13. Convert the binary number -111.111111_2 to decimal.

14. What decimal numbers do the following represent using the six-digit floating-point format?

    a. 479323
    b. 499323
    c. 509323
    d. 549323

15. Given their implied precision, what range of numbers do the following represent using the six-digit floating-point format?

    a. 521234
    b. 522345

16. Represent the following numbers in six-digit floating-point format:

    a. Square root of two (≈ 1.414213562)
    b. One million (1000000)
    c. $-e^{-10} \approx -0.00004539992976$

17. Convert the decimal number 1/8 to binary double format.
18. Convert the hexadecimal double format number c01d600000000000 to binary and to decimal.
19. Convert the following binary double format numbers to decimal:

    a. 0100000001100011001011111100000000000000000000000000000000000000
    b. 0011111111101000100000000000000000000000000000000000000000000000

20. Add the following two hexadecimal double format numbers: 3fe8000000000000 and 4011000000000000.
21. Using the six-digit floating-point format:

    a. What is the largest value which can be added to 3.523 which will result in a sum of 3.523 and why?
    b. What is the largest float which may be added to 722.4 which will result in a sum of 722.4 and why?
    c. What is the largest float which may be added to 722.3 which will result in a sum of 722.3 and why?

22. How would you calculate the sum of n^{-2} for $n = 1, 2, \ldots, 100{,}000$ and why?

Chapter 3
Iteration

3.1 Introduction

This chapter and the following four chapters introduce five basic mathematical modelling tools: iteration, linear algebra, Taylor series, interpolation, and bracketing. While they can be used as simple modelling tools on their own, their main function is to provide the basic building blocks from which numerical methods and more complex models will be built.

One technique that will be used in almost every numerical method in this book consists in applying an algorithm to some initial value to compute an approximation of a modelled value, then to apply the algorithm to that approximation to compute an even better approximation and repeat this step until the approximation improves to a desired level. This process is called *iteration*, and it can be as simple as applying a mathematical formula over and over or complex enough to require conditional flow-control statements. It also doubles as a modelling tool for movement, both for physical movement in space and for the passage of time. In those cases, one iteration can represent a step along a path in space or the increment of a clock.

This chapter will introduce some basic notions and terminology related to the tool of iteration, including most notably the different halting conditions that can come into play in an iterating function.

3.2 Iteration and Convergence

Given a function $f(x)$ and an initial value x_0, it is possible to calculate the result of applying the function to the value as such:

$$x_1 = f(x_0) \tag{3.1}$$

© Springer International Publishing Switzerland 2016
R. Khoury, D.W. Harder, *Numerical Methods and Modelling for Engineering*,
DOI 10.1007/978-3-319-21176-3_3

This would be our first iteration. The second iteration would compute x_2 by applying the function to x_1, the third iteration would compute x_3 by applying the function to x_2, and so on. More generally, the ith iteration is given in Eq. (3.2):

$$\begin{aligned}
x_1 &= f(x_0) \\
x_2 &= f(x_1) \\
x_3 &= f(x_2) \\
&\cdots \\
x_i &= f(x_{i-1})
\end{aligned} \tag{3.2}$$

Each value of x will be different from the previous. However, for certain functions, each successive value of x will become more and more similar to the previous one, until they stop changing altogether. At that point, the function is said to have *converged* to the value x. The steps of a function starting at x_0 and converging to x_i are given in Eq. (3.3):

$$\begin{aligned}
x_1 &= f(x_0) \\
x_2 &= f(x_1) \\
x_3 &= f(x_2) \\
&\vdots \\
x_{i-1} &= f(x_{i-2}) \\
x_i &= f(x_{i-1}) \\
x_i &= f(x_i)
\end{aligned} \tag{3.3}$$

Convergence can be observed by doing a simple experiment using any scientific calculator. Set the calculator in degrees, compute the cosine of 0, and then iteratively compute the cosine of every result obtained. The first result will be 1, and then, depending on the number of digits of precision of the calculator, it can be 0.9998476951, then 0.9998477415, and so on. At every iteration, more decimals will stay constant and the decimals that change will be further away from the radix point; in other words, the difference between the two successive numbers is decreasing. Then at some point, the number will stop changing altogether and the function will have converged. Again, depending on the precision of the calculator, this point will change: the value has already converged for the ten digits of precision in this paragraph and can take another half-dozen iterations to converge to 30 digits of precision.

Running the same experiment with a calculator set in radians yields a different result. This time the sequence goes 0, 1, 0.5403, 0.8575, 0.6542, 0.79348, 0.7013, 0.76395, etc. It is still converging, but while in degrees the value was decreasing towards convergence, in radians it *oscillates* around the value the function converges to, jumping from one value that is greater than the convergence value to one that is lesser and back to greater. It can also be observed that convergence is a lot slower. While in degrees the function can converge in five to ten iterations, depending on the precision of the calculator, in degrees it takes six iterations only for the first decimal to converge, and it will take up to iteration 16 for the second decimal to converge to 0.73. This notion of the speed with which a function converges is called *convergence rate*, and in the numerical methods that will be introduced later on, it will be found that it is directly linked to the big O error rate of the functions themselves: a function with a lesser big O error rate will have less error remaining in the value computed at each successive iteration and will thus converge to the error-free value in fewer iterations than a function with a greater big O error rate.

Finally, it is easy to see that not all functions will converge. Using the x^2 function of the calculator and starting at any value greater than 1 will yield results that are both greater at each successive iteration and that increase more between each successive iteration, until the maximum value or the precision of the calculator is exceeded. Functions with that behaviour are said to *diverge*. It is worth noting that some functions display both behaviours: they can converge when iterating from certain initial values, but diverge when iterating from others. The x^2 function is in fact an example of such a function: it diverges for any initial value greater than 1 or lesser than -1, but converges to 0 using any initial value between -1 and 1.

Example 3.1

Starting from any positive or negative non-zero value, compute ten iterations of the function:

$$f(x) = \frac{x}{2} + \frac{1}{x}$$

Solution

This function has been known since antiquity to converge to $\pm\sqrt{2}$. The exact convergence sequence will depend on the initial value, but positive and negative examples are given in the table below.

Iteration	Positive sequence	Negative sequence
x_0	44.0	−88.0
x_1	22.0227272727	−44.0113636364
x_2	11.0567712731	−22.0284032228
x_3	5.6188279524	−11.0595975482
x_4	2.9873870354	−5.6202179774
x_5	1.8284342085	−2.9880380306
x_6	1.4611331460	−1.8286867771
x_7	1.4149668980	−1.4611838931
x_8	1.4142137629	−1.4149685022
x_9	1.4142135623	−1.4142137637
x_{10}	1.4142135623	−1.4142135623

(continued)

Example 3.1 (continued)

The convergence can be further observed by plotting the value of the sequence at each iteration, as is done in the figure below (in blue for the positive sequence and in red for the negative sequence). It becomes clear to see in such a graphic that not only the sequence converges quickly, but it forms a decreasing logarithmic curve as it does so. This sequence is said to converge logarithmically.

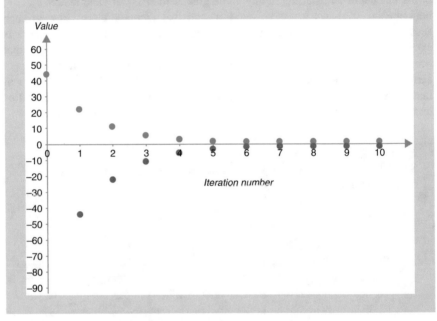

3.3   Halting Conditions

As explained in the previous section, a function might converge quickly to an exact value, or it might converge at a very slow rate, or even diverge altogether. And even in the cases where the function converges, the function might reach an approximate result quickly or spend endless iterations computing more and more digits of precision. For these reasons, it is important to give any iterating algorithm a set of *halting conditions* at which it should stop iterating. There are two types of halting conditions: success conditions, which indicate that the last iteration yielded a converged value of sufficient precision for a given model or application, and failure conditions, which indicate that the function has failed to converge for some reason. The exact details of these conditions and the specific conditions they test for will vary based on the numerical method that is being implemented and the situation that is being modelled. However, in general terms, a success condition will test the relative error between two iterations, and a failure condition will test the number of iterations computed.

An iterative function succeeds when it converges to a value. However, this converged value might be a rational number with an endless number of non-zero decimals (such as 1/3) or an irrational number, which would require endless iterations to compute completely. Moreover, in engineering practice, it seldom makes sense to compute all possible digits of a value, since after a certain point, these digits represent values too small to have a practical impact or even smaller than the precision of the instruments used to measure them. For these reasons, iterating until the exact equality condition of Eq. (3.3) is usually neither possible nor desirable. It is preferable to iterate until two successive values are approximately equal within a relative error, the value of which is problem dependent. Recall that relative error was defined in Eq. (1.2) to be the difference between an approximation and a real value as a ratio of the real value. In that definition, it is possible for an engineer to set a threshold relative error, below which an approximation is close enough to the real value to be considered equal for all intents and purposes. However, Eq. (1.2) requires knowledge of the real value, which is unknown in this current situation; the iterating function is being used to refine an approximation of the value and would be entirely unnecessary if the exact value was known to begin with! Nonetheless, if the function is converging, then each successive value is more exact than the previous one, and it is possible to rewrite Eq. (1.2) to use that fact as follows:

$$E_{rel} = \left| \frac{\text{previous value} - \text{current value}}{\text{current value}} \right| \tag{3.4}$$

$$E_i = \left| \frac{x_{i-1} - x_i}{x_i} \right| \tag{3.5}$$

In that definition, when the values computed in two successive iterations have a relative error E_i less than a preset threshold, they are close enough to be considered equal for all intents and purposes, and the function has successfully converged.

If the algorithm iterates to discover the value of a vector instead of a scalar, a better test for convergence is to compute the *Euclidean distance* between two successive vectors. This distance is the square root of the sum of squared differences between each pair of values in the two vectors. More formally, given two successive $n \times 1$ vectors $\mathbf{x}_i = [x_{i,0}, \ldots, x_{i,n-1}]^T$ and $\mathbf{x}_{i-1} = [x_{i-1,0}, \ldots, x_{i-1,n-1}]^T$, the Euclidean distance is defined as

$$E_i = \|\mathbf{x}_{i-1} - \mathbf{x}_i\| = \sqrt{(x_{i-1,0} - x_{i,0})^2 + \cdots + (x_{i-1,n-1} - x_{i,n-1})^2} \tag{3.6}$$

Once again, when two successive vectors have a distance E_i less than a preset threshold, they are close enough to be considered equal for all intents and purposes, and the function has successfully converged.

An iterative function fails to converge if it does not reach a success condition in a reasonable amount of time. A simple catch-all failure condition would be to set a maximum number of iterations and to terminate the function if that number is reached. Note however that this does not provide any feedback on why the function failed to converge. There are in fact several possible reasons why that could have happened. It could be that the function cannot possibly converge, that it is a diverging function, and that a solution is impossible to find. Or the problem might be the initial value chosen; it might be in a diverging region of the function or even on a singularity (a value that causes a division by zero). It could even be the case that the function was converging, but too slowly to reach the success condition in the set maximum number of iterations. In that case, it would be necessary to increase the maximum number of iterations allowed or to redesign the mathematical formulas to increase the convergence rate somehow. The important conclusion to retain is that a function failing to converge after a set number of iterations is not synonymous with a function never converging and no solution existing for the problem. This is important to remember because this failure condition, of setting a maximum number of iterations, is one of the most commonly used ones in practice.

Putting it all together, the pseudo-code of an iterative algorithm to compute the iterations of Eq. (3.2) using a threshold on the relative error of Eq. (3.5) as a success condition and a maximum number of iterations as a failure condition is given in Fig. 3.1.

```
CurrentValue ← Initial Value
IterationCounter ← 0
IterationMaximum ← Maximum Iteration Threshold
ErrorMinimum ← Minimum Relative Error Threshold

WHILE (TRUE):
    PreviousValue ← CurrentValue
    CurrentValue ← CALL F(CurrentValue)
    IterationCounter ← IterationCounter + 1
    CurrentError ← absolute((PreviousValue - CurrentValue)/ CurrentValue)

    IF (CurrentError <= ErrorMinimum)
        RETURN Success
    ELSE IF(IterationCounter = IterationMaximum)
        RETURN Failure
    END IF
END WHILE

FUNCTION F(x)
    RETURN compute a result given an input variable x
END FUNCTION
```

Fig. 3.1 Pseudo-code of an iterative software

Example 3.2
Iterate $f(x) = \frac{x - (\sin(x) + 0.5)}{\cos(x)}$ in radians using $x_0 = 0.5$, until the relative error is 10^{-5} or up to a maximum of ten iterations.

Solution
Since the target relative error is 10^{-5}, it is necessary to keep six digits of precision in the values. Any less would make it impossible to compute the error, while more digits would be unnecessary.

The results of the iterations, presented in the table below, show that the threshold relative error has been surpassed after four iterations. The function has converged and reached the success condition. It is unnecessary to compute the remaining six iterations.

Iteration	Value	Relative error
x_0	0.5	N/A
x_1	−0.616049	1.811624
x_2	−0.520707	0.183101
x_3	−0.523596	0.005518
x_4	−0.523599	0.000006

3.4   Summary

This chapter has introduced the concept of iteration, the first of the five mathematical tools that will underlie the numerical methods and modelling algorithms of later chapters. While iteration is a simple tool, it is also a very versatile technique and will be used in most algorithms in coming chapters. This chapter has introduced notions related to iterations, such as the notion of convergence, divergence, and divergence rate and the notions of success and failure halting conditions.

3.5   Exercises

1. What value does the function $f(x) = 2.4x(x - 1)$ converge to? How many iterations do you need to get the equality condition of Eq. (3.3)?
2. Starting with $x_0 = 0.5$, compute iterations of $f(x) = \sin(x)$ and $f(x) = \cos(x)$. Which converges faster? Is the difference significant?
3. Consider $f(x) = x^3$. For which range of values will the function converge or diverge?
4. Consider the function $f(x) = x + \sin(x)$, where the sin function is in radians (Beeler et al. 1972). Starting from $x_0 = 0.5$, compute the value of x_i and its relative error as an approximation of π over five iterations.

5. Consider the function $f(x) = (3x^4 + 10x^3 - 20x^2 - 24)/(4x^3 + 15x^2 - 40x)$. Starting from $x_0 = 5$, compute the value of x_i and its relative error as an approximation of 2 over five iterations.

6. Consider the following functions. How many values can each one converge to? What are they?

   (a) $f(x) = 1.2x^{-1} + 0.8$.
   (b) $f(x) = (x - 8)/3$.
   (c) $f(x) = (0.5x^2 - 10)/(x-6)$.
   (d) $f(x) = (x^2 + 0.4)/(2x + 0.6)$.
   (e) $f(x) = (2x^3 - 10.3x^2 - 36.5)/(3x^2 - 20.6x + 9.7)$.

Chapter 4
Linear Algebra

4.1 Introduction

The first of the five mathematical modelling tools, introduced in the previous chapter, is iteration. The second is solving systems of linear algebraic equations and is the topic of this chapter. A system of linear algebraic equations is any set of n equations with n unknown variables $x_0, \ldots, x_{n-1}$:

$$
\begin{aligned}
m_{0,0}x_0 + m_{0,1}x_1 + \cdots + m_{0,n-1}x_{n-1} &= b_0 \\
m_{1,0}x_0 + m_{1,1}x_1 + \cdots + m_{1,n-1}x_{n-1} &= b_1 \\
&\vdots \\
m_{n-1,0}x_0 + m_{n-1,1}x_1 + \cdots + m_{n-1,n-1}x_{n-1} &= b_{n-1}
\end{aligned}
\tag{4.1}
$$

where the $n \times n$ values $m_{0,0}, \ldots, m_{n-1,n-1}$ are known coefficient values that multiply the variables, and the n values $b_0, \ldots, b_{n-1}$ are the known result of each equation. A system of that form can arise in many ways in engineering practice. For example, it would be the result of taking measurements of a dynamic system at n different times. It also results from taking measurements of a static system at n different internal points. Consider, for example, the simple electrical circuit in Fig. 4.1. Four internal nodes have been identified in it. If one wants to model this circuit, for example, to be able to predict the voltage flowing between two nodes, the corresponding energy losses, or other of its properties, it is first necessary to model the voltages at each node using Kirchhoff's current law. Removing units and with appropriate scaling, this gives the following set of four equations and four unknown variables:

© Springer International Publishing Switzerland 2016
R. Khoury, D.W. Harder, *Numerical Methods and Modelling for Engineering*,
DOI 10.1007/978-3-319-21176-3_4

Fig. 4.1 Example electrical
circuit

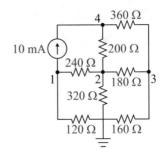

Node 1 : $\dfrac{v_1 - 0}{120} + \dfrac{v_1 - v_2}{240} = -0.01$

Node 2 : $\dfrac{v_2 - 0}{320} + \dfrac{v_2 - v_1}{240} + \dfrac{v_2 - v_3}{180} + \dfrac{v_2 - v_4}{200} = 0$

Node 3 : $\dfrac{v_3 - 0}{160} + \dfrac{v_3 - v_2}{180} + \dfrac{v_3 - v_4}{360} = 0$ (4.2)

Node 4 : $\dfrac{v_4 - v_2}{200} + \dfrac{v_4 - v_3}{360} = 0.01$

which is a system of four linear algebraic equations of the same form as Eq. (4.1).
The linear system of Eq. (4.1) can be written as a classic matrix-vector problem:

$$\mathbf{Mx=b} \qquad (4.3)$$

where $\mathbf{M}$ is the $n \times n$ matrix of known coefficients, $\mathbf{x}$ is an $n \times 1$ column vector of
unknowns, and $\mathbf{b}$ is an $n \times 1$ column vector of known equation result values. The
sample system given in Eq. (4.2), for example, would be written as

$$
\begin{bmatrix}
\dfrac{1}{120} + \dfrac{1}{240} & -\dfrac{1}{240} & 0 & 0 \\[3mm]
-\dfrac{1}{240} & \dfrac{1}{320} + \dfrac{1}{240} + \dfrac{1}{180} + \dfrac{1}{200} & -\dfrac{1}{180} & -\dfrac{1}{200} \\[3mm]
0 & -\dfrac{1}{180} & \dfrac{1}{160} + \dfrac{1}{180} + \dfrac{1}{360} & -\dfrac{1}{360} \\[3mm]
0 & -\dfrac{1}{200} & -\dfrac{1}{360} & \dfrac{1}{200} + \dfrac{1}{360}
\end{bmatrix}
\begin{bmatrix}
v_1 \\ v_2 \\ v_3 \\ v_4
\end{bmatrix}
$$

$$
=
\begin{bmatrix}
-0.01 \\ 0 \\ 0 \\ 0.01
\end{bmatrix}
$$

$$(4.4)$$

and each of the node equations can be recovered by multiplying the corresponding row of the matrix by the vector of unknowns and keeping the matching result value. Writing the system in matrix-vector form makes it easier to solve and discover the value of each unknown variable.

Some readers may have learned how to solve a system of linear equations using Gaussian elimination together with backward substitution, possibly in a previous course on linear algebra. However, there are two problems with Gaussian elimination. The first is its lack of efficiency. Even an optimal implementation of a Gaussian elimination and backward substitution algorithm for solving a system of n linear equations will require $(n^3/3 - n/3)$ multiplications and additions and $(n^2/2 - n/2)$ divisions and negations for the Gaussian elimination step in addition to $(n^2/2 - n/2)$ multiplications and subtractions and n divisions for the backward substitution step. In other words, it is an algorithm with $O(n^3)$ time complexity, which is very inefficient: the computation time required to solve a linear system will grow proportionally to the cube of the size of the system! Doubling the size of a system will require eight times more computations to solve, and trying to solve a system with 10 times more equations and unknowns will take 1000 times as long. The second problem with Gaussian elimination is related to the step that requires adding a multiple of one row to the others. In any situation where the coefficients of one row are a lot bigger in magnitude than those of another, and given a finite number of digits of precision, the algorithm will suffer from the problem of non-associativity of addition explained in Chap. 2. When that happens, the values computed for the variables by Gaussian elimination will end up having a very high relative error compared to the real values that should have been obtained.

This chapter will introduce better methods for solving the $\mathbf{Mx} = \mathbf{b}$ system, both for general matrices and for some special cases. The ability to solve linear systems will be important for other mathematical tools and for numerical methods, as it will make it possible to easily, efficiently, and accurately solve complex systems.

4.2 PLU Decomposition

The *PLU decomposition*, also called the *LUP decomposition* or the *PLU factorization*, is an improvement on the Gaussian elimination technique. It addresses the two problems highlighted in the introduction: by decomposing the matrix $\mathbf{M}$ into a lower triangular matrix $\mathbf{L}$ and an upper triangular matrix $\mathbf{U}$, it can solve the system in $O(n^2)$ instead of $O(n^3)$. And by doing a permutation of the rows so that the element with the maximum absolute value is always in the diagonal and keeping track of these changes in a permutation matrix $\mathbf{P}$, it can avoid the non-associativity problem.

The PLU decomposition technique thus works in two steps: first, decomposing the matrix $\mathbf{M}$ into three equivalent matrices:

```
M ← Input n×n matrix
P ← n×n identity matrix
L ← n×n zero matrix
U ← M

ColumnIndex ← 0
WHILE (ColumnIndex < n-1)

    ColumnVector ← columnColumnIndex, rows ColumnIndex to n of U
    IndexOfMaximum ← index of maximum absolute value in ColumnVector

    P ← swap row ColumnIndex and row IndexOfMaximum in P
    L ← swap row ColumnIndex and row IndexOfMaximum in L
    U ← swap row ColumnIndex and row IndexOfMaximum in U

    RowIndex ← ColumnIndex + 1
    WHILE (RowIndex < n)
        s ← -1 ×(element at row RowIndex, column ColumnIndex of U) /
              (element at row RowIndex, column ColumnIndex of U)
        row Update of U ← (row RowIndex of U) + s ×(row ColumnIndex of U)
        element at row RowIndex, column ColumnIndex of L ← -1 × s
        RowIndex ← RowIndex + 1
    END WHILE

    ColumnIndex ← ColumnIndex + 1
END WHILE
L ← L + (n×n identity matrix)

RETURN P, L, U
```

Fig. 4.2 Pseudo-code of the PLU decomposition

$$\mathbf{M} = \mathbf{PLU} \tag{4.5}$$

then solving the $\mathbf{PLUx} = \mathbf{b}$ system using simple forward and backward substitutions.

There is a simple step-by-step algorithm to decompose the matrix $\mathbf{M}$ into $\mathbf{L}$, $\mathbf{U}$, and $\mathbf{P}^{\mathrm{T}}$. Note that the algorithm decomposes into the transpose of the permutation matrix $\mathbf{P}$; later the forward and backward substitution operations will need this transposed matrix, so this actually saves a bit of time in the overall process. A pseudo-code version of this algorithm is presented in Fig. 4.2.

Step 1: Initialization. Each of the three decomposition matrices will have initial values. The matrix $\mathbf{U}$ is initially equal to $\mathbf{M}$, the matrix $\mathbf{L}$ is initially an $n \times n$ zero matrix, and the matrix $\mathbf{P}^{\mathrm{T}}$ is initially an $n \times n$ identity matrix.

Step 2: Decomposition. The algorithm considers each column in turn, working left to right from column 0 to the penultimate column $n - 2$. For the current column i, find the row j in the matrix $\mathbf{U}$ such that the element $u_{j,i}$ has the greatest absolute value in the column and $j \geq i$, meaning that the element is on or below the diagonal element of the column. If that element is zero, then the matrix is singular and cannot be decomposed with this method, and the algorithm terminates. Next, swap rows i and j in all three matrices $\mathbf{U}$, $\mathbf{L}$, and $\mathbf{P}^{\mathrm{T}}$. This will bring the greatest-valued element found on the diagonal of matrix $\mathbf{U}$ as element $u_{i,i}$. Finally, for every row k below the diagonal, calculate a scalar value $s = u_{k,i}/u_{i,i}$. Save the value of s at element $l_{k,i}$ in matrix $\mathbf{L}$ in order to fill the entries below the diagonal in that matrix, and add $-s$ times row i to row k in matrix $\mathbf{U}$. This addition will cause every value in column i under the diagonal of $\mathbf{U}$ to become 0.

Step 3: Finalization. Once the decomposition step has been done for every column except the last right-most one, the algorithm ends by adding an $n \times n$ identity matrix to the lower-diagonal matrix $\mathbf{L}$.

Once the matrix $\mathbf{M}$ has been decomposed, the $\mathbf{PLUx} = \mathbf{b}$ system can be solved in two steps. The matrix-vector $\mathbf{Ux}$ is replaced with a vector $\mathbf{y}$ in the equation, which can be computed with a forward substitution step:

$$\mathbf{Ly} = \mathbf{P}^{\mathrm{T}}\mathbf{b} \tag{4.6}$$

then knowing $\mathbf{y}$, a backward substitution step can compute $\mathbf{x}$:

$$\mathbf{Ux} = \mathbf{y} \tag{4.7}$$

Example 4.1
Use PLU decomposition to solve the following matrix-vector problem. Keep two decimals of precision:

$$\begin{bmatrix} 0.7 & 4 & -7.4 & 4.3 \\ -1.4 & 8 & 2.8 & -0.6 \\ 7 & 0 & 1 & -2 \\ 1.4 & 0.8 & 2.75 & 6.25 \end{bmatrix} \begin{bmatrix} x_0 \\ x_1 \\ x_2 \\ x_3 \end{bmatrix} = \begin{bmatrix} 7 \\ 4 \\ 8 \\ 3 \end{bmatrix}$$

Solution
Initialize the values of the matrices $\mathbf{P}^{\mathrm{T}}$, $\mathbf{L}$, and $\mathbf{U}$:

$$\mathbf{P}^{\mathrm{T}} = \begin{bmatrix} 1 & 0 & 0 & 0 \\ 0 & 1 & 0 & 0 \\ 0 & 0 & 1 & 0 \\ 0 & 0 & 0 & 1 \end{bmatrix} \quad \mathbf{L} = \begin{bmatrix} 0 & 0 & 0 & 0 \\ 0 & 0 & 0 & 0 \\ 0 & 0 & 0 & 0 \\ 0 & 0 & 0 & 0 \end{bmatrix} \quad \mathbf{U} = \begin{bmatrix} 0.7 & 4 & -7.4 & 4.3 \\ -1.4 & 8 & 2.8 & -0.6 \\ 7 & 0 & 1 & -2 \\ 1.4 & 0.8 & 2.75 & 6.25 \end{bmatrix}$$

Find the element on or under the diagonal in column 0 that has the largest absolute value. In this case, it is element $u_{2,0}$, so rows 0 and 2 need to be swapped in all three matrices.

(continued)

Example 4.1 (continued)

$$\mathbf{P}^T = \begin{bmatrix} 0 & 0 & 1 & 0 \\ 0 & 1 & 0 & 0 \\ 1 & 0 & 0 & 0 \\ 0 & 0 & 0 & 1 \end{bmatrix} \quad \mathbf{L} = \begin{bmatrix} 0 & 0 & 0 & 0 \\ 0 & 0 & 0 & 0 \\ 0 & 0 & 0 & 0 \\ 0 & 0 & 0 & 0 \end{bmatrix} \quad \mathbf{U} = \begin{bmatrix} 7 & 0 & 1 & -2 \\ -1.4 & 8 & 2.8 & -0.6 \\ 0.7 & 4 & -7.4 & 4.3 \\ 1.4 & 0.8 & 2.75 & 6.25 \end{bmatrix}$$

Then, for each row under the diagonal, compute the scalar value s. For row 1, the value is $s = -1.4/7 = -0.2$. That value will take position $l_{1,0}$ in matrix $\mathbf{L}$. Multiplying $-s$ times row 0 gives $[1.4, 0, 0.2, -0.4]$, and adding that to row 1 gives $[0, 8, 3, -1]$. For row 2 the scalar value is $s = 0.7/7 = 0.1$, which is saved as $l_{2,0}$; multiplying $-s$ by row 0 gives $[-0.7, 0, -0.1, 0.2]$, and adding that to row 2 gives $[0, 4, -7.5, 4.5]$. Finally, for row 3, $s = 1.4/7 = 0.2$, which is saved as $l_{3,0}$; multiplying $-s$ by row 0 gives $[-1.4, 0, -0.2, 0.4]$, and adding that to row 3 gives $[0, 0.8, 2.55, 6.65]$. The resulting matrices are

$$\mathbf{P}^T = \begin{bmatrix} 0 & 0 & 1 & 0 \\ 0 & 1 & 0 & 0 \\ 1 & 0 & 0 & 0 \\ 0 & 0 & 0 & 1 \end{bmatrix} \quad \mathbf{L} = \begin{bmatrix} 0 & 0 & 0 & 0 \\ -0.2 & 0 & 0 & 0 \\ 0.1 & 0 & 0 & 0 \\ 0.2 & 0 & 0 & 0 \end{bmatrix} \quad \mathbf{U} = \begin{bmatrix} 7 & 0 & 1 & -2 \\ 0 & 8 & 3 & -1 \\ 0 & 4 & -7.5 & 4.5 \\ 0 & 0.8 & 2.55 & 6.65 \end{bmatrix}$$

Moving on to column 1, the largest absolute value element of that column in $\mathbf{U}$ is already on the diagonal, so no swapping takes place. For row 2, the scalar computed is $s = 4/8 = 0.5$, and for row 3 it is $s = 0.8/8 = 0.1$. These values are saved in matrix $\mathbf{L}$ as $l_{2,1}$ and $l_{3,1}$, respectively, and after adding multiplied versions of row 1 in $\mathbf{U}$, the resulting matrices are

$$\mathbf{P}^T = \begin{bmatrix} 0 & 0 & 1 & 0 \\ 0 & 1 & 0 & 0 \\ 1 & 0 & 0 & 0 \\ 0 & 0 & 0 & 1 \end{bmatrix} \quad \mathbf{L} = \begin{bmatrix} 0 & 0 & 0 & 0 \\ -0.2 & 0 & 0 & 0 \\ 0.1 & 0.5 & 0 & 0 \\ 0.2 & 0.1 & 0 & 0 \end{bmatrix}$$

$$\mathbf{U} = \begin{bmatrix} 7 & 0 & 1 & -2 \\ 0 & 8 & 3 & -1 \\ 0 & 0 & -9 & 5 \\ 0 & 0 & 2.25 & 6.75 \end{bmatrix}$$

Moving on to column 2, the largest absolute value element of that column in $\mathbf{U}$ is again on the diagonal, so no swapping takes place. The scalar computed for row 3 is $s = 2.25/(-9) = -0.25$, which is saved in $\mathbf{L}$ as element $l_{3,2}$. Multiplying row 2 by $-s$ gives $[0, 0, -2.25, 1.25]$ and adding that to row 3 gives the following matrices:

(continued)

Example 4.1 (continued)

$$\mathbf{P}^{\mathrm{T}} = \begin{bmatrix} 0 & 0 & 1 & 0 \\ 0 & 1 & 0 & 0 \\ 1 & 0 & 0 & 0 \\ 0 & 0 & 0 & 1 \end{bmatrix} \quad \mathbf{L} = \begin{bmatrix} 0 & 0 & 0 & 0 \\ -0.2 & 0 & 0 & 0 \\ 0.1 & 0.5 & 0 & 0 \\ 0.2 & 0.1 & -0.25 & 0 \end{bmatrix} \quad \mathbf{U} = \begin{bmatrix} 7 & 0 & 1 & -2 \\ 0 & 8 & 3 & -1 \\ 0 & 0 & -9 & 5 \\ 0 & 0 & 0 & 8 \end{bmatrix}$$

Since this operation is not done on the final row, the decomposition step is over. The final step is to add a 4×4 identity matrix to $\mathbf{L}$, to get the final matrices:

$$\mathbf{P}^{\mathrm{T}} = \begin{bmatrix} 0 & 0 & 1 & 0 \\ 0 & 1 & 0 & 0 \\ 1 & 0 & 0 & 0 \\ 0 & 0 & 0 & 1 \end{bmatrix} \quad \mathbf{L} = \begin{bmatrix} 1 & 0 & 0 & 0 \\ -0.2 & 1 & 0 & 0 \\ 0.1 & 0.5 & 1 & 0 \\ 0.2 & 0.1 & -0.25 & 1 \end{bmatrix} \quad \mathbf{U} = \begin{bmatrix} 7 & 0 & 1 & -2 \\ 0 & 8 & 3 & -1 \\ 0 & 0 & -9 & 5 \\ 0 & 0 & 0 & 8 \end{bmatrix}$$

Now it is possible to solve the original system by substituting the matrix $\mathbf{M}$ for $\mathbf{PLU}$, then replacing $\mathbf{Ux}$ with a vector $\mathbf{y}$:

$$\mathbf{Mx} = \mathbf{b}$$

$$\mathbf{PLUx} = \mathbf{b}$$

$$\mathbf{PL(Ux)} = \mathbf{b}$$

$$\mathbf{PLy} = \mathbf{b}$$

$$\mathbf{Ly} = \mathbf{P}^{\mathrm{T}}\mathbf{b}$$

$$\begin{bmatrix} 1 & 0 & 0 & 0 \\ -0.2 & 1 & 0 & 0 \\ 0.1 & 0.5 & 1 & 0 \\ 0.2 & 0.1 & -0.25 & 1 \end{bmatrix} \begin{bmatrix} y_0 \\ y_1 \\ y_2 \\ y_3 \end{bmatrix} = \begin{bmatrix} 0 & 0 & 1 & 0 \\ 0 & 1 & 0 & 0 \\ 1 & 0 & 0 & 0 \\ 0 & 0 & 0 & 1 \end{bmatrix} \begin{bmatrix} 7 \\ 4 \\ 8 \\ 3 \end{bmatrix}$$

By forward substitution, it is easy to find the values of $\mathbf{y}$:

$$y_0 = 8$$
$$-0.2y_0 + y_1 = 4$$
$$0.1y_0 + 0.5y_1 + y_2 = 7$$
$$0.2y_0 + 0.1y_1 - 0.25y_2 + y_3 = 3$$

$$\begin{bmatrix} y_0 \\ y_1 \\ y_2 \\ y_3 \end{bmatrix} = \begin{bmatrix} 8.0 \\ 5.6 \\ 3.4 \\ 1.7 \end{bmatrix}$$

Then the system $\mathbf{Ux} = \mathbf{y}$ can be solved by backward substitution to get the value of $\mathbf{x}$:

(continued)

Example 4.1 (continued)

$$\begin{bmatrix} 7 & 0 & 1 & -2 \\ 0 & 8 & 3 & -1 \\ 0 & 0 & -9 & 5 \\ 0 & 0 & 0 & 8 \end{bmatrix} \begin{bmatrix} x_0 \\ x_1 \\ x_2 \\ x_3 \end{bmatrix} = \begin{bmatrix} 8.0 \\ 5.6 \\ 3.4 \\ 1.7 \end{bmatrix}$$

$$8x_3 = 1.7$$
$$-9x_2 + 5x_3 = 3.4$$
$$8x_1 + 3x_2 - x_3 = 5.6$$
$$7x_0 + x_2 - 2x_3 = 8.0$$

$$\begin{bmatrix} x_0 \\ x_1 \\ x_2 \\ x_3 \end{bmatrix} = \begin{bmatrix} 1.24 \\ 0.82 \\ -0.26 \\ 0.21 \end{bmatrix}$$

4.3   Cholesky Decomposition

Under certain circumstances, it is possible to decompose a matrix $\mathbf{M}$ into the form $\mathbf{LL}^T$. Such a decomposition is called a *Cholesky decomposition*. It requires half the memory and half the number operations of a PLU decomposition, since there is only one matrix $\mathbf{L}$ to compute. However, this technique can only be applied in specific circumstances, namely, when the matrix $\mathbf{M}$ is real, symmetric, and positive definite. The criterion of being real simply means that the matrix contains no complex numbers, and symmetric means that for all non-diagonal entries $m_{i,j} = m_{j,i}$. A positive-definite matrix that is also symmetric is one where all diagonal entries are positive, and each one is equal or greater than the sum of absolute values of all other entries in the row. While these criteria may seem restrictive, matrices of this form often arise in engineering practice. The matrix of Eq. (4.4), for example, is a real, symmetric, positive-definite matrix.

The easiest way to understand the Cholesky decomposition is to visualize the equality $\mathbf{M} = \mathbf{LL}^T$. This is presented in the case of a 4×4 matrix in Eq. (4.8). The matrix $\mathbf{M}$ is symmetric, but so is the product $\mathbf{LL}^T$. By inspection, a direct equality can be seen between each element $m_{i,j}$ in $\mathbf{M}$ and the corresponding element in $\mathbf{LL}^T$, which gives a simple equation of elements of $\mathbf{L}$.

$$\begin{bmatrix} m_{0,0} & m_{1,0} & m_{2,0} & m_{3,0} \\ m_{1,0} & m_{1,1} & m_{2,1} & m_{3,1} \\ m_{2,0} & m_{2,1} & m_{2,2} & m_{3,2} \\ m_{3,0} & m_{3,1} & m_{3,2} & m_{3,3} \end{bmatrix} = \begin{bmatrix} l_{0,0} & 0 & 0 & 0 \\ l_{1,0} & l_{1,1} & 0 & 0 \\ l_{2,0} & l_{2,1} & l_{2,2} & 0 \\ l_{3,0} & l_{3,1} & l_{3,2} & l_{3,3} \end{bmatrix} \begin{bmatrix} l_{0,0} & l_{1,0} & l_{2,0} & l_{3,0} \\ 0 & l_{1,1} & l_{2,1} & l_{3,1} \\ 0 & 0 & l_{2,2} & l_{3,2} \\ 0 & 0 & 0 & l_{3,3} \end{bmatrix}$$

$$
\begin{bmatrix}
m_{0,0} & m_{1,0} & m_{2,0} & m_{3,0} \\
m_{1,0} & m_{1,1} & m_{2,1} & m_{3,1} \\
m_{2,0} & m_{2,1} & m_{2,2} & m_{3,2} \\
m_{3,0} & m_{3,1} & m_{3,2} & m_{3,3}
\end{bmatrix}
$$

$$
=
\begin{bmatrix}
l_{0,0}^2 & l_{0,0}l_{1,0} & l_{0,0}l_{2,0} & l_{0,0}l_{3,0} \\
l_{0,0}l_{1,0} & l_{1,0}^2 + l_{1,1}^2 & l_{1,0}l_{2,0} + l_{1,1}l_{2,1} & l_{1,0}l_{3,0} + l_{1,1}l_{3,1} \\
l_{0,0}l_{2,0} & l_{1,0}l_{2,0} + l_{1,1}l_{2,1} & l_{2,0}^2 + l_{2,1}^2 + l_{2,2}^2 & l_{2,0}l_{3,0} + l_{2,1}l_{3,1} + l_{2,2}l_{3,2} \\
l_{0,0}l_{3,0} & l_{1,0}l_{3,0} + l_{1,1}l_{3,1} & l_{2,0}l_{3,0} + l_{2,1}l_{3,1} + l_{2,2}l_{3,2} & l_{3,0}^2 + l_{3,1}^2 + l_{3,2}^2 + l_{3,3}^2
\end{bmatrix}
\tag{4.8}
$$

Furthermore, looking at the matrix $\mathbf{LL}^{\mathrm{T}}$ column by column (or row by row, since it is symmetric), it can be seen that each element $l_{i,j}$ can be discovered by forward substitution in order, starting from element $l_{0,0}$. Column 0 gives the set of equations:

$$
l_{0,0} = \sqrt{m_{0,0}} \tag{4.9}
$$

$$
l_{1,0} = \frac{m_{1,0}}{l_{0,0}} \tag{4.10}
$$

$$
l_{2,0} = \frac{m_{2,0}}{l_{0,0}} \tag{4.11}
$$

$$
l_{3,0} = \frac{m_{3,0}}{l_{0,0}} \tag{4.12}
$$

Then column 1 gives

$$
l_{1,1} = \sqrt{m_{1,1} - l_{1,0}^2} \tag{4.13}
$$

$$
l_{2,1} = \frac{m_{2,1} - l_{1,0}l_{2,0}}{l_{1,1}} \tag{4.14}
$$

$$
l_{3,1} = \frac{m_{3,1} - l_{1,0}l_{3,0}}{l_{1,1}} \tag{4.15}
$$

And columns 2 and 3 give

$$
l_{2,2} = \sqrt{m_{2,2} - l_{2,0}^2 - l_{2,1}^2} \tag{4.16}
$$

$$
l_{3,2} = \frac{m_{3,2} - l_{2,0}l_{3,0} - l_{2,1}l_{3,1}}{l_{2,2}} \tag{4.17}
$$

$$
l_{3,3} = \sqrt{m_{3,3} - l_{3,0}^2 - l_{3,1}^2 - l_{3,2}^2} \tag{4.18}
$$

The step-by-step algorithm to construct the **L** matrix is thus apparent. For an $n \times n$ matrix **M**, take each column j in order from 0 to $n-1$. Then for each element i where $j \leq i \leq n-1$:

$$l_{i,j} = \begin{cases} \sqrt{m_{i,j} - \sum_{k=0}^{j-1} l_{i,k}^2} & \text{if } i = j \\ \dfrac{m_{i,j} - \sum_{k=0}^{j-1} l_{j,k} l_{i,k}}{l_{j,j}} & \text{if } i \neq j \end{cases} \qquad (4.19)$$

In other words, if the current element being computed is on the diagonal of **L**, subtract from the corresponding diagonal element of **M** the dot product of the current row of **L** (as constructed so far) with itself and take the square root of this result. If the current element being computed is below the diagonal of **L**, subtract from the corresponding element of **M** the dot product of the current row and current column of **L** (as constructed so far), and divide this result by the column's diagonal entry. The pseudo-code for an algorithm to compute this decomposition is given in Fig. 4.3.

Once the matrix **M** has been decomposed, the system can be solved in a manner very similar to the one used in the PLU decomposition technique. The matrix **M** is replaced by its decomposition in the **Mx** = **b** system to get

```
M ← Input nxn matrix
L ← nxn zero matrix

Index ← 0
WHILE (Index < n)

    element at row Index column Index of L ← square root of [(element at
            row Index column Index of M ) - (sum of: square of element at
            row 0 to Index column Index of L) ]

    BelowIndex ← Index + 1
    WHILE (BelowIndex < n)

        element at row BelowIndex column Index of L ← [(element at row
            BelowIndex column Index of M ) - (sum of: element at row
            0 to Index column Index of L × element at row 0 to Index
            column BelowIndex of L) ] / (element at row Index column
            Index of L)
        BelowIndex ← BelowIndex + 1

    END WHILE

    Index ← Index + 1
END WHILE

RETURN L
```

Fig. 4.3 Pseudo-code of the Cholesky decomposition

$$LL^Tx=b \qquad (4.20)$$

which can then be solved by replacing the matrix-vector L^Tx with a vector y and performing a forward substitution step:

$$Ly=b \qquad (4.21)$$

followed by a backward substitution step to compute x:

$$L^Tx=y \qquad (4.22)$$

Example 4.2
Use Cholesky decomposition to solve the system of Eq. (4.4). Keep four decimals of precision.

Solution
Begin by writing the matrix M with four digits of precision, to make it easier to work with:

$$M = \begin{bmatrix} 0.0125 & -0.0042 & 0 & 0 \\ -0.0042 & 0.0178 & -0.0056 & -0.0050 \\ 0 & -0.0056 & 0.0146 & -0.0028 \\ 0 & -0.0050 & -0.0028 & 0.0078 \end{bmatrix}$$

Since it is a 4×4 matrix, it could be decomposed using the general formula of Eq. (4.19), or the expanded set of equations from Eqs. (4.9) to (4.18) which is obtained by applying Eq. (4.19) to each of the ten entries of L. The values computed are

$$l_{0,0} = \sqrt{0.0125} = 0.1118$$

$$l_{1,0} = \frac{-0.0042}{0.1118} = -0.0373$$

$$l_{2,0} = \frac{0}{0.1118} = 0$$

$$l_{3,0} = \frac{0}{0.1118} = 0$$

$$l_{1,1} = \sqrt{0.0178 - (-0.0373)^2} = 0.1283$$

$$l_{2,1} = \frac{-0.0056 - (-0.0373)(0)}{0.1283} = -0.0433$$

(continued)

Example 4.2 (continued)

$$l_{3,1} = \frac{-0.0050 - (-0.0373)(0)}{0.1283} = -0.0390$$

$$l_{2,2} = \sqrt{0.0146 - (0)^2 - (-0.0433)^2} = 0.1127$$

$$l_{3,2} = \frac{-0.0028 - (0)(0) - (-0.0433)(-0.0390)}{0.1127} = -0.0396$$

$$l_{3,3} = \sqrt{0.0078 - (0)^2 - (-0.0390)^2 - (-0.0369)^2} = 0.0685$$

The matrix **M** thus decomposes into the following matrix **L**:

$$\mathbf{L} = \begin{bmatrix} 0.1118 & 0 & 0 & 0 \\ -0.0373 & 0.1283 & 0 & 0 \\ 0 & -0.0433 & 0.1127 & 0 \\ 0 & -0.0390 & -0.0396 & 0.0685 \end{bmatrix}$$

The forward substitution step to compute the vector **y** is the following:

$$\begin{bmatrix} 0.1118 & 0 & 0 & 0 \\ -0.0373 & 0.1283 & 0 & 0 \\ 0 & -0.0433 & 0.1127 & 0 \\ 0 & -0.0390 & -0.0396 & 0.0685 \end{bmatrix} \begin{bmatrix} y_0 \\ y_1 \\ y_2 \\ y_3 \end{bmatrix} = \begin{bmatrix} -0.01 \\ 0 \\ 0 \\ 0.01 \end{bmatrix}$$

and the vector **y** obtained is $[-0.0894, -0.0260, -0.0100, 0.1255]^T$. This vector is then used for the backward substitution step (keeping the vector **v** from Eq. (4.4) instead of the vector **x** from Eq. (4.22)):

$$\begin{bmatrix} 0.1118 & -0.0373 & 0 & 0 \\ 0 & 0.1283 & -0.0433 & -0.0390 \\ 0 & 0 & 0.1127 & -0.0396 \\ 0 & 0 & 0 & 0.0685 \end{bmatrix} \begin{bmatrix} v_1 \\ v_2 \\ v_3 \\ v_4 \end{bmatrix} = \begin{bmatrix} -0.0894 \\ -0.0260 \\ -0.0100 \\ 0.1255 \end{bmatrix}$$

to get the values of the voltage vector $\mathbf{v} = [-0.6195, 0.5415, 0.5553, 1.8321]^T$, the values that solve the system and model the circuit of Fig. 4.1.

4.4   Jacobi Method

The PLU decomposition and Cholesky decomposition methods make it possible to solve $\mathbf{Mx} = \mathbf{b}$ systems provided no information on what **x** may be. However, if information on **x** is available, for instance, if a similar system has been solved in the past, it could be beneficial to make use of it. The *Jacobi method* is an iterative algorithm for solving $\mathbf{Mx} = \mathbf{b}$ systems provided some initial estimate of the value of **x**.

It is important to note though that the Jacobi method can be used even without prior knowledge of $\mathbf{x}$, by using a random vector or a zero vector. It will take longer to converge to a solution in that case, but may still be more efficient than the PLU decomposition, especially for very large systems, and it is not restricted to positive-definite matrices like the Cholesky decomposition. The only requirement to use the Jacobi method is that the matrix $\mathbf{M}$ must have non-zero diagonal entries.

The Jacobi method begins by decomposing the matrix $\mathbf{M}$ into the sum of two matrices $\mathbf{D}$ and $\mathbf{E}$, where $\mathbf{D}$ contains the diagonal entries of $\mathbf{M}$ and zeros everywhere else, and $\mathbf{E}$ contains the off-diagonal entries of $\mathbf{M}$ and zeros on the diagonal. For a simple 3×3 example:

$$\mathbf{M} = \mathbf{D} + \mathbf{E}$$

$$\begin{bmatrix} a & b & c \\ d & e & f \\ g & h & i \end{bmatrix} = \begin{bmatrix} a & 0 & 0 \\ 0 & e & 0 \\ 0 & 0 & i \end{bmatrix} + \begin{bmatrix} 0 & b & c \\ d & 0 & f \\ g & h & 0 \end{bmatrix} \tag{4.23}$$

This makes it possible to rewrite the system in this way:

$$\begin{aligned} \mathbf{Mx} &= \mathbf{b} \\ (\mathbf{D} + \mathbf{E})\mathbf{x} &= \mathbf{b} \\ \mathbf{Dx} + \mathbf{Ex} &= \mathbf{b} \\ \mathbf{x} &= \mathbf{D}^{-1}(\mathbf{b} - \mathbf{Ex}) \end{aligned} \tag{4.24}$$

where the inverse matrix $\mathbf{D}^{-1}$ is the *reciprocal matrix* of $\mathbf{D}$, defined as

$$\mathbf{DD}^{-1} = \mathbf{I} \tag{4.25}$$

Moreover, in the special case where $\mathbf{D}$ is already a diagonal matrix, the reciprocal $\mathbf{D}^{-1}$ is simply a diagonal matrix containing the inverse of each diagonal scalar value:

$$\begin{bmatrix} a & 0 & 0 \\ 0 & e & 0 \\ 0 & 0 & i \end{bmatrix} \begin{bmatrix} \dfrac{1}{a} & 0 & 0 \\ 0 & \dfrac{1}{e} & 0 \\ 0 & 0 & \dfrac{1}{i} \end{bmatrix} = \begin{bmatrix} 1 & 0 & 0 \\ 0 & 1 & 0 \\ 0 & 0 & 1 \end{bmatrix} \tag{4.26}$$

The Jacobi method implements an iterative algorithm using Eq. (4.24) to refine the estimate of $\mathbf{x}$ at every iteration k as follows:

$$\mathbf{x}_{k+1} = \mathbf{D}^{-1}(\mathbf{b} - \mathbf{Ex}_k) \tag{4.27}$$

```
x ← Input initial value vector of length n
b ← Input solution vector of length n
M ← Input n×n matrix
IterationMaximum ← Input maximum number of iterations
ErrorMinimum ← Input minimum relative error

D ← n×n zero matrix
diagonal elements of D ← 1 / (diagonal elements of M)
E ← M
Diagonal elements of E ← 0

IterationCounter ← 0
WHILE (TRUE)

    PreviousValue ← x
    x ← D × (b - E × x)
    CurrentError ← Euclidean distance between x and PreviousValue
    IterationCounter ← IterationCounter + 1

    IF (CurrentError <= ErrorMinimum)
        RETURN Success, x
    ELSE IF (IterationCounter = IterationMaximum)
        RETURN Failure
    END IF

END WHILE
```

Fig. 4.4 Pseudo-code of the Jacobi method

The iterations continue until one of two halting conditions is reached: either the Euclidean distance (see Chap. 3) between two successive iterations of x_k is less than a target error, in which case the algorithm has converged to a good estimate of **x**, or k increments to a preset maximum number of iterations, in which case the method has failed to converge. The pseudo-code for this algorithm is presented in Fig. 4.4.

Example 4.3
Use the Jacobi method to solve the following system to an accuracy of 0.1, keeping two decimals of precision and starting with a zero vector.

$$\begin{bmatrix} 5 & 2 & -1 \\ 3 & 7 & 3 \\ 1 & -4 & 6 \end{bmatrix} \begin{bmatrix} x_0 \\ x_1 \\ x_2 \end{bmatrix} = \begin{bmatrix} 2 \\ -1 \\ 1 \end{bmatrix}$$

Solution
Begin by decomposing the matrix into diagonal and off-diagonal matrices:

(continued)

Example 4.3 (continued)

$$\begin{bmatrix} 5 & 2 & -1 \\ 3 & 7 & 3 \\ 1 & -4 & 6 \end{bmatrix} = \begin{bmatrix} 5 & 0 & 0 \\ 0 & 7 & 0 \\ 0 & 0 & 6 \end{bmatrix} + \begin{bmatrix} 0 & 2 & -1 \\ 3 & 0 & 3 \\ 1 & -4 & 0 \end{bmatrix}$$

Then build the iterative equation:

$$\begin{bmatrix} x_{k+1,0} \\ x_{k+1,1} \\ x_{k+1,2} \end{bmatrix} = \begin{bmatrix} \frac{1}{5} & 0 & 0 \\ 0 & \frac{1}{7} & 0 \\ 0 & 0 & \frac{1}{6} \end{bmatrix} \left(\begin{bmatrix} 2 \\ -1 \\ 1 \end{bmatrix} - \begin{bmatrix} 0 & 2 & -1 \\ 3 & 0 & 3 \\ 1 & -4 & 0 \end{bmatrix} \begin{bmatrix} x_{k,0} \\ x_{k,1} \\ x_{k,2} \end{bmatrix} \right)$$

It can be seen that the zeros off-diagonal in $\mathbf{D}^{-1}$ will simplify the equations a lot. In fact, the three values of $\mathbf{x}_{k+1}$ will be computed by these simple equations:

$$x_{k+1,0} = \frac{1}{5}(2 - (2x_{k,1} - 1x_{k,2}))$$

$$x_{k+1,1} = \frac{1}{7}(-1 - (3x_{k,0} + 3x_{k,2}))$$

$$x_{k+1,2} = \frac{1}{6}(1 - (1x_{k,0} - 4x_{k,1}))$$

Starting with $\mathbf{x}_0 = [0\,0\,0]^{\mathrm{T}}$, the first iteration will give $\mathbf{x}_1 = [0.40\,-0.14\,0.17]^{\mathrm{T}}$. The Euclidean distance between these two iterations is

$$E_1 = \|\mathbf{x}_1 - \mathbf{x}_0\| = \sqrt{(0.40 - 0)^2 + (-0.14 - 0)^2 + (0.17 - 0)^2} \approx 0.46$$

The following iterations will be

$$\mathbf{x}_2 = [0.49 - 0.39\,0.00]^{\mathrm{T}} \quad \|\mathbf{x}_2 - \mathbf{x}_1\| = 0.31$$

$$\mathbf{x}_3 = [0.56 - 0.36 - 0.17]^{\mathrm{T}} \quad \|\mathbf{x}_3 - \mathbf{x}_2\| = 0.19$$

$$\mathbf{x}_4 = [0.51 - 0.31 - 0.16]^{\mathrm{T}} \quad \|\mathbf{x}_4 - \mathbf{x}_3\| = 0.07$$

The target accuracy has been reached after the fourth iteration. For reference, the correct answer to this system is $\mathbf{x} = [0.50\,-0.31\,-0.12]^{\mathrm{T}}$, which the method approximated well. It could be noted that the value x_{i2} actually started off in the wrong direction, starting at 0 and increasing to 0.17, before dropping towards its correct negative value in subsequent iterations.

4.5 Gauss-Seidel Method

The *Gauss-Seidel method* is a technical improvement on the Jacobi method. The iterative formula of Eq. (4.27) computes the vector x_{k+1} based on the vector x_k, using the matrix-vector multiplication Ex_k. This means that every element $x_{k+1,i}$ of x_{k+1} will be computed using some combination of the elements x_{kj} of x_k, including the elements where $j < i$ for which better estimates have already been computed in x_{k+1}. But if better values of these elements are already available, why continue using the older values in the computations? The Gauss-Seidel method proposes the simple improvement of using each new value of x_{k+1} in the computation of Eq. (4.27) as soon as it is available, rather than waiting to compute the entire vector x_{k+1} using x_k only. This modifies the iterative Eq. (4.27) into

$$x_{k+1} = D^{-1}(b - Ex_{k+1}) \tag{4.28}$$

and adds the step to being each iteration by setting $x_{k+1} = x_k$. Since the value of x_k is converging, this simple improvement of using the updated values earlier in the iterations actually allows it to converge faster. This change to the pseudo-code of the Jacobi method is shown in Fig. 4.5. Notice that the single line computing the new value of the vector x in the code of Fig. 4.4 has been replaced by a loop that computes each new element of vector x one at a time and uses all new values from previous iterations of the loop to compute subsequent ones.

Example 4.4
Use the Gauss-Seidel method to solve the following system to an accuracy of 0.1, keeping two decimals of precision and starting with a zero vector.

$$\begin{bmatrix} 5 & 2 & -1 \\ 3 & 7 & 3 \\ 1 & -4 & 6 \end{bmatrix} \begin{bmatrix} x_0 \\ x_1 \\ x_2 \end{bmatrix} = \begin{bmatrix} 2 \\ -1 \\ 1 \end{bmatrix}$$

Solution
This is the same system as in Example 4.3, and using Eq. (4.28) will build an almost identical iterative equation, with the only difference that it uses x_{k+1} instead of x_k in its computations:

$$\begin{bmatrix} x_{k+1,0} \\ x_{k+1,1} \\ x_{k+1,2} \end{bmatrix} = \begin{bmatrix} \frac{1}{5} & 0 & 0 \\ 0 & \frac{1}{7} & 0 \\ 0 & 0 & \frac{1}{6} \end{bmatrix} \left(\begin{bmatrix} 2 \\ -1 \\ 1 \end{bmatrix} - \begin{bmatrix} 0 & 2 & -1 \\ 3 & 0 & 3 \\ 1 & -4 & 0 \end{bmatrix} \begin{bmatrix} x_{k+1,0} \\ x_{k+1,1} \\ x_{k+1,2} \end{bmatrix} \right)$$

(continued)

Example 4.4 (continued)

$$x_{k+1,0} = \frac{1}{5}(2 - (2x_{k+1,1} - 1x_{k+1,2}))$$

$$x_{k+1,1} = \frac{1}{7}(-1 - (3x_{k+1,0} + 3x_{k+1,2}))$$

$$x_{k+1,2} = \frac{1}{6}(1 - (1x_{k+1,0} - 4x_{k+1,1}))$$

Starting with $\mathbf{x}_1 = \mathbf{x}_0 = [0\ 0\ 0]^T$, the first element will remain the same as was computed with the Jacobi method, $x_{10} = 0.40$. However, that value is immediately part of $\mathbf{x}_1$ and used in the computation of the next element. Using $\mathbf{x}_1 = [0.40\ 0\ 0]^T$, the next element is $x_{11} = -0.31$. Then using $\mathbf{x}_1 = [0.40\ -0.31\ 0]^T$, the last element is $x_{12} = -0.11$. This gives a final result for the first iteration of $\mathbf{x}_1 = [0.40\ -0.31\ -0.11]^T$. The Euclidean distance between the two iterations is

$$E_1 = \|\mathbf{x}_1 - \mathbf{x}_0\| = \sqrt{(0.40 - 0)^2 + (-0.31 - 0)^2 + (-0.11 - 0)^2} = 0.52$$

The following iterations will be

$$\mathbf{x}_2 = [0.50\ -0.31\ -0.13]^T \quad \|\mathbf{x}_2 - \mathbf{x}_1\| = 0.11$$

$$\mathbf{x}_3 = [0.50\ -0.30\ -0.12]^T \quad \|\mathbf{x}_3 - \mathbf{x}_2\| = 0.01$$

The target accuracy has been reached after the third iteration, one sooner than the Jacobi method achieved in Example 4.3. Comparing the results of the first iteration obtained with the two methods demonstrates how beneficial using the updated values is. While the element computed will always be the same with both methods, the second element computed here was $x_{1,1} = -0.31$ (or more precisely $x_{1,1} = -0.314$), almost exactly correct compared to the correct answer of -0.307 and a much better estimate than the $x_{1,1} = -0.14$ computed by the Jacobi method. And the third element computed here was $x_{1,2} = -0.11$, a very good step towards the correct answer of -0.12, and a definite improvement compared to the Jacobi method which had started off with a step the wrong direction entirely, at $x_{1,2} = 0.17$.

```
x ← Input initial value vector of length n
b ← Input solution vector of length n
M ← Input nxn matrix
IterationMaximum ← Input maximum number of iterations
ErrorMinimum ← Input minimum relative error

D ← nxn zero matrix
diagonal elements of D ← 1 / (diagonal elements of M)
E ← M
Diagonal elements of E ← 0

IterationCounter ← 0
WHILE (TRUE)

    PreviousValue ← x

    Index ← 0
    WHILE (Index < n)
        element Index of x ← D × (b - E × x)
        Index ← Index + 1
    END WHILE

    CurrentError ← Euclidean distance between x and PreviousValue
    IterationCounter ← IterationCounter + 1

    IF (CurrentError <= ErrorMinimum)
        RETURN Success, x
    ELSE IF (IterationCounter = IterationMaximum)
        RETURN Failure
    END IF

END WHILE
```

Fig. 4.5 Pseudo-code of the Gauss-Seidel method

4.6 Error Analysis

When solving an $Mx = b$ system, one must be mindful to account for the error both in the matrix M and the vector b, as well as for the propagation of that error to the solution vector x. The error on the values in M and b will be dependent on the method used to obtain those values; it could be, for example, the error on the instruments used to measure these values or the error of the mathematical models used to estimate the values. Since both M and b are used to compute x, it should be no surprise that the error on x will be dependent on those of M and b. Unfortunately, the error propagation is not linear: a 10 % relative error in either M or b does not translate to a 10 % relative error in x, but could in fact be a lot more.

The error propagation involves a property of the matrix $\mathbf{M}$ called the *condition number* of the matrix, which is defined as

$$\text{cond}(\mathbf{M}) = \|\mathbf{M}\|\|\mathbf{M}^{-1}\| \tag{4.29}$$

where the double bar represents the *Euclidean* norm (or 2-norm) of the matrix defined as the maximum that the matrix stretches the length of any vector, that is, the largest value of $\frac{\|\mathbf{Mv}\|}{\|\mathbf{v}\|}$ for any vector $\mathbf{v}$. This value may be calculated by finding the square root of the maximum eigenvalue of the matrix multiplied by its transpose:

$$\|\mathbf{M}\| = \sqrt{\lambda_{\max}\left(\mathbf{MM}^{\mathsf{T}}\right)} \tag{4.30}$$

and the *eigenvalue*, in turn, is any scalar value λ that is a solution to the matrix-vector problem:

$$\mathbf{Mv} = \lambda\mathbf{v} \tag{4.31}$$

where $\mathbf{v}$ is the *eigenvector* corresponding to λ. An algorithm to compute the maximum eigenvalue will be discussed later in this section. The inverse matrix $\mathbf{M}^{-1}$ in Eq. (4.29) is the *reciprocal matrix* of $\mathbf{M}$, which will also be discussed later in this section.

It is also necessary to know the relative error on the solution vector $\mathbf{b}$. If the vector of absolute errors $\mathbf{e_b}$ giving the error on each value in $\mathbf{b}$ is known, then the relative error on $\mathbf{b}$ will be the scalar Δb computed as

$$\Delta b = \frac{\|\mathbf{e_b}\|}{\|\mathbf{b}\|} \tag{4.32}$$

where the double-bar operator is the *Euclidean norm* of the vector or the square root of sum of squares of the elements of $\mathbf{b}$. More formally, if $\mathbf{b}$ is an $n \times 1$ vector $[b_0, \ldots, b_{n-1}]^{\mathsf{T}}$:

$$\|\mathbf{b}\| = \sqrt{\sum_{i=0}^{n-1} b_i^2} \tag{4.33}$$

Finally, the relative error on the solution of the system $\mathbf{x}$ will be bounded by the relative error on $\mathbf{b}$ and the condition number of $\mathbf{M}$:

$$E_{\mathbf{x}} = \Delta b \cdot \text{cond}(\mathbf{M}) \tag{4.34}$$

Example 4.5
Consider the following $\mathbf{Mx} = \mathbf{b}$ system:

$$\begin{bmatrix} 5 & 1 \\ -2 & 10 \end{bmatrix} \begin{bmatrix} x_0 \\ x_1 \end{bmatrix} = \begin{bmatrix} 0.005 \\ 0.901 \end{bmatrix}$$

Compute the value of $\mathbf{x}$. If the value of $\mathbf{b}$ was perturbed to $[0.095, 0.901]^T$, compute the relative error on $\mathbf{b}$, then compute the perturbed $\mathbf{x}$ and the relative error on $\mathbf{x}$.

Solution
A two-equation two-variable system can be solved easily to find $\mathbf{x} = [-0.016, 0.087]^T$. Using the perturbed value of $\mathbf{b}$, which we will note $\mathbf{b_p}$, the perturbed solution vector $\mathbf{x_p} = [0.001, 0.090]^T$.

The absolute error caused by the perturbation in each case can be computed by taking the absolute-valued difference between the correct vector and its perturbed version. This gives $\mathbf{e_b} = [0.090, 0]^T$ and $\mathbf{e_x} = [0.015, 0.003]^T$. The relative error on each vector is

$$\Delta b = \frac{\|\mathbf{e_b}\|}{\|\mathbf{b}\|} = \frac{\sqrt{(0.090^2 + 0^2)}}{\sqrt{(0.005^2 + 0.901^2)}} = 0.100$$

$$\Delta x = \frac{\|\mathbf{e_x}\|}{\|\mathbf{x}\|} = \frac{\sqrt{(0.015^2 + 0.003^2)}}{\sqrt{(-0.016^2 + 0.087^2)}} = 0.173$$

Thus, a perturbation causing a relative error of 10 % on $\mathbf{b}$ has caused a corresponding relative error of over 17 % on the solution $\mathbf{x}$.

Example 4.6
An $\mathbf{Mx} = \mathbf{b}$ system has the following matrix of coefficients:

$$\mathbf{M} = \begin{bmatrix} 3 & 1 & 1 \\ 2 & 7 & 3 \\ 4 & 2 & 9 \end{bmatrix}$$

and the value of $\mathbf{b}$ is being measured by instruments that have a relative error of 5 %. What relative error can be expected on the value of $\mathbf{x}$?

(continued)

Example 4.6 (continued)
Solution
Answering this question requires knowing the condition number of the matrix
M. The first step to computing Eq. (4.29) is to multiply

$$\mathbf{MM}^\mathsf{T} = \begin{bmatrix} 11 & 16 & 23 \\ 16 & 62 & 49 \\ 23 & 49 & 101 \end{bmatrix}$$

The maximum eigenvalue of that matrix is 140.31 (see Example 4.8), and the
square root of it is 11.85. Next, compute the inverse matrix (see Example 4.7):

$$\mathbf{M}^{-1} = \begin{bmatrix} 0.40 & -0.05 & -0.03 \\ -0.04 & 0.16 & -0.05 \\ -0.17 & -0.01 & 0.13 \end{bmatrix}$$

multiply it by its transpose and find the largest eigenvalue to be 0.20, the
square root of which is 0.45. Finally, we can use these values in Eq. (4.29):

$$\mathrm{cond}(\mathbf{M}) = 11.85 \times 0.45 \approx 5.33$$

The relative error on the vector **b** is given to be 5%, or 0.05. Using
Eq. (4.29), the relative error on **x** will be

$$E_\mathbf{x} = 0.05 \times 5.33 = 0.267$$

which means the relative error on **x** will be bounded to 26.7 % at a maximum.

4.6.1   Reciprocal Matrix

The *inverse matrix* or *reciprocal matrix* $\mathbf{M}^{-1}$ of a matrix **M** is a matrix such that

$$\mathbf{MM}^{-1} = \mathbf{I} \tag{4.35}$$

and is computed as the *adjoint matrix* (the transpose of the cofactor matrix) of
M divided by the determinant of **M**.

   The *cofactor* of a matrix given an element $m_{i,j}$ is the determinant of the
submatrix obtained by deleting row i and column j. The *cofactor matrix* of an
$n \times n$ matrix **M** is the $n \times n$ matrix obtained by computing the cofactor given each
element $m_{i,j}$ in the corresponding position of **M** and alternating + and − signs, with
the initial cofactor given $m_{0,0}$ having a positive sign.

The *determinant* of a square matrix **M**, written $|\mathbf{M}|$, is a scalar value obtained by computing the *expansion by cofactors*, or the sum of each cofactor given $m_{1,j}$ multiplied by $m_{1,j}$ for the first row of the matrix, alternating $+$ and $-$ signs, and starting with a positive sign. Since the cofactor given $m_{1,j}$ is itself the determinant of the submatrix with one less row and column, this will compute determinants recursively, until the scalar case where the determinant is the scalar number itself.

Two common cases of the reciprocal matrix can help clarify notions. If the matrix **M** is 2×2,

$$\mathbf{M} = \begin{bmatrix} a & b \\ c & d \end{bmatrix} \tag{4.36}$$

its cofactor matrix will be

$$\begin{bmatrix} |d| & -|c| \\ -|b| & |a| \end{bmatrix} = \begin{bmatrix} d & -c \\ -b & a \end{bmatrix} \tag{4.37}$$

then its adjoint will be the transpose of the cofactor matrix:

$$\begin{bmatrix} d & -c \\ -b & a \end{bmatrix}^{\mathrm{T}} = \begin{bmatrix} d & -b \\ -c & a \end{bmatrix} \tag{4.38}$$

and finally its reciprocal will be the adjoint divided by the determinant:

$$\mathbf{M}^{-1} = \frac{1}{\begin{vmatrix} a & b \\ c & d \end{vmatrix}} \begin{bmatrix} d & -c \\ -b & a \end{bmatrix}^{\mathrm{T}} \tag{4.39}$$

$$= \frac{1}{ad - bc} \begin{bmatrix} d & -b \\ -c & a \end{bmatrix}$$

Likewise, if **M** is a 3×3 matrix,

$$\mathbf{M} = \begin{bmatrix} a & b & c \\ d & e & f \\ g & h & i \end{bmatrix} \tag{4.40}$$

its reciprocal will be

$$\mathbf{M}^{-1} = \frac{1}{\begin{vmatrix} a & b & c \\ d & e & f \\ g & h & i \end{vmatrix}} \begin{bmatrix} \begin{vmatrix} e & f \\ h & i \end{vmatrix} & -\begin{vmatrix} d & f \\ g & i \end{vmatrix} & \begin{vmatrix} d & e \\ g & h \end{vmatrix} \\ -\begin{vmatrix} b & c \\ h & i \end{vmatrix} & \begin{vmatrix} a & c \\ g & i \end{vmatrix} & -\begin{vmatrix} a & b \\ g & h \end{vmatrix} \\ \begin{vmatrix} b & c \\ e & f \end{vmatrix} & -\begin{vmatrix} a & c \\ d & f \end{vmatrix} & \begin{vmatrix} a & b \\ d & e \end{vmatrix} \end{bmatrix}^{\mathrm{T}}$$

$$= \frac{1}{a(ei - hf) - b(di - gf) + c(dh - ge)} \begin{bmatrix} \begin{vmatrix} e & f \\ h & i \end{vmatrix} & \begin{vmatrix} c & b \\ i & h \end{vmatrix} & \begin{vmatrix} b & c \\ e & f \end{vmatrix} \\ \begin{vmatrix} f & d \\ i & g \end{vmatrix} & \begin{vmatrix} a & c \\ g & i \end{vmatrix} & \begin{vmatrix} c & a \\ f & d \end{vmatrix} \\ \begin{vmatrix} d & e \\ g & h \end{vmatrix} & \begin{vmatrix} b & a \\ h & g \end{vmatrix} & \begin{vmatrix} a & b \\ d & e \end{vmatrix} \end{bmatrix}$$

$$(4.41)$$

Example 4.7
Compute the inverse of this matrix, using two decimals of precision:

$$\mathbf{M} = \begin{bmatrix} 3 & 1 & 1 \\ 2 & 7 & 3 \\ 4 & 2 & 9 \end{bmatrix}$$

Solution
The cofactor matrix is obtained by computing, at each position (i,j), the determiner of the submatrix without row i and column j. So, for example, at position (0,0),

$$\begin{vmatrix} 7 & 3 \\ 2 & 9 \end{vmatrix} = 7 \times 9 - 2 \times 3 = 57$$

and at position (1,2),

$$\begin{vmatrix} 3 & 1 \\ 4 & 2 \end{vmatrix} = 3 \times 2 - 4 \times 1 = 2$$

The cofactors are assembled into the matrix with alternating signs, to create the cofactor matrix:

(continued)

Example 4.7 (continued)

$$\begin{bmatrix} 57 & -6 & -24 \\ -7 & 23 & -2 \\ -4 & -7 & 19 \end{bmatrix}$$

and the adjoint matrix is simply the transpose of that one.

Next, the determinant can be computed as the sum of cofactor of each element in the first row multiplied by the corresponding element, with alternating signs:

$$\begin{vmatrix} 3 & 1 & 1 \\ 2 & 7 & 3 \\ 4 & 2 & 9 \end{vmatrix} = 3 \begin{vmatrix} 7 & 3 \\ 2 & 9 \end{vmatrix} - 1 \begin{vmatrix} 2 & 3 \\ 4 & 9 \end{vmatrix} + 1 \begin{vmatrix} 2 & 7 \\ 4 & 2 \end{vmatrix} = 141$$

Finally, the inverse matrix is the adjoint matrix divided by the determinant:

$$\mathbf{M}^{-1} = \frac{1}{141} \begin{bmatrix} 57 & -7 & -4 \\ -6 & 23 & -7 \\ -24 & -2 & 19 \end{bmatrix} = \begin{bmatrix} 0.40 & -0.05 & -0.03 \\ -0.04 & 0.16 & -0.05 \\ -0.17 & -0.01 & 0.13 \end{bmatrix}$$

4.6.2   Maximum Eigenvalue

A simple iterative algorithm can be used to find the maximum eigenvalue of an $n \times n$ matrix $\mathbf{M}$.

1. Begin by defining an $n \times 1$ vector of random values $\mathbf{x}_0$.
2. At each iteration step k, compute:

$$\mathbf{x}_{k+1} = \frac{\mathbf{M}\mathbf{x}_k}{\|\mathbf{x}_k\|} \tag{4.42}$$

   where $\|\mathbf{x}_k\|$ is the Euclidean norm of the vector.
3. End the iterations when a predefined maximum number of steps has been reached (failure condition) or when the relative error between the Euclidean norm of $\mathbf{x}_k$ and of $\mathbf{x}_{k+1}$ is less than a target error threshold (success condition).

If the iterative algorithm ends in success, then the maximum eigenvalue of $\mathbf{M}$ is the Euclidean norm $\|\mathbf{x}_{k+1}\|$. That is why the error used in the iterative algorithm is the relative error between the norms, rather than the Euclidean distance between the vectors, which is what would normally be used in an algorithm like this one as explained in Chap. 3. The pseudo-code for this algorithm is presented in Fig. 4.6.

```
x ← Input initial value vector of length n
M ← Input n×n matrix
IterationMaximum ← Input maximum number of iterations
ErrorMinimum ← Input minimum relative error

IterationCounter ← 0
WHILE (TRUE)

    PreviousValue ← x
    x ← M × x / (Euclidean norm of x)
    Eigenvalue ← square root of (sum of square of elements of x)

    CurrentError ← Euclidean distance between x and PreviousValue
    IterationCounter ← IterationCounter + 1

    IF (CurrentError <= ErrorMinimum)
        RETURN Success, Eigenvalue
    ELSE IF (IterationCounter = IterationMaximum)
        RETURN Failure
    END IF

END WHILE
```

Fig. 4.6 Pseudo-code of the maximum eigenvalue algorithm

Example 4.8
Compute the maximum eigenvalue of this matrix with two decimals to a relative error of less than 0.01.

$$M = \begin{bmatrix} 11 & 16 & 23 \\ 16 & 62 & 49 \\ 23 & 49 & 101 \end{bmatrix}$$

Solution
Start with a vector of random values, such as $x_0 = [0.1, 0.2, 0.3]^T$. Then compute the iterations, keeping track of the error at each one.

$$x_1 = \frac{Mx_0}{\|x_0\|} = \frac{[11.20 \quad 28.70 \quad 42.40]^T}{\sqrt{0.1^2 + 0.2^2 + 0.3^2}} = [29.93 \quad 76.70 \quad 113.32]^T$$

$$e_1 = \frac{\|x_1\| - \|x_0\|}{\|x_1\|} = \frac{140.07 - 0.37}{140.07} = 1.00$$

$$x_2 = [29.72 \quad 77.01 \quad 113.46]^T$$

$$e_2 = 0.002$$

(continued)

Example 4.8 (continued)

The target error has been surpassed after the second iteration. The maximum eigenvalue is the norm:

$$\|\mathbf{x}_2\| = \sqrt{29.72^2 + 77.01^2 + 113.46^2} = 140.31$$

4.7   Summary

Systems of linear algebraic equations linking n variables in n equations arise frequently in engineering practice, and so modelling and solving them accurately and efficiently is of great importance. This chapter has introduced four methods to that end. The first method, the PLU decomposition method, can work on any non-singular matrix. The second method, the Cholesky decomposition, is more limited and only works on symmetric positive-definite matrices; however, such matrices often model engineering systems, and when it can be used, it is much more efficient than the PLU decomposition in terms of time and space complexity. The Jacobi method and its improved version the Gauss-Seidel method are both iterative methods whose only requirement is that the matrix must have non-zero diagonal entries. These methods converge most quickly when an initial estimate of the solution is available to begin the iterations, but like any iterative algorithm, they carry the risk of not converging to a solution at all. Whichever method is used, the fact remains that errors in the matrix and the result vector of the system will propagate to the solution vector computed. The chapter thus ended by presenting the mathematical tools needed to estimate the upper bound of that error.

4.8   Exercises

1. Solve the following $\mathbf{Mx} = \mathbf{b}$ systems using the PLU decomposition.

(a) $\mathbf{M} = \begin{bmatrix} 0.6 & -2.6 & -5.8 \\ 6.0 & 2.0 & 1.0 \\ 1.2 & -8.4 & -2.8 \end{bmatrix} \quad \mathbf{b} = \begin{bmatrix} 16.4 \\ 15.4 \\ 12.0 \end{bmatrix}$

(b) $\mathbf{M} = \begin{bmatrix} 3.0 & 1.0 & -1.6 & 8.6 \\ 0.0 & -2.0 & 7.0 & -0.8 \\ 6.0 & 2.0 & 1.0 & 0.0 \\ 1.2 & -4.6 & 0.2 & 3.0 \end{bmatrix} \quad \mathbf{b} = \begin{bmatrix} -0.8 \\ 22.0 \\ -3.6 \\ 0.2 \end{bmatrix}$

(c) $\mathbf{M} = \begin{bmatrix} 10.0 & 3.0 & 2.0 & 3.0 \\ 1.0 & 0.3 & -11.8 & 3.3 \\ -2.0 & 3.4 & 1.3 & 8.1 \\ 2.0 & 8.6 & 1.4 & 2.6 \end{bmatrix} \quad \mathbf{b} = \begin{bmatrix} 14.0 \\ 12.9 \\ 5.8 \\ 19.4 \end{bmatrix}$

2. Solve the following $\mathbf{Mx} = \mathbf{b}$ systems using the Cholesky decomposition.

(a) $\mathbf{M} = \begin{bmatrix} 81 & 18 & 9 \\ 18 & 68 & -14 \\ 9 & -14 & 41 \end{bmatrix}$ $\mathbf{b} = \begin{bmatrix} 909 \\ 506 \\ 61 \end{bmatrix}$

(b) $\mathbf{M} = \begin{bmatrix} 49 & -14 & 28 \\ -14 & 85 & 46 \\ 28 & 46 & 173 \end{bmatrix}$ $\mathbf{b} = \begin{bmatrix} 196 \\ 25 \\ 529 \end{bmatrix}$

(c) $\mathbf{M} = \begin{bmatrix} 9.00 & 0.60 & -0.30 & 1.50 \\ 0.60 & 16.04 & 1.18 & -1.50 \\ -0.30 & 1.18 & 4.10 & -0.57 \\ 1.50 & -1.50 & -0.57 & 25.45 \end{bmatrix}$ $\mathbf{b} = \begin{bmatrix} 2.49 \\ 0.57 \\ 0.79 \\ -2.21 \end{bmatrix}$

(d) $\mathbf{M} = \begin{bmatrix} 49 & 14 & -7 & 7 \\ 14 & 29 & -12 & 2 \\ -7 & -12 & 41 & -19 \\ 7 & 2 & -19 & 35 \end{bmatrix}$ $\mathbf{b} = \begin{bmatrix} 14 \\ -46 \\ 36 \\ -32 \end{bmatrix}$

(e) $\mathbf{M} = \begin{bmatrix} 4.00 & 0.40 & 0.80 & -0.20 \\ 0.40 & 1.04 & -0.12 & 0.28 \\ 0.80 & -0.12 & 9.20 & 1.40 \\ -0.20 & 0.28 & 1.40 & 4.35 \end{bmatrix}$ $\mathbf{b} = \begin{bmatrix} -0.20 \\ -0.32 \\ 13.52 \\ 14.17 \end{bmatrix}$

3. Solve the following $\mathbf{Mx} = \mathbf{b}$ systems using the Jacobi method given the initial values of $\mathbf{x}$ and the target accuracy required.

(a) $\mathbf{M} = \begin{bmatrix} 2.00 & 1.00 \\ 1.00 & 10.00 \end{bmatrix}$ $\mathbf{b} = \begin{bmatrix} 1.00 \\ 2.00 \end{bmatrix}$ $\mathbf{x} = \begin{bmatrix} 0.00 \\ 0.00 \end{bmatrix}$ accuracy $= 0.1$

(b) $\mathbf{M} = \begin{bmatrix} 5.02 & 2.01 & -0.98 \\ 3.03 & 6.95 & 3.04 \\ 1.01 & -3.99 & 5.98 \end{bmatrix}$ $\mathbf{b} = \begin{bmatrix} 2.05 \\ -1.02 \\ 0.98 \end{bmatrix}$ $\mathbf{x} = \begin{bmatrix} 0.45 \\ -0.41 \\ -0.01 \end{bmatrix}$ accuracy $= 0.001$

4. Redo exercise 3 using the Gauss-Seidel method.

5. Given the matrix $\mathbf{M} = \begin{bmatrix} 5 & 0 \\ 0 & 2 \end{bmatrix}$ and a vector $\mathbf{b}$ with a relative error of 5%, what will the error on $\mathbf{x}$ be bounded to?

6. Given the matrix $\mathbf{M} = \begin{bmatrix} 2 & -3 \\ -1 & 2 \end{bmatrix}$ and a vector $\mathbf{b} = \begin{bmatrix} 2.0 \pm 0.2 \\ 4.0 \pm 0.1 \end{bmatrix}$, what will the error on $\mathbf{x}$ be bounded to?

7. What is the maximum eigenvalue of the matrix $\mathbf{M} = \begin{bmatrix} 0 & 1 \\ 1 & -1 \end{bmatrix}$?

Chapter 5
Taylor Series

5.1 Introduction

It is known that, zooming-in close enough to a curve, it will start to look like a straight line. This can be tested easily by using any graphic software to draw a curve, and then zooming into a smaller and smaller region of it. It is also the reason why the Earth appears flat to us; it is of course spherical, but humans on its surface see a small portion up close so that it appears like a plane. This leads to the intuition for the third mathematical and modelling tool in this book: it is possible to represent a high-order polynomial (such as a curve or a sphere) with a lower-order polynomial (such as a line or a plain), at least over a small region. The mathematical tool that allows this is called the Taylor series. And, since the straight line mentioned in the first intuitive example is actually the tangent (or first derivative) of the curve, it should come as no surprise that this Taylor series will make heavy use of derivatives of the functions being modelled.

5.2 Taylor Series and nth-Order Approximation

Assume a function $f(x)$ which has infinite continuous derivatives (which can be zero after a point). Assume furthermore than the function has a known value at a point x_i, and that an approximation of the function's value is needed at another point x (which will generally be near x_i). That approximation can be obtained by expanding the derivates of the function around x_i in this manner:

© Springer International Publishing Switzerland 2016
R. Khoury, D.W. Harder, *Numerical Methods and Modelling for Engineering*,
DOI 10.1007/978-3-319-21176-3_5

$$f(x) = f(x_i) + f^{(1)}(x_i)(x - x_i) + \frac{f^{(2)}(x_i)}{2!}(x - x_i)^2 + \frac{f^{(3)}(x_i)}{3!}(x - x_i)^3 + \cdots$$
$$+ \frac{f^{(n)}(x_i)}{n!}(x - x_i)^n + \cdots \tag{5.1}$$

Equation (5.1) can be written in an equivalent but more compact form as:

$$f(x) = \sum_{k=0}^{\infty} \frac{f^{(k)}(x_i)}{k!}(x - x_i)^k \tag{5.2}$$

This expansion is called the *Taylor series*. Note that Eq. (5.2) makes explicit the divisions by 0! and 1!, which were not written in Eq. (5.1) because both those factorials are equal to 1, as well as a multiplication by $(x - x_i)^0$ also excluded from Eq. (5.1) for the same reason. In the special case where $x_i = 0$, the series simplifies into Eq. (5.3), which is also called the *Maclaurin series*.

$$f(x) = \sum_{k=0}^{\infty} \frac{f^{(k)}(0)}{k!}x^k \tag{5.3}$$

Instead of speaking of a general point x in the vicinity of x_i, it can be useful to formally define a point x_{i+1} exactly one step h distant from x_i. This updates Eq. (5.2) as:

$$f(x_{i+1}) = \sum_{k=0}^{\infty} \frac{f^{(k)}(x_i)}{k!}(x_{i+1} - x_i)^k$$
$$f(x_i + h) = \sum_{k=0}^{\infty} \frac{f^{(k)}(x_i)}{k!}h^k \tag{5.4}$$

This form of the Taylor series given in Eq. (5.4) is the one that will be most useful in this textbook. One problem with it, however, is that it requires computing an infinite summation, something that is never practical to do in engineering! Consequently, it is usually truncated at some value n as such:

$$f(x_i + h) \approx \sum_{k=0}^{n} \frac{f^{(k)}(x_i)}{k!}h^k \tag{5.5}$$

In this truncated form, it is an *nth-order Taylor series approximation*. As indicated by the name and by the fact it eliminates an infinite number of terms from the summation, in this form the series only computes an approximation of the value of the function at x_{i+1}, not the actual exact value.

Example 5.1

Approximate the value of the following function $f(x)$ at a step $h=1$ after $x_0 = 0$, using a 0th, 1st, 2nd, and 3rd-order Taylor series approximation.

$$f(x) = 1 + x + \frac{x^2}{2!} + \frac{x^3}{3!} + \frac{x^4}{4!} + \cdots$$

Solution

To begin, note that the summation is actually the expansion of $f(x) = e^x$. This means that the value $f(1)$ that is being approximated is the constant $e = 2.7183\ldots$ Note as well that the summation is infinitely derivable and that each derivative is equal to the original summation, as is the case for $f(x) = e^x$:

$$f(x) = f^{(1)}(x) = f^{(2)}(x) = f^{(3)}(x) = \cdots$$

The Taylor series expansion of the function at any step h after $x_0 = 0$ is:

$$f(0+h) = f(h) = \sum_{k=0}^{\infty} \frac{f^{(k)}(0)}{k!} h^k$$

$$= f(0) + f^{(1)}(0)h + \frac{f^{(2)}(0)}{2!}h^2 + \frac{f^{(3)}(0)}{3!}h^3 + \frac{f^{(4)}(0)}{4!}h^4 + \cdots$$

and the 0th, 1st, 2nd, and 3rd-order approximations truncate the series after the first, second, third, and fourth terms respectively. In other words, the 0th-order approximation is:

$$f(h) \approx f(0)$$
$$f(1) \approx 1$$

The 1st-order approximation is:

$$f(h) \approx f(0) + f^{(1)}(0)h = 1 + h$$
$$f(1) \approx 2$$

The 2nd-order approximation is:

$$f(h) \approx f(0) + f^{(1)}(0)h + \frac{f^{(2)}(0)}{2!}h^2 = 1 + h + \frac{h^2}{2!}$$
$$f(1) \approx 2.5$$

And the 3rd-order approximation is:

(continued)

Example 5.1 (continued)

$$f(h) \approx f(0) + f^{(1)}(0)h + \frac{f^{(2)}(0)}{2!}h^2 + \frac{f^{(3)}(0)}{3!}h^3 = 1 + h + \frac{h^2}{2!} + \frac{h^3}{3!} = 2.67$$

$$f(1) \approx 2.67$$

Comparing these approximations to the original equation in the example, it can be seen that every additional order makes the expansion $f(h)$ match the original function $f(x)$ more closely. Likewise, knowing that $f(1) = 2.7183\ldots$, it can be seen that with every additional order, the approximation gets closer to the real value.

To further illustrate the situation, the following figure compares $f(x)$ in black to the 0th-order approximation $f(h)$ in red, the 1st-order approximation $f(h)$ in dashed blue, the 2nd-order approximation $f(h)$ in green, and the 3rd-order approximation $f(h)$ in dashed orange. Once again, it can be seen that the higher the order, the more closely the approximation matches the real function.

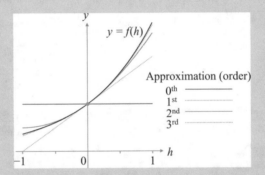

Alternatively, it can be seen that all five functions go through the same point $f(0)$. However, the 0th approximation in red then diverges immediately from the correct function in black, while the 1st approximation in blue matches the correct function over a short step to about 0.1, the 2nd approximation in green follows the function in black over a longer step to approximately 0.3, and the 3rd approximation in orange has the longest overlaps with the function in black, to a step of almost 0.7. This highlights another understanding of the Taylor series approximation: the greater the approximation order, the greater the step for which it will give an accurate approximation.

5.3   Error Analysis

When expanded to infinity, the Taylor series of Eq. (5.4) is exactly equivalent to the original function. That is to say, the error in that case is null. Problems arise however when the series is truncated into the nth-order approximation of Eq. (5.5). Clearly, the truncated series is not equivalent to the complete series nor to the original equation, and there is an approximation error to account for.

Comparing Eqs. (5.4) and (5.5), it is clear to see that the error will be exactly equal to the truncated portion of the series:

$$f(x_i + h) = \sum_{k=0}^{n} \frac{f^{(k)}(x_i)}{k!} h^k + \sum_{k=n+1}^{\infty} \frac{f^{(k)}(x_i)}{k!} h^k \tag{5.6}$$

Unfortunately, this brings back the summation to infinity that the nth-order approximation was meant to eliminate. Fortunately, there is a way out of this, by noting that the terms of the Taylor series are ordered in decreased absolute value. That is to say, each term contributes less than the previous but more than the next to the total summation. This phenomenon could be observed in Example 5.1: note that the 0th and 1st-order approximation add a value of 1 to the total, the 2nd-order approximation adds a value of 0.5, and the 3rd-order approximation a value of 0.17. Likewise, graphically in that example, it can be seen that while each step brings the approximation closer to the real value, it also leaves much less room for further improvement with the infinite number of remaining terms. This observation can be formalized by writing:

$$\sum_{k=n+1}^{\infty} \frac{f^{(k)}(x_i)}{k!} h^k \approx \frac{f^{(n+1)}(x_i)}{(n+1)!} h^{n+1} \tag{5.7}$$

In other words, the error on an nth-order Taylor series approximation will be proportional to the $(n+1)$th term of the series. Using the big O notation introduced in Chap. 1, Eqs. (5.6) and (5.7) can also be written as:

$$f(x_i + h) = \sum_{k=0}^{n} \frac{f^{(k)}(x_i)}{k!} h^k + O(h^{n+1}) \tag{5.8}$$

Special care should be taken with Eq. (5.7) when dealing with series that alternate zero and non-zero terms (such as trigonometric functions). If the $(n+1)$th term happens to be one of the zero terms of the series, it should not be mistaken for the approximation having no error! Rather, in that case, the $(n+1)$th term and all subsequent zero terms should be skipped, and the error will be proportional to the next non-zero term.

Example 5.2
What is the error of the 1st-order Taylor series approximation of the following function at a step $h = 1$ after $x_0 = 0$?

$$f(x) = 1 - 0.2x + 0.6x^2 - 0.3x^3 + 0.5x^4 - 0.1x^5$$

Solution
The 1st-order approximation is:

$$f(x_0 + h) = f(x_0) + f^{(1)}(x_0)h + E(x_0)$$

where the error term $E(x)$ is the 2nd-order term:

$$E(x_0) = \frac{f^{(2)}(x_0)}{2!}h^2$$

and the relevant derivatives are:

$$f^{(1)}(x) = -0.2 + 1.2x - 0.9x^2 + 2x^3 - 0.5x^4$$
$$f^{(2)}(x) = 1.2 - 1.8x + 6x^2 - 2x^3$$

The series can now be evaluated at step $h = 1$ after $x_0 = 0$:

$$f(x_0) = 1$$
$$f^{(1)}(x_0) = -0.2$$
$$f^{(2)}(x_0) = -1.2$$
$$f(x_0 + h) = 1 + (-0.2)(1) + E(x_0) = 0.8 + E(x_0)$$
$$E(x_0) = \frac{-1.2}{2}(1)^2 = -0.6$$

The function is thus approximated as 0.8 with an error on the order of 0.6. To verify, the actual value of $f(1)$ can be computed from the equation as 1.5, so the approximation has an absolute error of 0.7, which is indeed on the same order as 0.6.

Example 5.3
What is the error of the 2nd-order Taylor series approximation of $\cos(x)$ in radians at a step $h = 0.01$ after $x_0 = 0$?

Solution
The derivatives of the cos and sin functions are:

(continued)

Example 5.3 (continued)

$$\frac{d}{dx}\cos(x) = -\sin(x)$$

$$\frac{d}{dx}\sin(x) = \cos(x)$$

and so the 2nd-order approximation will be:

$$f(x_0 + h) = f(x_0) + f^{(1)}(x_0)h + \frac{f^{(2)}(x_0)}{2!}h^2 + E(x_0)$$

$$= \cos(x_0) - \sin(x_0)h - \frac{\cos(x_0)}{2!}h^2 + E(x_0)$$

$$f(0 + 0.01) = \cos(0) - \sin(0)(0.01) - \frac{\cos(0)}{2}(0.01)^2 + E(0)$$

$$= 1 + \frac{1}{2}(0.01)^2 + E(0)$$

$$= 0.99995 + E(0)$$

The error term $E(x)$ is the 3rd-order term:

$$E(x_0) = \frac{f^{(3)}(x_0)}{3!}h^3 = \frac{\sin(x_0)}{3!}h^3$$

However, since $\sin(0) = 0$, this term will be 0. That is clearly wrong, since 0.99995 is not a perfect approximation of the value of $\cos(0.01)$! In this case, the error is the next non-zero term, which is the 4th-order term:

$$E(x_0) = \frac{f^{(4)}(x_0)}{4!}h^4 = \frac{\cos(x_0)}{4!}h^4 = \frac{1}{24}(0.01)^4 = 4.16 \times 10^{-10}$$

Now that it is possible to measure the error of a Taylor series approximation, the natural next question is, how can this information be used to create better approximations of real systems? Given both the discussion in this section and Eq. (5.8) specifically, it can be seen that there are two ways to reduce the error term $O(h^{n+1})$: by using smaller values of h or greater values of n. Using smaller values of h means taking smaller steps, or evaluating the approximation nearer to the known point. Indeed, it has been established and clearly illustrated in Example 5.1 that the approximation diverges more from the real function the further it gets from the evaluated point; conversely, even a low-order approximation is accurate for a small step around the point. It makes sense, then, that reducing the step size h will lead to a smaller approximation error. The second option is to increase the approximation order n, which means adding more terms to the series. This will make the approximation more complete and similar to the original function, and therefore reduce the error.

5.4   Modelling with the Taylor Series

The Taylor series is a powerful modelling approximation tool. It can be used to model even the most complex, infinite, high-degree mathematical functions with a simpler, finite, lower-degree polynomial, provided only that the original function is derivable. Most notably, the 1st-order Taylor series approximation can be used to linearize a complex system into a linear degree-1 polynomial (a.k.a. a straight line) which will be a simple but accurate local model over a small neighborhood.

Additionally, the Taylor series is a useful tool to perform a mathematical analysis of other more complex mathematical formula, such as the numerical methods presented later in this book. By modelling these methods using Taylor series approximations, it will become possible to define and measure the upper bound of their errors.

5.5   Summary

For many engineering applications, it can be useful to model a complex function as a simpler low-order polynomial. This chapter has introduced the infinite Taylor series, which gives an exact equivalent of the function, and the finite nth-order Taylor series approximation, which gives an nth-order polynomial model of the function with a predictable $O(h^{n+1})$ error.

5.6   Exercises

1. Rewrite the Taylor series

$$f(x) = f(x_0) + f^{(1)}(x_0)(x - x_0) + \frac{f^{(2)}(x_0)}{2!}(x - x_0)^2 + \frac{f^{(3)}(x_0)}{3!}(x - x_0)^3$$

   in the form of $f(x_0 + h)$ where $h = x - x_0$.
2. Approximate $\sin(1.1)$ (in radians) using a 1st-order Taylor series approximation expanded around $x_0 = 1$. What is the relative error of this answer?
3. Approximate $\sin(1.1)$ (in radians) using a 2nd-order Taylor series approximation expanded around $x_0 = 1$.
4. What is the bound on the error of using a 1st-order Taylor series approximation expanded around $x_0 = 0.5$ for the function $f(x) = e^x$ when computing the approximation for:

   (a) $x = 0$
   (b) $x = 1$

5. Compute the 0th to 2nd Taylor series approximation of the following functions for $x_0 = 1$ and $h = 0.5$. For each one, use the Taylor series to estimate the approximation error and compute the absolute error to the real value.

   (a) $f(x) = x^2 - 4x + 3$
   (b) $f(x) = 3x^3 + x^2 - 4x + 3$
   (c) $f(x) = -2x^5 + 3x^3 + x^2 - 4x + 3$

Chapter 6
Interpolation, Regression, and Extrapolation

6.1 Introduction

Oftentimes, practicing engineers are required to develop new models of existing undocumented systems they need to understand. These could be, for example, man-made legacy systems for which documentation is outdated or missing, or natural systems that have never been properly studied. In all cases, there are no design documents or theoretical resources available to guide the modelling. The only option available is to take discrete measurements of the system and to discover the underlying mathematical function that generates these points. This chapter will present a set of mathematical and modelling tools that can perform this task.

The tools presented in this chapter can be divided into three variations of this challenge. In the first variation, an exact and error-free set of measurements is available, and the mathematical function computed must have each and every one of these points as an exact result, as illustrated for a 2D case in Fig. 6.1 (left). This is the challenge of *interpolation*. In the second variation, the measurements have errors, and as a result the mathematical function computed does not need to exactly account for all the measurements (or even any of the measurements) but is rather the function with the minimal average error to the set of measurements, as illustrated in Fig. 6.1 (right). This challenge is called *linear regression*. Finally, given a set of measurements, it may be necessary to find not the function that exactly fits or best approximates them, but the function that can best predict future (or past) behavior of the system beyond the measurements. That is the challenge of *extrapolation*. In both graphs of Fig. 6.1, the portions left and right of the first and last points, marked in a dashed line, are extrapolated.

The tools presented in this chapter are an important addition to the toolbox developed over the last three chapters. While iteration, linear algebra, and the Taylor series, are all useful mathematical modelling tools, they all assume that a mathematical function of the system is known and available to iterate, solve, or derive. The tools of iteration, linear regression, and extrapolation make it possible to discover new functions where none are known.

© Springer International Publishing Switzerland 2016
R. Khoury, D.W. Harder, *Numerical Methods and Modelling for Engineering*,
DOI 10.1007/978-3-319-21176-3_6

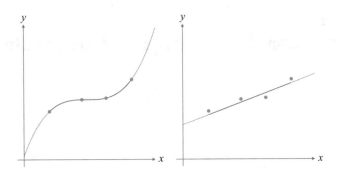

Fig. 6.1 *Left*: A set of exact measurement points in 2D and the interpolated mathematical function (*solid line*) and extrapolated function (*dashed line*). *Right*: The same points as inexact measurements in 2D and the regressed mathematical function (*solid line*) and extrapolated function (*dashed line*)

6.2   Interpolation

Given a set of n measurement points, the challenge of *interpolation* consists in discovering the lowest-degree polynomial that has those points as exact solutions. It is said that polynomial *fits* the points. To be sure, many polynomials of higher degrees could fit the points as well; however, given no other information on the function that generated the data (such as which degree it should be) it makes sense to prefer the simplest, lowest-degree function that could be found. Moreover, picking the lowest-degree function solves a problem of ambiguity: there could be multiple different functions with the same higher degree that fit the points and no way to prefer one over another, but the function of the lowest possible degree will always be unique.

The lowest degree of the function depends on the number of points being used for interpolation. With a set of n points, the lowest-degree polynomial it is possible to interpolate is $n - 1$, meaning the highest power any variable will have is $n - 1$. In the simplest 2D case, which is the one most of this chapter will focus on and the most common one in engineering practice, the equation will thus follow this form:

$$y = f(x) = c_0 + c_1 x + c_2 x^2 + c_3 x^3 + \ldots + c_{n-1} x^{n-1} \qquad (6.1)$$

where every value c_k is a coefficient of the polynomial whose value must be discovered by the interpolation technique.

How is it known that only one unique polynomial of degree $n - 1$ can fit the n points? To understand this, observe two facts:

- The sum or difference of two polynomials of degree $n - 1$ must be a polynomial of degree $n - 1$ or less. It will be less than degree $n - 1$ if the degree-$n - 1$ terms in the two original polynomials cancel each other out, and it will be degree $n - 1$ if they do not, it cannot possibly contain terms of a degree higher than $n - 1$ since those did not exist in the original polynomials being summed or subtracted.

- A polynomial with $n-1$ roots must be of degree $n-1$ or more. The roots are the values for which the polynomial's solution is 0. Each root is a value for which one of the terms of the polynomial becomes exactly equal but of opposite sign to all the other terms combined, which is why a polynomial with $n-1$ roots must have at least $n-1$ terms and be at least of degree $n-1$. Graphically, when plotted, a polynomial of degree $n-1$ will have $n-2$ local optima where the curve changes directions. In order for the curve to intersect the zero axis $n-1$ times by crossing it, changing directions, and crossing it again, it will need to perform at least $n-2$ changes of directions, and thus be at least of degree $n-1$. The one exception to this rule is the zero polynomial $f(x) = 0$, which has more roots (an infinite number of them) than its degree of 0.

Now assume there exists two polynomials of degree $n-1$ $p_1(x)$ and $p_2(x)$ that both interpolate a set of n points. Define a third polynomial as the difference of these two:

$$r(x) = p_1(x) - p_2(x) \tag{6.2}$$

By the first of the two observations above, it is clear that $r(x)$ will be a polynomial of degree $n-1$. Moreover, the polynomials interpolate the same set of n points, which means they will both have the same value at those points and their difference will be zero. These points will therefore be the roots of $r(x)$. And there will be n of them, one more than the degree of $r(x)$. By the second of the two observations above, the only polynomial $r(x)$ could be is the zero polynomial, which means $p_1(x)$ and $p_2(x)$ were the same polynomials in the first place. Q.E.D.

6.3 Vandermonde Method

6.3.1 Univariate Polynomials

The *Vandermonde method* is a very straightforward interpolation technique. It simply requires substituting each of the n points into a polynomial of the form of Eq. (6.1) to create a linear system of n equation, which can then be solved using any of the techniques learnt in Chap. 4. But despite its simplicity, the Vandermonde method is very powerful and can be generalized to use different non-polynomial functions, multidimensional points, and even to perform regressions, as will be presented later on.

Given a set of n points such as the list of Eq. (6.3), it has been proven in the previous section that it is possible to interpolate a unique polynomial of degree $n-1$ of the form of Eq. (6.1). Since each point is a solution of the polynomial, the set of points yields n discrete evaluations of the polynomial, written out in the set of equations (6.4).

$$(x_0, y_0), (x_1, y_1), \ldots, (x_i, y_i), \ldots, (x_{n-1}, y_{n-1}) \tag{6.3}$$

$$y_0 = f(x_0) = c_0 + c_1 x_0 + c_2 x_0^2 + c_3 x_0^3 + \ldots + c_{n-1} x_0^{n-1}$$
$$y_1 = f(x_1) = c_0 + c_1 x_1 + c_2 x_1^2 + c_3 x_1^3 + \ldots + c_{n-1} x_1^{n-1}$$
$$\ldots$$
$$y_i = f(x_i) = c_0 + c_1 x_i + c_2 x_i^2 + c_3 x_i^3 + \ldots + c_{n-1} x_i^{n-1} \tag{6.4}$$
$$\ldots$$
$$y_{n-1} = f(x_{n-1}) = c_0 + c_1 x_{n-1} + c_2 x_{n-1}^2 + c_3 x_{n-1}^3 + \ldots + c_{n-1} x_{n-1}^{n-1}$$

This should be immediately recognizable as a system of equations identical to the one in Eq. (4.1), except with different notation, and with the coefficients of the equations being the unknowns and the variables being known instead of the other way around. Writing this system of equations into matrix–vector form gives:

$$\begin{bmatrix} 1 & x_0 & \cdots & x_0^i & \cdots & x_0^{n-1} \\ 1 & x_1 & \cdots & x_1^i & \cdots & x_1^{n-1} \\ \vdots & \vdots & \ddots & \vdots & & \vdots \\ 1 & x_i & \cdots & x_i^i & \cdots & x_i^{n-1} \\ \vdots & \vdots & & \vdots & \ddots & \vdots \\ 1 & x_{n-1} & \cdots & x_{n-1}^i & \cdots & x_{n-1}^{n-1} \end{bmatrix} \begin{bmatrix} c_0 \\ c_1 \\ \vdots \\ c_i \\ \vdots \\ c_{n-1} \end{bmatrix} = \begin{bmatrix} y_0 \\ y_1 \\ \vdots \\ y_i \\ \vdots \\ y_{n-1} \end{bmatrix} \tag{6.5}$$

Here, the matrix containing the values of the variables of the polynomial is called the Vandermonde Matrix and is written $\mathbf{V}$, the vector of unknown coefficients is written $\mathbf{c}$, and the vector of solutions of the polynomial, or the evaluations of the points, is $\mathbf{y}$. This gives a $\mathbf{Vc} = \mathbf{y}$ system. This system can then be solved using any of the techniques learnt in Chap. 4, or any other decomposition technique, to discover the values of the coefficients and thus the polynomial interpolating the points.

Example 6.1
Four measurements of an electrical system were taken. At time 0 s the output is 1 V, at time 1 s it is 2 V, at time 2 s it is 9 V, and at time 3 s it is 28 V. Find a mathematical model for this system.

Solution
There are four 2D points: (0, 1), (1, 2), (2, 9), and (3, 28). Four points can interpolate a polynomial of degree 3 of the form:

$$y = f(x) = c_0 + c_1 x + c_2 x^2 + c_3 x^3$$

Writing this into a Vandermonde system of the form of Eq. (6.5) gives:

(continued)

Example 6.1 (continued)

$$\begin{bmatrix} 1 & 0 & 0 & 0 \\ 1 & 1 & 1 & 1 \\ 1 & 2 & 4 & 8 \\ 1 & 3 & 9 & 27 \end{bmatrix} \begin{bmatrix} c_0 \\ c_1 \\ c_2 \\ c_3 \end{bmatrix} = \begin{bmatrix} 1 \\ 2 \\ 9 \\ 28 \end{bmatrix}$$

This system can then be solved using any technique to find the solution:

$$\mathbf{c} = \begin{bmatrix} 1 & 0 & 0 & 1 \end{bmatrix}^{\mathrm{T}}$$

Meaning that the polynomial is:

$$y = f(x) = 1 + x^3$$

To illustrate, the four data points and the interpolated polynomial are presented in the figure below.

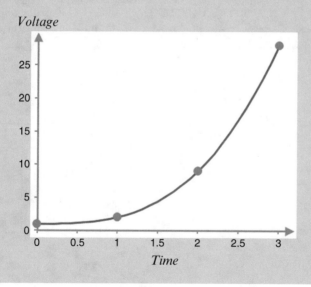

6.3.2   *Univariate General Functions*

One of the advantages of the Vandermonde method is its flexibility: with little modification, it can be use to interpolate any function, not just a polynomial. If it is known that a non-polynomial function of x, such as a trigonometric function for example, is part of the underlying model, the mathematical development can be adapted to include it. This is accomplished by rewriting Eq. (6.1) in a more general form, with each term having one of the desired functions of x multiplied by an unknown coefficient:

$$y = f(x) = c_0 + c_1 f_1(x) + c_2 f_2(x) + c_3 f_3(x) + \ldots + c_{n-1} f_{n-1}(x) \qquad (6.6)$$

It can be seen that the original polynomial equation was simply a special case of this equation with $f_i(x) = x^i$. In the more general case, any function of x can be used, its result evaluated for each given sample of x and stored in the matrix $\mathbf{V}$, and then used to solve the vector of coefficients.

Example 6.2
Four measurements of an electrical system were taken. At time 0 s the output is 1 V, at time 1 s it is 2 V, at time 2 s it is 9 V, and at time 3 s it is 28 V. Find a mathematical model for this system, knowing that the system handles a sine and a cosine wave. Work in radians.

Solution
There are four 2D points: (0,1), (1,2), (2,9), and (3,28). Given the information in the question, they interpolate a trigonometric function of the form:

$$y = f(x) = c_0 + c_1 \sin(x) + c_2 \cos(x) + c_3 \sin(x) \cos(x)$$

Writing this into a Vandermonde system gives:

$$\begin{bmatrix} 1 & 0 & 1 & 0 \\ 1 & 0.84 & 0.54 & 0.45 \\ 1 & 0.91 & -0.42 & -0.38 \\ 1 & 0.14 & -0.99 & -0.14 \end{bmatrix} \begin{bmatrix} c_0 \\ c_1 \\ c_2 \\ c_3 \end{bmatrix} = \begin{bmatrix} 1 \\ 2 \\ 9 \\ 28 \end{bmatrix}$$

This system can then be solved using any technique to find the solution:

$$\mathbf{c} = \begin{bmatrix} 15.91 & -11.19 & -14.91 & 7.87 \end{bmatrix}^{\mathrm{T}}$$

Meaning that the polynomial is:

$$y = f(x) = 15.91 - 11.19 \sin(x) - 14.91 \cos(x) + 7.87 \sin(x) \cos(x)$$

To illustrate, the four data points and the interpolated polynomial are presented in the figure below. Comparing to the solution of Example 6.1, it can be seen that the polynomial fits the measurements just as well, but behaves a bit differently in-between the data points.

(continued)

Example 6.2 (continued)

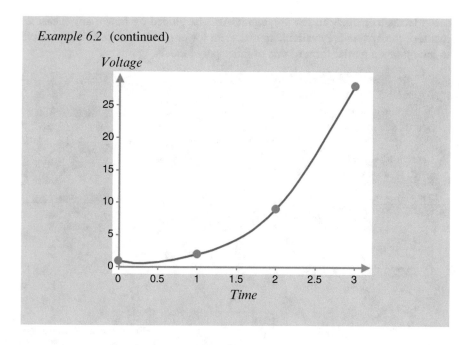

6.3.3 *Multidimensional Polynomial*

The same basic technique can be used to generalize the Vandermonde method to interpolate multivariate or multidimensional functions. In this case, the single-variable functions $f_i(x)$ in Eq. (6.6) become multivariate functions of k variables $f_i(x_0, \ldots, x_{k-1})$. Each function $f_i(x_0, \ldots, x_{k-1})$ is a different product of the k variables, and the entire polynomial would exhaustively list all such products, starting with the constant term (all variables exponent 0), then the single-variable products (one variable at exponent 1 multiplying all others at exponent 0), then the two-variable products (two variables at exponent 1 multiplying all others at exponent 0), and so on. After the last degree-1 term (multiplying all variables together), the exhaustive list continues with one variable at exponent 2.

The most common multidimensional case in engineering is the three-dimensional case, with measurement points being of the form $(x, y, z = f(x, y))$. In this case, the polynomial exhaustively listing all terms at degree 1 and at degree 2 are given in Eqs. (6.7) and (6.8) respectively.

$$z = f(x, y) = c_0 + c_1 x + c_2 y + c_3 xy \tag{6.7}$$

$$z = f(x, y) = c_0 + c_1 x + c_2 y + c_3 xy + c_4 x^2 + c_5 y^2 + c_6 x^2 y + c_7 xy^2 + c_8 x^2 y^2 \tag{6.8}$$

Given the number of coefficients to compute, Eq. (6.7) requires four points to interpolate, and Eq. (6.8) requires nine points to interpolate. The choice of how many terms to include in the polynomial will be guided by how many measurements of the system are available to use. With fewer points available, it is possible to interpolate a partial form of one of those polynomials.

Example 6.3

Four measurements of the height of a structure were taken. At position (3 km, 3 km) the height is 5 km, at position (3 km, 4 km) it is 6 km, at position (4 km, 3 km) it is 7 km, and at position (4 km, 4 km) it is 9 km in height. Find a mathematical model for this structure.

Solution

There are four 3D points: (3, 3, 5), (3, 4, 6), (4, 3, 7), (4, 4, 9). Given the information in the question, they interpolate a 3D function of the form:

$$z = f(x, y) = c_0 + c_1 x + c_2 y + c_3 xy$$

Writing this into a Vandermonde system gives:

$$\begin{bmatrix} 1 & 3 & 3 & 9 \\ 1 & 3 & 4 & 12 \\ 1 & 4 & 3 & 12 \\ 1 & 4 & 4 & 16 \end{bmatrix} \begin{bmatrix} c_0 \\ c_1 \\ c_2 \\ c_3 \end{bmatrix} = \begin{bmatrix} 5 \\ 6 \\ 7 \\ 9 \end{bmatrix}$$

This system can then be solved using any technique to find the solution:

$$\mathbf{c} = \begin{bmatrix} 5 & -1 & -2 & 1 \end{bmatrix}^T$$

Meaning that the polynomial is:

$$z = f(x, y) = 5 - x - 2y + xy$$

6.4 Lagrange Polynomials

The Vandermonde method gives a simple and flexible technique to interpolate polynomials. However, it requires solving a linear system, which in turn requires the use of a matrix calculator (or a lot of patience to get through long and tedious mathematical equations). The method of *Lagrange polynomials* is not as flexible as Vandermonde and is limited to the 2D case, but it has the benefit of being intuitive for humans and to be computable by hand, at least for a small number of points.

The Lagrange polynomials technique works in two simple steps. Given a set of n measurement points (x_0, y_0), ..., (x_i, y_i), ..., (x_{n-1}, y_{n-1}), the first step computes

n separate polynomials, with each one being equal to 1 for one of the n points and equal to 0 for all others. These polynomials are actually quite simple to define; they will each have the form:

$$L_i(x) = \frac{(x - x_0) \ldots (x - x_{i-1})(x - x_{i+1}) \ldots (x - x_{n-1})}{(x_i - x_0) \ldots (x_i - x_{i-1})(x_i - x_{i+1}) \ldots (x_i - x_{n-1})} \tag{6.9}$$

Notice that this polynomial, developed for point x_i, skips over the $(x - x_i)$ term in the numerator and the $(x_i - x_i)$ term in the denominator, but has $n - 1$ terms for the other $n - 1$ points in the set. When $x = x_i$, the denominator and numerator will be equal and the polynomial will evaluate to 1. At any of the other points, one of the subtractions in the numerator will give 0, as will the entire polynomial. The second point then multiplies each polynomial $L_i(x)$ with the value y_i of the measurement at that point, and sums them all together.

$$y = f(x) = \sum_{i=0}^{n-1} y_i L_i(x) \tag{6.10}$$

The final polynomial can then optionally be simplified.

Example 6.4
Four measurements of an electrical system were taken. At time 0 s the output is 1 V, at time 1 s it is 2 V, at time 2 s it is 9 V, and at time 3 s it is 28 V. Find a mathematical model for this system.

Solution
There are four 2D points: $(0, 1)$, $(1, 2)$, $(2, 9)$, and $(3, 28)$. Write a polynomial of the form of Eq. (6.9) for each of the points:

$$L_0(x) = \frac{(x - 1)(x - 2)(x - 3)}{(0 - 1)(0 - 2)(0 - 3)} = \frac{x^3 - 6x^2 + 11x - 6}{-6}$$

$$L_1(x) = \frac{(x - 0)(x - 2)(x - 3)}{(1 - 0)(1 - 2)(1 - 3)} = \frac{x^3 - 5x^2 + 6x}{2}$$

$$L_2(x) = \frac{(x - 0)(x - 1)(x - 3)}{(2 - 0)(2 - 1)(2 - 3)} = \frac{x^3 - 4x^2 + 3x}{-2}$$

(continued)

Example 6.4 (continued)

$$L_3(x) = \frac{(x-0)(x-1)(x-2)}{(3-0)(3-1)(3-2)} = \frac{x^3 - 3x^2 + 2x}{6}$$

Next, each polynomial is multiplied by its matching value, and they are all summed up together to get:

$$y = f(x) = y_0 L_0(x) + y_1 L_1(x) + y_2 L_2(x) + y_3 L_3(x)$$

$$= 1\frac{x^3 - 6x^2 + 11x - 6}{-6} + 2\frac{x^3 - 5x^2 + 6x}{2} + 9\frac{x^3 - 4x^2 + 3x}{-2}$$

$$+ 28\frac{x^3 - 3x^2 + 2x}{6}$$

which is the polynomial interpolating the points. It can further be simplified into:

$$y = f(x)$$

$$= x^3\left(\frac{-1}{6} + \frac{6}{6} + \frac{-27}{6} + \frac{28}{6}\right) + x^2\left(\frac{6}{6} + \frac{-30}{6} + \frac{108}{6} + \frac{-84}{6}\right)$$

$$+ x\left(\frac{-11}{6} + \frac{36}{6} + \frac{-81}{6} + \frac{56}{6}\right) + \left(\frac{6}{6} + \frac{0}{6} + \frac{0}{6} + \frac{0}{6}\right)$$

$$= x^3 + 1$$

which is the same polynomial that was found in Example 6.1.

While the Lagrange polynomials method is the easiest interpolation method for humans to understand and use, it is also the most complicated one to implement in software, as can be seen from the pseudocode in Fig. 6.2. It also suffers from the problem that interpolating a polynomial for a set of n points with this method gives no information whatsoever on the polynomial that could be interpolated with $n + 1$ points including the same n points. In practical terms, this means that if a polynomial has been interpolated for a set of n points and new measurements of the system are made subsequently, the computations have to be done over from scratch. To be sure, that was also the case with the Vandermonde method. However, since with Lagrange polynomials the computations are also made by hand, this can become a major limitation of this method.

```
Points ← Input list of n 2D points (x,y)
Function ← empty function

FunctionIndex ← 0
WHILE (FunctionIndex < n)

      PointIndex ← 0
      Numerator ← 1
      Denominator ← 1

      WHILE (PointIndex < n)

           IF (PointIndex different from FunctionIndex)
                Numerator ← Numerator × [(x variable) - (x coordinate of point
                            number PointIndex)]
                Denominator ← Denominator × [(x coordinate of point number
                              FunctionIndex) - (x coordinate of point number
                              PointIndex)]
           END IF

           PointIndex ← PointIndex + 1
      END WHILE

      Function ← Function  + (y coordinate of point number FunctionIndex) ×
                 Numerator / Denominator
      FunctionIndex ← FunctionIndex + 1
END WHILE

RETURN Function
```

Fig. 6.2 Pseudocode of Lagrange polynomial

6.5 Newton Polynomials

The *Newton polynomials* method discovers the polynomial that interpolates a set of n points under the form of a sum of polynomials going from degree 0 to degree $n - 1$, in the form given in Eq. (6.11). That equation may look long, but it is actually quite straightforward: each individual term i is composed of a coefficient c_i multiplied by a series of subtractions of x by every measurement point from x_0 to x_{i-1}.

$$
\begin{aligned}
y &= f(x) \\
&= c_0 + c_1(x - x_0) + c_2(x - x_0)(x - x_1) + \dots \\
&\quad + c_{n-1}(x - x_0)(x - x_1) \dots (x - x_{n-2})
\end{aligned}
\tag{6.11}
$$

Unlike the Vandermonde method and Lagrange polynomials, the Newton polynomials method can be used to incrementally add points to the interpolation set. A new measurement point (x_n, y_n) will simply add the term $c_n(x - x_0)\cdots(x - x_{n-1})$ to the sum of Eq. (6.11). This new term will be a polynomial of degree n, as will the entire polynomial (as it should be since it now interpolates a set of $n + 1$ points). Moreover, it can be seen that this new term will not have any effect on the terms computed previously: since it is multiplied by $(x - x_0)\cdots(x - x_{n-1})$, it was 0 at all previous interpolated points. The polynomial of Eq. (6.11) was correct for n points, and the newly added $n + 1$ point makes it possible to compute a refinement to that equation without requiring recomputing of the entire interpolation.

The biggest challenge in Newton polynomials is to compute the set of coefficients. There is actually a simple method for computing them, but to understand where the equations come from, it is best to learn the underlying logic by computing the first few coefficients.

Much like Eq. (6.11) makes it possible to incrementally add new points into the interpolation, the coefficients are computed by incrementally adding new points into the set considered. The first coefficient, c_0, will be computed using only the first point (x_0, y_0). Evaluating Eq. (6.11) at that first point reduces it to the straight-line polynomial $y = f(x) = c_0$ since, when the polynomial is evaluated at x_0, all other terms are multiplied by $(x_0 - x_0)$ and become 0. The value of the coefficient is thus clear:

$$y_0 = f(x_0) = c_0 \tag{6.12}$$

Taking the second point into consideration and evaluating Eq. (6.11) at that coordinate while including the result of Eq. (6.12) gives a polynomial the degree 1:

$$y_1 = f(x_1) = y_0 + c_1(x_1 - x_0) \tag{6.13}$$

The value of the coefficient c_1 in the newly added term of the equation is the only unknown in that equation, and can be discovered simply by isolating it in that equation:

$$c_1 = \frac{y_1 - y_0}{x_1 - x_0} = \frac{f(x_1) - f(x_0)}{x_1 - x_0} \tag{6.14}$$

The right-hand side of Eq. (6.14) can be written in a more general form:

$$f(x_i, x_{i+1}) = \frac{f(x_{i+1}) - f(x_i)}{x_{i+1} - x_i} \tag{6.15}$$

in which case the coefficient c_1 of Eq. (6.14) becomes:

$$c_1 = f(x_0, x_1) \tag{6.16}$$

Next, a third measurement point (x_2, y_2) is observed. Evaluating Eq. (6.11) with that new point gives:

$$y_2 = f(x_2) = f(x_0) + f(x_0, x_1)(x_2 - x_0) + c_2(x_2 - x_0)(x_2 - x_1) \tag{6.17}$$

Once again, the value of the coefficient in the newly added term of the equation is the only unknown in that equation, and its value can be discovered simply by isolating it in that equation:

$$c_2 = \frac{f(x_2) - f(x_0) - f(x_0, x_1)(x_2 - x_0)}{(x_2 - x_0)(x_2 - x_2)} = \frac{f(x_1, x_2) - f(x_0, x_1)}{x_2 - x_0} \tag{6.18}$$

And once again that result can be written in a more compact function form:

$$f(x_i, x_{i+1}, x_{i+2}) = \frac{f(x_{i+1}, x_{i+2}) - f(x_i, x_{i+1})}{x_{i+2} - x_i} \tag{6.19}$$

$$c_2 = f(x_0, x_1, x_2) \tag{6.20}$$

A general rule should be apparent from these examples. For any new point (x_k, y_k) added to the interpolation set, a new function can be written as:

$$f(x_i, x_{i+1}, \ldots, x_{i+k}) = \frac{f(x_{i+1}, \ldots, x_{i+k}) - f(x_i, \ldots, x_{i+k-1})}{x_{i+k} - x_i} \tag{6.21}$$

and the coefficient of the new term added to the polynomial is the evaluation of that new function from x_0 to x_k:

$$c_k = f(x_0, x_1, \ldots, x_k) \tag{6.22}$$

This gives the interpolated polynomial of Eq. (6.11) the form:

$$\begin{aligned} y &= f(x) \\ &= f(x_0) + f(x_0, x_1)(x - x_0) + f(x_0, x_1, x_2)(x - x_0)(x - x_1) + \ldots \\ &\quad + f(x_0, \ldots, x_{n-1})(x - x_0)(x - x_1) \ldots (x - x_{n-2}) \end{aligned} \tag{6.23}$$

One thing that should be evident from the examples and from the general formula of Eq. (6.21) is that calculating one level of the function $f(x_i, \ldots, x_{i+k})$ requires knowledge of the previous level of the function $f(x_i, \ldots, x_{i+k-1})$ and, recursively, knowledge of all previous levels of the function down to $f(x_i)$. There is in fact a simple method of systematically computing all these values, by building what is called a *table of divided differences*. One such table combining the information of the sample computations from Eqs. (6.12) to (6.20) is given in Table 6.1. Each column of this table is filled in by computing one level of the function f. The first column simply contains the measurement values x_i, and the second column the corresponding values $f(x_i)$. The third column then has the values $f(x_i, x_{i+1})$, which are computed from the first two columns. Moreover, following Eq. (6.15), each individual value is computed by subtracting the two immediately adjacent values in the previous column, divided by the subtraction of highest and lowest value of x_i. That column will also have one less value than the previous one, since there are fewer combinations possible at that level. The fourth column has the values of $f(x_i, x_{i+1}, x_{i+2})$, which are computed from the third and first column. Once again, each individual value of the new column is computed by subtracting the two immediately adjacent values in the previous column divided by the subtraction of highest and lowest value of x_i, as per Eq. (6.19). And once again, there will be one less value in the new column than there was in the previous one. This process goes on until the

Table 6.1 Sample table of divided differences

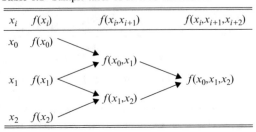

```
Points ← Input list of n 2D points (x,y)
DividedDifferences ← table of n rows and n + 1 columns

RowIndex ← 0
WHILE (RowIndex < n)
    element at column 0, row RowIndex of DividedDifferences ← x coordinate
            of point number RowIndex
    element at column 1, row RowIndex of DividedDifferences ← y coordinate
            of point number RowIndex
    RowIndex ← RowIndex + 1
END WHILE

ColumnIndex ← 2
WHILE (ColumnIndex < n + 1)

    RowIndex ← 0
    WHILE (RowIndex < n - ColumnIndex + 1)

        element at column ColumnIndex, row RowIndex of DividedDifferences
            ← [ (element at column ColumnIndex - 1,
                  row RowIndex + 1 of DividedDifferences)
              - (element at column ColumnIndex - 1,
                  row RowIndex of DividedDifferences)]
          / [ (element at column 0,
                  row RowIndex + ColumnIndex - 1 of DividedDifferences)
              - (element at column 0,
                  row RowIndex of DividedDifferences)]

        RowIndex ← RowIndex + 1
    END WHILE

    ColumnIndex ← ColumnIndex + 1
END WHILE

RETURN Row 0 of DividedDifferences
```

Fig. 6.3 Pseudocode of Newton polynomial

last column has only one value. The coefficients of the polynomial in Eq. (6.23) are immediately available in the final table, as the first value of each column.

The pseudocode for an algorithm to compute the table of divided differences is presented in Fig. 6.3. This algorithm will return the coefficients needed to build a polynomial of the form of Eq. (6.23). An additional step would be needed to recover coefficients for a simpler but equivalent polynomial of the form of Eq. (6.1); this step would be to multiply the Newton coefficients with subsets of x coordinates of the interpolation points and adding all products of the same degree together. This additional step is not included here.

Example 6.5
Four measurements of an electrical system were taken. At time 0 s the output is 1 V, at time 1 s it is 2 V, at time 2 s it is 9 V, and at time 3 s it is 28 V. Find a mathematical model for this system.

Solution
There are four 2D points: (0,1), (1,2), (2,9), and (3,28). Build the table of divided differences. The first two columns are immediately available.

x_i	$f(x_i)$	$f(x_i,x_{i+1})$	$f(x_i,x_{i+1},x_{i+2})$	$f(x_i,x_{i+1},x_{i+2},x_{i+3})$
0	1			
1	2			
2	9			
3	28			

Values in the third column are computed using Eq. (6.15), combining values from the previous two columns. Then, values in the fourth column will be computed using Eq. (6.19) and the values of the third column and the first column.

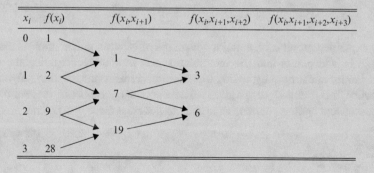

(continued)

Example 6.5 (continued)

The equation to compute the value in the last column can be derived from Eq. (6.21) as:

$$f(x_i, x_{i+1}, x_{i+2}, x_{i+3}) = \frac{f(x_{i+1}, x_{i+2}, x_{i+3}) - f(x_i, x_{i+1}, x_{i+2})}{x_{i+3} - x_i}$$

and the values needed to compute it are the two values in column four and the largest and smallest values of x_i. This completes the table:

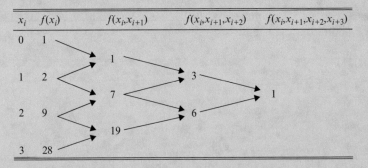

x_i	$f(x_i)$	$f(x_i, x_{i+1})$	$f(x_i, x_{i+1}, x_{i+2})$	$f(x_i, x_{i+1}, x_{i+2}, x_{i+3})$
0	1			
		1		
1	2		3	
		7		1
2	9		6	
		19		
3	28			

Finally, the polynomial of Eq. (6.23) can be constructed by using the first entry of each column as the coefficients.

$$y = f(x) = 1 + 1(x - 0) + 3(x - 0)(x - 1) + 1(x - 0)(x - 1)(x - 2)$$
$$= x^3 + 1$$

Again, the final simplified polynomial is the same one that was computed in Examples 6.1 and 6.4.

As explained previously, a major advantage of Newton polynomials is that it is possible to add points into the interpolation set without recomputing the entire interpolation, but simply by adding higher-order terms to the existing polynomial. In practice, this is done by appending the new points to the existing table of divided differences and adding columns as needed to generate more coefficients.

Example 6.6

A fifth measurement of the electrical system of Example 6.5 has been taken. At time 5 s, the measurement is 54 V. Update the mathematical model for this system.

(continued)

Example 6.6 (continued)
Solution
Append the table of divided differences by adding 5 and 54 to columns one and two respectively. Then, using the equations already known, the functions f can be computed. In the end, a new column must be added at the right-hand side of the table, for the new function (derived from Eq. (6.21)):

$$f(x_i, x_{i+1}, x_{i+2}, x_{i+3}, x_{i+4}) = \frac{f(x_{i+1}, x_{i+2}, x_{i+3}, x_{i+4}) - f(x_i, x_{i+1}, x_{i+2}, x_{i+3})}{x_{i+4} - x_i}$$

The complete table is:

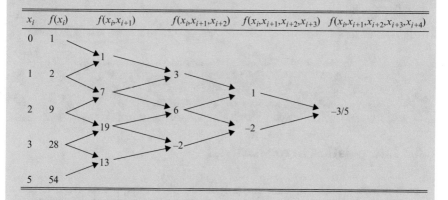

and the polynomial is:

$$y = f(x) = 1 + 1(x - 0) + 3(x - 0)(x - 1) + 1(x - 0)(x - 1)(x - 2)$$
$$- \frac{3}{5}(x - 0)(x - 1)(x - 2)(x - 3)$$
$$= 1 + \frac{18}{5}x - \frac{33}{5}x^2 + \frac{23}{5}x^3 - \frac{3}{5}x^4$$

which is the same as for Example 6.5 with one additional term added to account for the new measurement point. The five measurements are presented in the figure below, along with the original interpolated function from Example 6.5 in red and the updated interpolated function above in purple.

(continued)

Example 6.6 (continued)

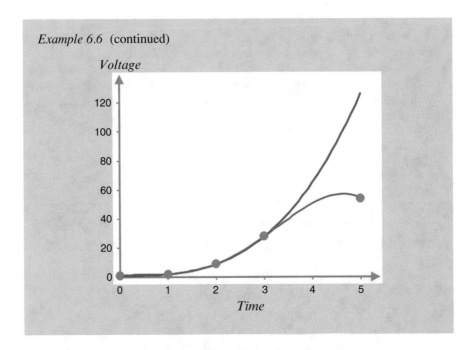

6.6   Interpolation Error Analysis

It is worth remembering that the polynomial $f(x)$ interpolated from a set of n measurement points is, by design, the unique lowest-degree polynomial that can exactly fit the points. However, there is no guarantee that the real polynomial $p(x)$ that generated those measurements is that polynomial; it could be, for example, a much higher-degree polynomial that cannot be uniquely determined from the limited set of points available. This observation is illustrated in Fig. 6.4: the interpolated parabola $f(x)$ in red fits the three data points perfectly, but is inaccurate compared to the real polynomial $p(x)$ in blue elsewhere in the interpolation interval. Formally, for any point x in the interval $[x_0, x_{n-1}]$, the relationship between reality and interpolation is:

Fig. 6.4 Three points on a polynomial (*blue*) and the interpolated parabola (*red*)

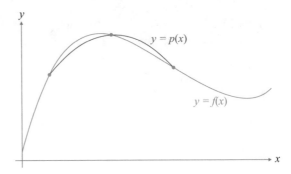

$$p(x) = f(x) + E(x) \tag{6.24}$$

where $E(x)$ is an interpolation error term, which is equal to 0 at the interpolated measurement points.

Chapter 5 has already introduced the Taylor series as an error modelling tool. While it may be difficult to see how it could be applied to the error of a polynomial in the form of Eq. (6.1), writing it in the equivalent Newton polynomial form of Eq. (6.11) makes things a lot clearer. Indeed, Eq. (6.11) can be seen as an $(n-1)^{\text{th}}$-order Taylor series approximation with:

$$c_i = \frac{p^{(i)}(x)}{i!} \tag{6.25}$$

Consequently, the error term will be the n-order term of the series evaluated at a point x_ε in the interval $[x_0, x_{n-1}]$:

$$E(x) = \frac{p^{(n)}(x_\varepsilon)}{n!}(x - x_0)(x - x_1)\ldots(x - x_{n-1}) \tag{6.26}$$

Unfortunately, Eq. (6.26) cannot be used to compute the error term, for the same reason Eq. (6.25) could not be used to compute the coefficients: the polynomial $p(x)$ is unknown. It is, after all, the very polynomial that is being modelled by interpolation. However, an alternative is immediately available from Eq. (6.25): using the coefficient c_n, which can be computed from Eqs. (6.21) and (6.22). The error term then becomes:

$$E(x) = f(x_0, x_1, \ldots, x_{n-1} x_\varepsilon)(x - x_0)(x - x_1)\ldots(x - x_{n-1}) \tag{6.27}$$

The coefficient of the error term can thus be computed using Newton polynomials and the table of divided differences learnt in the previous section, provided an additional point x_ε not used in the interpolation is available.

It is worth noting that, while the development of the error term above uses explicitly Newton polynomials, the error term will be the same for any interpolation method, including Vandermonde and Lagrange polynomials. It is also worth remembering again that this error term is only valid within the interpolation interval.

Example 6.7

Given the interpolated model of the electrical system from Example 6.5, estimate the modelling error on a point computed at time 2.5 s. Use the additional measurement of 4 V at 1.5 s to compute the coefficient.

(continued)

Example 6.7 (continued)
Solution
Append the table of divided differences of Example 6.5 by adding 1.5 and 4 to
columns one and two respectively and compute the new coefficients:

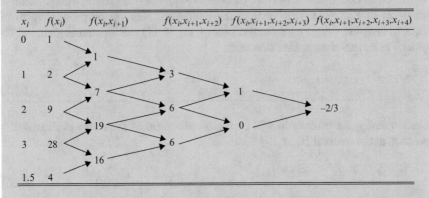

x_i	$f(x_i)$	$f(x_i,x_{i+1})$	$f(x_i,x_{i+1},x_{i+2})$	$f(x_i,x_{i+1},x_{i+2},x_{i+3})$	$f(x_i,x_{i+1},x_{i+2},x_{i+3},x_{i+4})$
0	1				
		1			
1	2		3		
		7		1	
2	9		6		−2/3
		19		0	
3	28		6		
		16			
1.5	4				

The error term is thus:

$$E(x) = -\frac{2}{3}(x-0)(x-1)(x-2)(x-3)$$

and the error on a computation at 2.5 s is $E(2.5) = 0.625$ V.

For comparison, the value computed at 2.5 s by the model of Example 6.5
is 16.625 V, while the value computed by the more accurate model of
Example 6.6 is 17.1875 V. The difference between these values is 0.5625 V,
very close to the predicted error of the model to the real polynomial.

6.7   Linear Regression

Given a set of n points, interpolation finds a polynomial of degree $n-1$ that fits
exactly all the points. Oftentimes, however, that is not the model required for a
problem. There could be a number of reasons for that. For example, it could be that
the number of measurements taken of the system is a lot greater than the expected
degree the model polynomial should be. It would be silly to ignore some (or most)
of the measurements in order to reach a target polynomial order. Moreover,
measurements taken in practice will normally be inexact, due for example to
noise, to observation inaccuracies, or to the inherent limits of the equipment used.
A model that fits these points exactly will therefore be the wrong model for the
system, since it fits erroneous data!

For example, it is well known that an ideal resistor is linear in its response.
However, the measured response of a resistor in practice might not be linear,

Fig. 6.5 20 measurements with errors of a linear system

because of measuring equipment error, fluctuations in the system, defects in the resistor itself, or many other reasons. Simply taking two measurements of the resistor and interpolating a straight line will thus lead to an incorrect model. Using multiple readings to approximate the straight-line response of the resistor will make it possible to create a much more accurate model of the resistor.

Figure 6.5 illustrates the problem. The 20 measurements are clearly pointing to a linear system. However, because of measurement errors, the points do not line up properly. With 20 points, an interpolation method would generate a polynomial of degree 19, which is much, much more complex than what the data is pointing to. On the other hand, selecting two points could lead to the interpolation of very different lines, depending on which pair of points are selected, whereas the correct line can be determined less ambiguously when the entire set of points is considered.

The solution to this problem is *regression*, or the process of discovering the polynomial $y \approx f(x)$ that best approximates (as opposed to fits) the measurement data. When this polynomial is linear, such as the one of Eq. (6.1), the process is called *linear regression* (the term "linear" here has nothing to do with whether the polynomial is for a straight line or not). When the regression is actually looking for a straight line (a polynomial of degree 1), it is called a *simple linear regression*.

One question remains however: how to define the polynomial that "best" approximates a set of points? After all, several polynomials of the same degree could approximate a set of points, and be better approximations of certain points or regions of the set of points while being worse at others. For example, Fig. 6.6 shows three possible straight-line approximations (among countless others!) of the set of points of Fig. 6.5.

Indeed, some approximations could be a better fit for part of the set of points and minimize the errors over that region of the measurements, while others will be better fits for other parts of the measurements. The best approximation, in the context of regression, is not the one that will minimize the errors over part of the points, but the one that will lead to the lowest overall errors. The errors of the model on a measurement is the difference between the observed value y_i (which, while incorrect, is still the best information available about that point) and the value predicted by the model polynomial $f(x_i)$. Then, the overall error of the model on the set of measurements will be the *sum of square errors* (SSE) of each point:

Fig. 6.6 Three
approximations of the set of
points

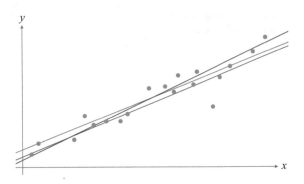

$$\text{SSE} = \sum_{i=0}^{n-1} (y_i - f(x_i))^2 \tag{6.28}$$

The best polynomial that can be regressed is the one that minimizes the value of
the SSE.

6.8   Method of Least Squares

The method of least squares computes the polynomial that minimizes the SSE
through a formal mathematical development. To understand the development,
consider the easiest case of a simple linear regression. In that case, Eq. (6.28)
becomes:

$$\text{SSE} = \sum_{i=0}^{n-1} (y_i - c_0 - c_1 x_i)^2 \tag{6.29}$$

The method is looking for the polynomial, or the values of c_0 and c_1, that minimize
the SSE. The minimum for each coefficient is found by computing the partial
derivative of the equation with respect to that coefficient and setting it equal to 0:

$$\frac{\partial \text{SSE}}{\partial c_0} = -2 \sum_{i=0}^{n-1} (y_i - c_0 - c_1 x_i) = 0 \tag{6.30}$$

$$\frac{\partial \text{SSE}}{\partial c_1} = -2 \sum_{i=0}^{n-1} (y_i - c_0 - c_1 x_i) x_i = 0 \tag{6.31}$$

The problem is now reduced to a system of two equations with two unknown
variables to solve together, which is trivial to do. It can be done by isolating c_0 and
c_1 in Eqs. (6.30) and (6.31) (note that the coefficients multiply the summation), or

by writing the equations into an $\mathbf{Mx} = \mathbf{b}$ form and solving the system using one of the decomposition techniques from Chap. 4:

$$
\begin{bmatrix} n & \sum_{i=0}^{n-1} x_i \\ \sum_{i=0}^{n-1} x_i & \sum_{i=0}^{n-1} x_i^2 \end{bmatrix} \begin{bmatrix} c_0 \\ c_1 \end{bmatrix} = \begin{bmatrix} \sum_{i=0}^{n-1} y_i \\ \sum_{i=0}^{n-1} y_i x_i \end{bmatrix}
\tag{6.32}
$$

The method of least squares can be applied to other cases of linear regression, to discover higher-order polynomials for the model. The only downside is that each additional term and coefficient in the polynomial to regress requires computing one more derivative and handling one more equation.

Example 6.8
An electrical system is measured at every second, starting at time 1 s, using noisy equipment. At time 1 s the initial output is 0.5 V, and the following measurements are 1.7 V at time 2 s, then 1.4, 2.8, 2.3, 3.6, 2.7, 4.1, 3.0 V, and finally 4.9 V at time 10 s. Find a linear model for this system.

Solution
Compute a simple linear regression by filling in the values into the matrix–vector system of Eq. (6.32):

$$
\begin{bmatrix} 10 & 55 \\ 55 & 385 \end{bmatrix} \begin{bmatrix} c_0 \\ c_1 \end{bmatrix} = \begin{bmatrix} 27.0 \\ 180.1 \end{bmatrix}
$$

Then solve the system to find $c_0 = 0.59$ and $c_1 = 0.38$. This means the model is:

$$
y \approx f(x) = 0.59 + 0.38x
$$

And the model's SSE, computed using Eq. (629), is 3.30.
    Graphically, the measurements and the regressed line are presented below. It can be seen that the data points are lined up in two uneven linear sets. While the model does not actually go through any of the points, it is nonetheless the best approximation, as it goes roughly in-between the two sets, a bit closer to the larger one. Since the errors are squared in the SSE, attempting to reduce the error by moving the line closer to one of the sets of points would cause a much larger increase from the error to the other set.

(continued)

Example 6.8 (continued)

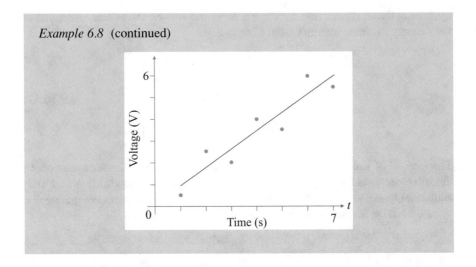

6.9   Vandermonde Method

The Vandermonde method learned for interpolation in Sect. 6.3 can be used for linear regression as well. Much like before, this is done first by writing out the polynomial of the model, then filling in a $\mathbf{Vc} = \mathbf{y}$ system using the values of the measurements, and finally solving for the coefficient vector. The main difference with the previous version of the method is that there are a lot more points than coefficients, so the matrix–vector system does not balance out. This can be simply solved by multiplying both sides of the system by the transpose of $\mathbf{V}$. The correct system for regression is thus:

$$\mathbf{V^T V c} = \mathbf{V^T y} \tag{6.33}$$

It is worth noting that the Vandermonde method is equivalent to the method of least squares. The multiplications $\mathbf{V^T V}$ and $\mathbf{V^T y}$ yield the values computed by deriving and expanding the SSE equations. The main advantage of the Vandermonde method is its simplicity. The matrix $\mathbf{V}$ and vector $\mathbf{y}$ are straightforward to build from the observations without having to derive equations or remember sets of summations, then the system can be built from two multiplications only.

An important benefit of having a fast and simple way to compute regressions is to make it possible to easily compute multiple regressions of a set of measurements. This is a benefit when the degree of the polynomial required to model a system is unknown, and must be discovered through trial-and-error. Generating multiple regressions at different degrees and finding which one gives the best trade-off between low SSE and simplicity is one modelling approach that can work when there is no other information available.

Example 6.9

The following measurements of an electrical system were taken with noisy equipment. At time 1 s the output is −0.3 V, at time 2 s it is −0.2 V, at time 3 s it is 0.5 V, at time 4 s it is 2.0 V, at time 5 s it is 4.0 V, at time 6 s it is 6.0 V, at time 7 s it is 9.0 V, at time 8 s it is 13.0 V, at time 9 s it is 17.0 V, and at time 10 s it is 22.0 V. Find a model for this system.

Solution

Since the degree of the model is unknown, use a trial-and-error approach to find the correct one. Begin by computing three regressions for polynomials of degree 1, 2, and 3, to see if one of those can approximate the data well enough. If none of them are appropriate, higher-degree regressions might be required. The three polynomials are:

$$f_1(x) = c_0 + c_1 x$$
$$f_2(x) = c_0 + c_1 x + c_2 x^2$$
$$f_3(x) = c_0 + c_1 x + c_2 x^2 + c_3 x^3$$

The corresponding Vandermonde systems are:

$$\mathbf{V}_1^T \mathbf{V}_1 \mathbf{c}_1 = \mathbf{V}_1^T \mathbf{y}$$
$$\mathbf{V}_2^T \mathbf{V}_2 \mathbf{c}_2 = \mathbf{V}_2^T \mathbf{y}$$
$$\mathbf{V}_3^T \mathbf{V}_3 \mathbf{c}_3 = \mathbf{V}_3^T \mathbf{y}$$

where the matrices $\mathbf{V}_i$ and the vectors of coefficients $\mathbf{c}_i$ will have two, three, or four columns or rows, respectively, depending on the polynomial being computed. Expanded, the seven vectors and matrices used in the above equations are:

$$\mathbf{V}_1 = \begin{bmatrix} 1 & 1 \\ 1 & 2 \\ 1 & 3 \\ 1 & 4 \\ 1 & 5 \\ 1 & 6 \\ 1 & 7 \\ 1 & 8 \\ 1 & 9 \\ 1 & 10 \end{bmatrix} \quad \mathbf{V}_2 = \begin{bmatrix} 1 & 1 & 1 \\ 1 & 2 & 4 \\ 1 & 3 & 9 \\ 1 & 4 & 16 \\ 1 & 5 & 25 \\ 1 & 6 & 36 \\ 1 & 7 & 49 \\ 1 & 8 & 64 \\ 1 & 9 & 81 \\ 1 & 10 & 100 \end{bmatrix} \quad \mathbf{V}_3 = \begin{bmatrix} 1 & 1 & 1 & 1 \\ 1 & 2 & 4 & 8 \\ 1 & 3 & 9 & 27 \\ 1 & 4 & 16 & 64 \\ 1 & 5 & 25 & 125 \\ 1 & 6 & 36 & 216 \\ 1 & 7 & 49 & 343 \\ 1 & 8 & 64 & 512 \\ 1 & 9 & 81 & 729 \\ 1 & 10 & 100 & 1000 \end{bmatrix} \quad \mathbf{y} = \begin{bmatrix} -0.3 \\ -0.2 \\ 0.5 \\ 2 \\ 4 \\ 6 \\ 9 \\ 13 \\ 17 \\ 22 \end{bmatrix}$$

(continued)

Example 6.9 (continued)

$$\mathbf{c_1} = \begin{bmatrix} c_0 \\ c_1 \end{bmatrix} \mathbf{c_2} = \begin{bmatrix} c_0 \\ c_1 \\ c_2 \end{bmatrix} \mathbf{c_3} = \begin{bmatrix} c_0 \\ c_1 \\ c_2 \\ c_3 \end{bmatrix}$$

Solving each of the three vector–matrix systems finds the coefficients of the corresponding polynomial. Those polynomials are:

$$f_1(x) = -6.25 + 2.46x$$
$$f_2(x) = 0.19 - 0.76x + 0.29x^2$$
$$f_3(x) = -0.10 - 0.50x + 0.24x^2 + 0.003x^3$$

The final challenge is to decide which of these three polynomials, if any, is the best approximation of the system, to use in a model. To make this decision, consider the SSE values. For $f_1(x)$ it is 45.70, for $f_2(x)$ it is 0.23, and for $f_3(x)$ it is 0.19. These error values clearly indicate that a polynomial of degree 1 is a very wrong approximation of the data, while a polynomial of degree 3 gives very little improvement compared to the one of degree 2. The polynomial of degree 2 is the best model in this situation.

Alternatively, looking at the situation graphically can help shed some light on it. The data points are presented in the following figure, along with the approximations $f_1(x)$ in solid red, $f_2(x)$ in dashed red, and $f_3(x)$ in dashed brown A visual inspection makes it clear that the measurements are following a parabola curve and that the straight-line regression is a very poor approximation. Meanwhile, the degree-3 approximation overlaps very much with the degree-2 approximation and does not offer a better approximation.

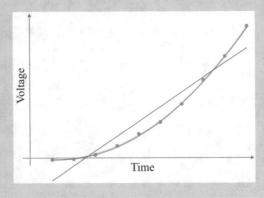

(continued)

Example 6.9 (continued)

Finally, consider the values of the coefficient of the highest-degree term in each equation. For $f_1(x)$ and $f_2(x)$, the terms of degree 1 and degree 2, respectively, have coefficients that are of the same magnitude as those of other terms in the equation, indicating that they contribute as much as other terms in the equation and that degree 1 and degree 2 matter in this model. For $f_3(x)$, the coefficient of the degree-3 term is two orders of magnitude less than those of the other terms of the equation, making its contribution to the equation minimal and indicating that the system is probably not of degree 3.

6.9.1 Vandermonde Method for Multivariate Linear Regression

In interpolation, one of the major advantages of the Vandermonde method was that it made it possible to model multivariate cases and multidimensional problems easily. This is also true when the method is used for linear regression. Moreover, it is done in the same way, by computing the values of the matrix $\mathbf{V}$ using terms of a multivariate polynomial and solving the system to get the coefficients.

Example 6.10

The shape of a ski slope needs to be modelled. The elevation of various points on the slope has been measured, along with their GPS coordinates. Defining the low end of the ski lift at GPS coordinates (1,3) as elevation 0 m, the points measured are: (0,0) 5 m, (2,1) 10 m, (2.5,2) 9 m, (4,6) 3 m, and (7,2) 27 m. Knowing that the ski slope can be approximated as a plane, find the best model for it.

Solution

A plane is simply a linear polynomial in 3D, and its equation is:

$$z \approx f(x, y) = c_0 + c_1 x + c_2 y$$

The system to solve is $\mathbf{V}^T\mathbf{V}\mathbf{c} = \mathbf{V}^T\mathbf{z}$, where the matrix $\mathbf{V}$ will contain the values multiplying each coefficient, namely 1, x, and y respectively. The values of the three variables in the system are:

(continued)

Example 6.10 (continued)

$$V = \begin{bmatrix} 1 & 0 & 0 \\ 1 & 2 & 1 \\ 1 & 2.5 & 2 \\ 1 & 1 & 3 \\ 1 & 4 & 6 \\ 1 & 7 & 2 \end{bmatrix} \quad c = \begin{bmatrix} c_0 \\ c_1 \\ c_2 \end{bmatrix} \quad z = \begin{bmatrix} 5 \\ 10 \\ 9 \\ 0 \\ 3 \\ 27 \end{bmatrix}$$

Solving the system finds the coefficients $c_0 = 5$, $c_1 = 4$, and $c_2 = -3$, corresponding to the polynomial:

$$z \approx f(x, y) = 5 + 4x - 3y$$

which has an SSE of 0.0. Visually, the measurement points and the model plane look like this:

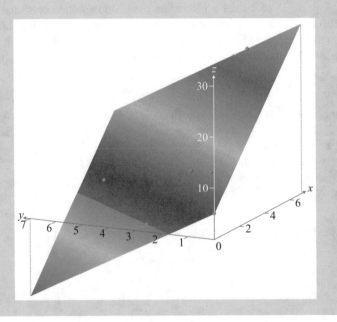

6.10   Transformations

The two regression methods seen so far are used specifically for linear regressions. However, many systems in engineering practice are not linear, but instead are logarithmic or exponential. Such systems cannot be modelled accurately by a linear polynomial, regardless of the order of the polynomial used. This is the case for example of models of population growths, and of capacitor charges in resistor-capacitor (RC) circuits.

In such a situation, the solution is to compute a *transformation* of the nonlinear equation into a linear one, compute the linear regression in that form to find the best approximation, then to reverse the transformation to find the real model. The transformation is whatever operation is needed to turn the polynomial into a linear function. For example, if the function is logarithmic, the transformation is to take its exponential, and the reverse transformation is to take the logarithm of the model.

Example 6.11
The following measures are taken of a discharging capacitor in an RC circuit: at time 0.25 s it registers 0.54 V, at 0.75 s it registers 0.25 V, at 1.25 s it registers 0.11 V, at 1.75 s it registers 0.06 V, and at 2.25 s it registers 0.04 V. Find the best model to approximate this capacitor.

Solution
Plotting the measurements graphically shows clearly that they follow an exponential relationship of the form:

$$y \approx f(x) = c_0 e^{c_1 x}$$

Such a function cannot be modelled using the linear regression tools seen so far. However, transforming by taking the natural log of each side of the equation yields a simple linear function:

$$\ln(y) \approx \ln(c_0 e^{c_1 x}) = \ln(c_0) + \ln(e^{c_1 x}) = c_0^{transform} + c_1 x$$

This simple linear regression problem can easily be solved using the method of least squares or the Vandermonde method to find the coefficients. The linear equation is:

$$\ln(y) \approx -0.25 - 1.66x$$

Finally, reverse the transformation by taking the exponential of each side of the equation:

$$y \approx e^{-0.25-1.66x} = e^{-0.25}e^{-1.66x} = 0.78e^{-1.66x}$$

That equation models the measured data with an SSE of 0.002. The data and the modelling exponential are illustrated below.

(continued)

Example 6.11 (continued)

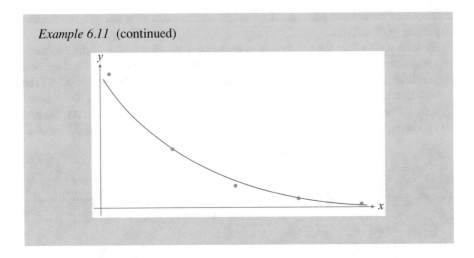

6.11   Linear Regression Error Analysis

Linear regression error is different from the interpolation error computed previously in some major respects. Interpolation methods compute a polynomial that fits the measured data exactly, and consequently constrain the error that can occur in-between those measures, since the error must always drop back to zero at the next interpolated measurement. Linear regression does not impose such a requirement; it computes a polynomial that approximates the measured data, and that polynomial might not actually fit any of the measurements with zero error. Consequently, the error in-between the measures are not constrained. It is instead probabilistic: the values in-between the approximated measurements are probably near the polynomial values (since it is the approximation with minimal SSE), but some of them might be far away from it. In fact, the same holds true for the measurements themselves. The situation could be understood visually by adding a third probability dimension on top of the two dimensions of the data, as in Fig. 6.7. That figure shows a polynomial $y = f(x)$ regressed from a set of points, and the probability of the position of measurements in the X–Y plane is illustrated in the third dimension. The

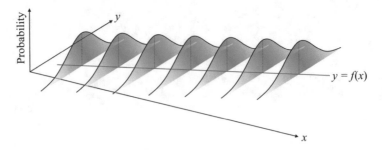

Fig. 6.7 Linear regression on the x- and y-axes, with the probability of the measurements on *top*

Table 6.2 Sample standard deviation and confidence interval

Range of s around $f(x)$	Confidence interval of observations
$\pm 1.00s$	0.6826895
$\pm 1.28s$	0.8000000
$\pm 1.64s$	0.9000000
$\pm 1.96s$	0.9500000
$\pm 2.00s$	0.9544997
$\pm 2.58s$	0.9900000
$\pm 2.81s$	0.9950000
$\pm 3.00s$	0.9973002
$\pm 3.29s$	0.9990000
$\pm 4.00s$	0.9999366
$\pm 5.00s$	0.9999994

error of the measurement is thus a normal distribution with the mean at the polynomial, which is the position with least error, and some standard deviation σ of unknown value.

The standard deviation may not be known, but given the set of measurement points it can be approximated as the sample standard deviation s:

$$\sigma \approx s = \sqrt{\frac{1}{n-1} \sum_{i=0}^{n-1} (y_i - f(x_i))^2} \tag{6.34}$$

This in turn makes it possible to compute the *confidence interval* (CI) of the approximation, or the area around the regressed polynomial that the measurements are likely to be found with a given probability. For a normal distribution, these intervals are well known: 68.3 % of the observed measurements y will be within ± 1 sample standard deviations of the regressed $f(x)$, 95.4 % of the measurements will be within $\pm 2\,s$, and 99.7 % of the measurements will be within $\pm 3\,s$ of $f(x)$. These points are also called the 0.683 CI, the 0.954 CI, and the 0.997 CI. Table 6.2 lists other common relationships between s and CI.

Example 6.12
The following measurements of an electrical system were taken with noisy equipment. At time 1 s the output is 2.6228 V, at time 2 s it is 2.9125 V, at time 3 s it is 3.1390 V, at time 4 s it is 4.2952 V, at time 5 s it is 4.9918 V, at time 6 s it is 4.6468 V, at time 7 s it is 5.4008 V, at time 8 s it is 6.3853 V, at time 9 s it is 6.7494 V, and at time 10 s it is 7.3864 V. Perform a simple linear regression to find a model of the system, and compute the 0.8 CI.

(continued)

Example 6.12 (continued)

Solution

Using any of the methods seen previously, the model of the system can be found to be the straight-line polynomial

$$y \approx f(x) = 1.889 + 0.539x$$

with an SSE of 0.745. Next, the sample standard deviation is given by Eq. (6.34) as:

$$s = \sqrt{\frac{1}{9}\,0.745} = 0.288$$

The 0.8 CI is the interval at ± 1.28 times s, as listed in Table 6.2. The requested function is thus

$$y \approx f(x) \pm 1.28s = (1.889 + 0.539x) \pm 0.369$$

Visually, the result is illustrated below. The ten data points are marked with dots, the regressed polynomial is the solid red line, and the upper and lower bounds of the confidence interval are the two dotted lines. It can be seen visually that this interval does include 8 of the 10 measurement points, as expected; only the points at 5 and 6 s fall outside the range.

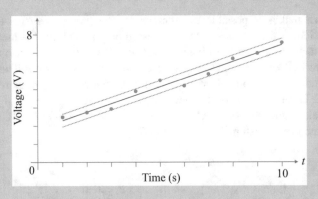

6.12 Extrapolation

The two techniques seen so far, interpolation and regression, have in common that they take in a set of measurements from x_0 to x_{n-1} and compute a model to represent the system within the interval covered by those n measurements in order to predict the value of new measurements with a predictable error. The model in question is

Fig. 6.8 Comparison of interpolation and extrapolation of a system

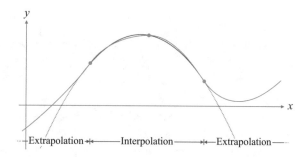

however not valid outside of that interval, and if used beyond those bounds it could lead to a massive misrepresentation of reality. The problem is illustrated graphically in Fig. 6.8, in the case of the interpolation of three points. The degree-2 polynomial interpolated (the solid red parabola in the figure) fits the measurements perfectly and is a good, low-error approximation of the real system (the blue line) in that interval. However, the real system is a degree-4 polynomial, and as a result, outside the interpolation region of the three measurements, the polynomial quickly becomes an inaccurate and high-error approximation of the system (the dashed red line), especially after the inflection points of the system that is not part of the model.

Nonetheless, being able to model and predict the values of a system beyond the confines of a measured interval is a common problem in engineering. It must be done, for example, in order to predict the future behavior of a natural system, in order to design structures that can withstand the likely natural conditions and variations they will be subjected to. It is also necessary to reconstruct historical data that has been lost or was never measured, for example to analyze the failure of a system after the fact and understand the conditions that caused it to go wrong. This challenge, of modelling a system beyond the limits of the measurements, is called *extrapolation*.

Performing an accurate extrapolation requires more information than interpolation and linear regression. Most notably, it requires knowledge of the nature of the system being modelled, and of the degree of the polynomial that can represent it. With that additional information, it becomes possible to compute a model that will have the correct number of inflection points and will avoid the error illustrated in Fig. 6.8. Then, by performing a linear regression over the set of measurements, it is possible to find the best polynomial of the required degree to approximate the data. That polynomial will also give the best extrapolation values.

Example 6.13
The following input/output measurements of a system were recorded:

$$(-0.73507, 0.17716), (-0.58236, 0.13734), (-0.22868, 0.00741),$$
$$(0.24253, -0.00397), (0.27129, 0.01410), (0.31244, 0.08215),$$
$$(0.51378, 0.04926), (0.59861, 0.14643), (0.63754, 0.08751)$$

(continued)

Example 6.13 (continued)

The system is believed to be either linear or quadratic. Predict the output of the system at $x = 1.5$ in each case, and determine which one is the best prediction.

Solution

Using any of the linear regression methods seen so far, the degree-1 and degree-2 polynomials can be found to be:

$$f_1(x) = 0.08270 - 0.04556x$$
$$f_2(x) = 0.00470 - 0.01834x + 0.30768x^2.$$

The value extrapolated with each one are $f_1(1.5) = 0.01436$ and $f_2(1.5) = 0.66947$. These are clearly two very different answers! However, a quick visual inspection of the data, presented below, shows that the data is quadratic rather than linear, and the correct extrapolated value is thus 0.66947. This illustrates the absolute necessity of knowing with certainty the correct degree of polynomial to use to model the data when doing extrapolation. An error of even one degree leads to a wrong inflection in the model and to extrapolated values that are potentially wildly divergent from reality.

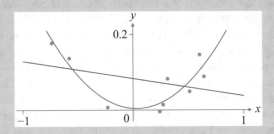

6.13   Summary

It is often necessary, in engineering practice, to develop a mathematical model of a system given only a set of observed measurements. This challenge can further be divided into three categories. When the measurements are exact and error-free and the model must account for them exactly, it is called interpolation. When the measurements are inexact and the model must approximate them, it is linear regression. And when values beyond the bounds of the measurements must be modelled and predicted, the challenge is extrapolation. This chapter has introduced several techniques to accomplish all three of these. The centerpiece is the Vandermonde method, a technique to reduce the challenge to a linear vector–matrix problem such as those studied in Chap. 4. This method benefits from very large flexibility: it can be adapted for interpolation or linear regression, to discover single-variable or multivariable polynomials, and even to handle nonlinear polynomials. The Lagrange polynomials method was also introduced as a more human-understandable technique for interpolation, while the Newton polynomials have the benefit of allowing the incremental improvement of the model when new measurements become available. For regression, in addition to the Vandermonde method, the equivalent mathematical development of the method of least squares was discussed in order to demonstrate the formal mathematical foundations of the techniques.

6.14   Exercises

1. Using the Vandermonde Method, find the polynomial which interpolates the following set of measurements:

   (a) (2,3), (5,7).
   (b) (0,2), (1,6), (2,12).
   (c) (−2,21), (0,1), (1,0), (3, −74).
   (d) (1,5), (2,7), (4,11), (6,15).
   (e) (−3.2,4.5), (−1.5,0.5), (0.3,0.6), (0.7,1.2), (2.5,3.5).
   (f) (1.3,0.51), (0.57,0.98), (−0.33,1.2), (−1.2,14), (2.1, −0.35), (0.36,0.52).

2. Must the x values be ordered from smallest to largest exponent for the Vandermonde method to work?

3. Using the Vandermonde Method, find the polynomial of the form $f(x) = c_1 \sin(x) + c_2 \cos(x)$ which interpolates the following set of measurements: (0.3,0.7), (1.9, −0.2).

4. Using the Vandermonde Method, find the polynomial of the form $f(x) = c_0 + c_1 \sin(x) + c_2 \cos(x)$ which interpolates the following set of measurements: (4,0.3), (5, 0.9), (6, −0.2).

5. Using Lagrange polynomials, find the polynomial which interpolates the following set of measurements:

   (a) (2,3), (5, −6).
   (b) (1,2), (3,4)
   (c) (2,9), (3, −14), (5, −24).
   (d) (0,1), (1,0), (3,4).
   (e) (5.3,4.6), (7.3,2.6).
   (f) (0,0), (1,2), (2,36)
   (g) (0,0), (1,2), (2,36), (3,252)

6. Using Newton polynomial, find the polynomial which interpolates the following set of measurements:

   (a) (2,3), (5,7).
   (b) (2,2), (3,1), (5,2).

    (c) $(-2, -39)$, $(0,3)$, $(1,6)$, $(3,36)$.

    (d) $(-2,21)$, $(0,1)$, $(1,0)$, $(3, -74)$.

    (e) $(1,5)$, $(2,7)$, $(4,11)$, $(6,15)$.

    (f) $(1.3, 0.51)$, $(0.57, 0.98)$, $(-0.33, 1.2)$, $(-1.2, 14)$, $(2.1, -0.35)$, $(0.36, 0.52)$.

7. Must the x values be ordered from smallest to largest for the method to find Newton polynomials to work?

8. Suppose you have computed the polynomial which interpolates the set of measurements $(1, 4)$, $(3, -2)$, $(4, 10)$, $(5, 16)$ using the following table of divided differences:

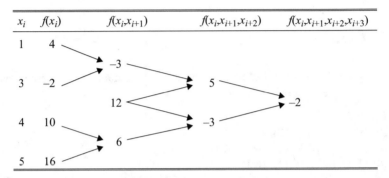

x_i	$f(x_i)$	$f(x_i,x_{i+1})$	$f(x_i,x_{i+1},x_{i+2})$	$f(x_i,x_{i+1},x_{i+2},x_{i+3})$
1	4			
		-3		
3	-2		5	
		12		-2
4	10		-3	
		6		
5	16			

    Use this result to compute the polynomial which interpolates the set of measurements $(3,-2)$, $(4,10)$, $(5,16)$, $(7,34)$.

9. Using the Vandermonde Method, find the polynomial which interpolates the following set of measurements:

    (a) $(0,0,5)$, $(0,1,4)$, $(1, 0,3)$, $(1,1,6)$.

    (b) $(2,4,12)$, $(2,5,11)$, $(3,4,10)$, $(3,5,14)$.

    (c) $(2,2,6)$, $(2,3,10)$, $(4,2,12)$, $(5,5,18)$.

    (d) $(2,3,16)$, $(2,5,15)$, $(4,3,14)$, $(6,6,17)$.

    (e) $(1,1,3.2)$, $(1,2,4.4)$, $(1,3,6.5)$, $(2,1,2.5)$, $(2,2,4.7)$, $(2,3,5.8)$, $(3,1,5.1)$, $(3,2,3.6)$, $(3,3,2.9)$.

    (f) $(1,2,3,4.9)$, $(3,5,2,2.6)$, $(5,4,2,3.7)$, $(4,1,4,7.8)$.

10. Compute a simple linear regression using the following set of measurements:

    (a) $(1,0)$, $(2,1)$, $(3,1)$, $(4,2)$.

    (b) $(0.282,0.685)$, $(0.555,0.563)$, $(0.089,0.733)$, $(0.157,0.722)$, $(0.357,0.662)$, $(0.572,0.588)$, $(0.222,0.693)$, $(0.800,0.530)$, $(0.266,0.650)$, $(0.056,0.713)$.

    (c) $(1, 2.6228)$, $(2, 2.9125)$, $(3, 3.1390)$, $(4,4.2952)$, $(5, 4.9918)$, $(6, 4.6468)$, $(7,5.4008)$, $(8, 6.3853)$, $(9, 6.7494)$, $(10, 7.3864)$.

    (d) $(0.350,2.909)$, $(0.406,2.987)$, $(0.597,3.259)$, $(1.022,3.645)$, $(1.357,4.212)$, $(1.507,4.295)$, $(2.228,5.277)$, $(2.475,5.574)$, $(2.974,6.293)$, $(2.975,6.259)$.

11. Consider the following set of measurements submitted for simple linear regression:

    $(1, 2.6228)$, $(2, 2.9125)$, $(3, 3.1390)$, $(4, 4.2952)$, $(5, 4.9918)$,

    $(6, 4.6468)$, $(7, 5.4008)$, $(8, 63.853)$, $(9, 6.7494)$, $(10, 7.3864)$

What would you consider to be problematic about it, and what would you consider a reasonable solution?

12. Compute a linear regression for a quadratic polynomial using the following set of measurements:

(a) $(-2,3)$, $(-1,1)$, $(0,0)$, $(1,1)$, $(2,5)$.
(b) $(1,0.5)$, $(2,1.7)$, $(3,3.4)$, $(4,5.7)$, $(5,8.4)$.
(c) $(0,2.1)$, $(1,7.7)$, $(2,13.6)$, $(3,27.2)$, $(4,40.9)$, $(5,61.1)$.

13. Compute a linear regression for an exponential polynomial using the following set of measurements:

(a) $(0.029,2.313)$, $(0.098, 2.235)$, $(0.213,2.094)$, $(0.352,1.949)$, $(0.376,1.924)$, $(0.393,1.907)$, $(0.473,1.828)$, $(0.639,1.674)$, $(0.855,1.493)$, $(0.909,1.451)$.
(b) $(0.228,0.239)$, $(0.266,0.196)$, $(0.268,0.218)$, $(0.345,0.173)$, $(0.351,0.188)$, $(0.543,0.090)$, $(0.667,0.057)$, $(0.942,0.022)$, $(0.959,0.026)$, $(0.991,0.019)$.
(c) $(0,0.71666)$, $(1,0.42591)$, $(2,0.25426)$, $(3,0.15122)$, $(4,0.08980)$, $(5,0.05336)$, $(6,0.03179)$, $(7,0.01889)$, $(8,0.01123)$, $(9,0.00666)$, $(10,0.00396)$.

14. Using the following set of measurements:
$(0,2.29)$, $(1,1.89)$, $(2,1.09)$, $(3,0.23)$, $(4, -0.80)$, $(5, -1.56)$, $(6, -2.18)$, $(7, -2.45)$, $(8, -2.29)$, $(9, -1.75)$, $(10, -1.01)$
compute a linear regression for a polynomial of the following form:

(a) $f(x) = c_1\sin(0.4x) + c_2\cos(0.4x)$.
(b) $f(x) = c_0 + c_1\sin(0.4x) + c_2\cos(0.4x)$.
(c) Comparing both polynomials, what conclusion can you reach about the constant term c_0?

15. Compute the requested value at the given following set of measurements, knowing that the polynomial is linear:

(a) $(-1, 7)$, $(0, 3)$, $(1, 0)$, $(2, -3)$, looking for $x = 3$.
(b) $(0.3,0.80)$, $(0.7,1.3)$, $(1.2,2.0)$, $(1.8,2.7)$, looking for $x = 2.3$.
(c) $(0.01559,0.73138)$, $(0.30748,0.91397)$, $(0.31205,0.83918)$, $(0.90105,1.05687)$, $(1.21687,1.18567)$, $(1.47891,1.23277)$, $(1.52135,1.25152)$, $(3.25427,1.79252)$, $(3.42342, 1.85110)$, $(3.84589,1.98475)$, looking for $x = 4.5$.

16. Compute the value at $x = 3$ given the following set of measurements, knowing that the polynomial is of the form $y(x) = c_1x^2$: $(-2, 5)$, $(-1, 1)$, $(0, 0)$, $(1, 2)$, $(2, 4)$.

17. The following measurements come from the exponential decrease of a discharging capacitor:
$(-1.5,1.11)$, $(-0.9,0.92)$, $(-0.7,0.85)$, $(0.7,0.57)$, $(1.2,0.49)$, $(1.4,0.45)$
At what time will the charge be half the value of the charge at time 0 s?

Chapter 7
Bracketing

7.1 Introduction

Consider the problem of searching for the word "lemniscate" in a dictionary of thousands of pages. It would be unthinkable to find the word by reading the dictionary systematically page by page. However, since words are sorted alphabetically from A to Z in the dictionary, it is easy to open the dictionary at a random page and determine whether the letter L is before or after that page. This single step will greatly reduce the number of pages to search through. Next, select a page at random in the portion of the dictionary the word is known to be in, and determine whether the word is before or after that second point. Once a point starting with the letter L is reached, the following letter E is considered the section kept is the one containing LE, then LEM, and so on until sufficient precision is achieved (namely that the page containing the word is found).

This type of search is called *bracketing*. It consists in defining brackets, or upper and lower bounds, on the value of the solution of a problem, then iteratively refining these bounds and reducing the interval the solution is found in, until the interval represents an acceptable error range on the solution.

7.2 Binary Search Algorithm

Possibly the simplest and most popular bracketing algorithm is the *binary search algorithm*. Assume that a solution to a problem is needed; its value is unknown but it is known to be somewhere between a lower bound value x_L and an upper bound value x_U. The algorithm words by iteratively dividing the interval into half and keeping the half that contains the solution. So, in the first iteration, the middle point would be:

© Springer International Publishing Switzerland 2016
R. Khoury, D.W. Harder, *Numerical Methods and Modelling for Engineering*,
DOI 10.1007/978-3-319-21176-3_7

$$x_L + \frac{x_U - x_L}{2} = x_i \tag{7.1}$$

and, assuming the solution is not exactly that point x_i (which it will rarely if ever be for any real-world problem except very simple ones), then one of the two intervals, either $[x_L, x_i]$ or $[x_i, x_U]$, will contain the solution. In either case, the search interval is now half the size it was before! In the next iteration, the remaining interval is again divided into half, then again and again. As with any iterative algorithm in Chap. 3, halting conditions must be defined, namely a maximum number of iterations and a target-relative error. The error can be defined as before, between two successive middle points (using for example x_{i+1}, the middle point of the interval $[x_L, x_i]$):

$$E_{\text{rel}} = \left| \frac{x_i - x_{i+1}}{x_{i+1}} \right| \tag{7.2}$$

The final result returned by the algorithm is not the solution to the problem, but an interval that the solution is found in. The solution can be approximated as the central point of that interval, with half the interval as absolute error:

$$x_{i+1} \pm \frac{|x_i - x_L|}{2} \tag{7.3}$$

Note that this definition of the absolute error could be substituted into Eq. (7.2) as well to compute the relative error:

$$E_{\text{rel}} = \left| \frac{(x_i - x_L)/2}{x_{i+1}} \right| \tag{7.4}$$

The pseudocode for the binary search algorithm is presented in Fig. 7.1. This code will serve as a foundation for the more sophisticated bracketing algorithms that will be presented in later chapters. Note that it requires a call to a function `SolutionInInterval(XL,XU)` which serves to determine if the solution is between the lower and upper bounds given in parameter. This function cannot be defined in pseudocode; it will necessarily be problem-specific. For example, in the dictionary lookup problem of the previous section it will consist in comparing the spelling of the target word to that of the bounds, while in more mathematical problems it can require evaluating a function and comparing the result to that of the bounds. Note as well that the first check in the `IF` uses the new point x as both bounds; it will return true if that point is the exact solution.

```
XL ← Input lower bound
XU ← Input upper bound
IterationMaximum ← Input maximum number of iterations
ErrorMinimum ← Input minimum relative error
x ← XL

IterationCounter ← 0
WHILE (TRUE)

    PreviousValue ← x

    x ← XL + (XU - XL) / 2
    IF ( CALL SolutionInInterval(x,x) )
        RETURN Success, x
    ELSE IF ( CALL SolutionInInterval(XL,x) )
        XU ← x
    ELSE IF ( CALL SolutionInInterval(x,XU) )
        XL ← x
    END IF

    CurrentError ← absolute value of [ (x - PreviousValue) / x ]
    IterationCounter ← IterationCounter+1

    IF (CurrentError <= ErrorMinimum)
        RETURN Success, x
    ELSE IF (IterationCounter = IterationMaximum)
        RETURN Failure
    END IF

END WHILE

FUNCTION SolutionInInterval(XL, XU)
    IF (solution to the problem is between XL and XU)
        RETURN TRUE
    ELSE
        RETURN FALSE
    END IF
END FUNCTION
```

Fig. 7.1 Pseudocode of the binary search algorithm

7.3 Advantages and Limitations

Bracketing is by far the simplest and most intuitive of the mathematical tools covered in this book. It is nonetheless quite powerful, since it makes it possible to zoom in on the solution to a problem given no information other than a way to evaluate bounds. The initial bounds for the algorithm can easily be selected by rule-of-thumb, rough estimation (or "guesstimation"), or visual inspection of the problem.

However, this tool does have important limitations as well. It is less efficient than the other four mathematical tools studied, both in terms of computational time and in terms of accuracy. As a result, numerical methods that make use of bracketing, despite being simpler, will also be the ones that converge to the solution in the longest time and to the least degree of accuracy. Bracketing methods also work best for one-dimensional $y = f(x)$ problems and scale up very poorly into n dimensions. Indeed, a system of n equations and n unknowns will require 2^n bounds to be properly bracketed.

7.4   Summary of the Five Tools

Chapters 3–7 have introduced five mathematical and modelling tools: iteration, linear algebra, Taylor series, interpolation, regression, and extrapolation, and now bracketing. Each of these tools is useful to solve a specific problem or to model a specific aspect of nature, and together they will form the building blocks of the more advanced numerical methods that will be presented in the next six chapters. It is worth finishing this section with a review of these important building blocks.

- Iteration is a tool useful to refine an approximation step by identical step. As a modelling tool, it can represent any naturally converging system. This tool will be the foundation of almost all numerical method algorithms coming up.
- Linear algebra is useful to model and solve systems with multiple independent variables that interact through a set of equations. This is often the case, especially when dealing with multivariate or multidimensional problems.
- Taylor series is an approximation tool that can give a good local representation of a system using its derivatives, and can do so with a measurable error. This will be useful mainly to develop algorithms to solve complex systems and to estimate the error on the solutions these algorithms provide.
- Interpolation, regression, and extrapolation tools are used to estimate a continuous model of a system given a set of discrete measurements. In many cases in engineering practice and in the upcoming chapters, discrete points are all that is available to work with, and these tools thus become cornerstones of the work.
- Bracketing, the topic of this chapter, is a search tool useful to converge on a solution when almost no other information is available. While the least efficient of the five tools, it remains nonetheless a useful last-resort tool to have available.

Chapter 8
Root-Finding

8.1 Introduction

The *root* of a continuous multidimensional function $f(\mathbf{x})$ is any point $\mathbf{x} = \mathbf{x}_r$ for which the function $f(\mathbf{r}) = 0$. Algorithms that discover the value of a function's root are called *root-finding* algorithms, and they constitute the first numerical method presented in this book.

Root-finding is an important engineering modelling skill. Indeed, many situations, once properly modelled by a set of mathematical equations, can be solved by finding the point where the multiple equations are equal to each other; or said differently, where the difference between the results of all the equations is zero. That is the root of the system of equations.

As an example, consider the simple circuit illustrated in Fig. 8.1. Suppose it is necessary to find the current running through this circuit. To model this system, recall that Kirchhoff's law states that the sum of voltages of the components in the circuit will be equal to the voltage of the source:

$$V_D + V_R = V_S \tag{8.1}$$

The voltage of the source V_S is known to be 0.5 V, while Ohm's law says that the voltage going through the resistor is:

$$V_R = RI \tag{8.2}$$

where R is also given in the circuit to be 10 Ω. The Shockley ideal diode equation gives the voltage going through the diode, V_D, as:

$$I = I_S \left(e^{\frac{V_D}{nV_T}} - 1 \right) \tag{8.3}$$

© Springer International Publishing Switzerland 2016
R. Khoury, D.W. Harder, *Numerical Methods and Modelling for Engineering*,
DOI 10.1007/978-3-319-21176-3_8

Fig. 8.1 A simple diode
circuit

where I_s is the saturation current, V_T is the thermal voltage, and n is the ideality
factor. Assume these values have been measured to be $I_s = 8.3 \times 10^{-10}$ A,
$V_T = 0.7$ V and $n = 2$.

Moving V_S to the left-hand side of Eq. (8.1) along with the other two voltage
values makes the model into a root-finding problem. Moreover, incorporating the
equations of V_R and V_D from Eqs. (8.2) and (8.3) respectively into Eq. (8.1) along
with the measured values given yields the following equation for the system:

$$-0.5 + 1.4\ln\left(1.20482 \times 10^9 I + 1\right) + 10I = 0 \tag{8.4}$$

The value of the current running through the circuit loop is thus the root of the
circuit's model. In the case of Eq. (8.4), that value is $3.562690102 \times 10^{-10}$ A.

8.2   Bisection Method

The bisection method is an iterative bracketing root-finding method. It follows the
bracketing algorithm steps presented in Chap. 7: it begins by setting an upper and
lower bound before and after the root, then at each iteration, it reduces the interval
between these bounds to zoom in on the root. The bisection method thus has the
same advantages and limitations that were highlighted in Chap. 7: it is not only a
simple, robust, and intuitive root-finding method, but also an inefficient one that
works best in one-dimensional problems such as the resistor example of the
previous section.

The algorithm for the bisection method is a simple three-step iterative process:
setup for the iterations, iterate, and terminate the iterations. The steps are explained
below, followed by pseudocode for the algorithm in Fig. 8.2.

Step 1: Set the initial lower bound x_L and upper bound x_U around the root of the
function $f(x)$. Since the root is a zero of the equation, the two bounds must be on
either side of a zero-crossing; in other words, they must be points where the
equation evaluates to values of opposite signs:

$$[x_L, x_U] \begin{cases} (f(x_L) < 0) \wedge (f(x_U) > 0) \\ (f(x_L) > 0) \wedge (f(x_U) < 0) \end{cases} \tag{8.5}$$

Step 2: Once points on either side of the zero-crossing have been selected, the
bisection method iteratively tightens them by picking the point exactly in-between
the two bounds:

```
XL ← Input lower bound
XU ← Input upper bound
IterationMaximum ← Input maximum number of iterations
ErrorMinimum ← Input minimum relative error
x ← XL

IterationCounter ← 0
WHILE (TRUE)

    PreviousValue ← x
    x ← (XL + XU) / 2
    Evaluation ← CALL F(x)

    IF ( Evaluation = 0 )
        RETURN Success, x
    ELSE IF ( Evaluation is of the same sign as CALL F(XL) )
        XL ← x
    ELSE IF ( Evaluation is of the same sign as CALL F(XU) )
        XU ← x
    END IF

    CurrentError ← absolute value of [ (x - PreviousValue) / x ]
    IterationCounter ← IterationCounter + 1

    IF (CurrentError <= ErrorMinimum)
        RETURN Success, x
    ELSE IF (IterationCounter = IterationMaximum)
        RETURN Failure
    END IF

END WHILE


FUNCTION F(x)
    RETURN evaluation of the target function at point x
END FUNCTION
```

Fig. 8.2 Pseudocode of the bisection method

$$x_i = \frac{x_L + x_U}{2} \tag{8.6}$$

The function is then evaluated at that point. In some rare cases, the function will evaluate to zero exactly, in which case the point x_i is the root and the algorithm can terminate. More generally however, the function will evaluate to a positive or negative value, and the middle point will be on one or the other side of the root. The middle point replaces the bound on the same side of the root as itself and becomes the new bound on that side of the root. The interval is thus reduced by half at each iteration, and the root remains bracketed between a lower and upper bound:

$$[x_L, x_U] \rightarrow \begin{cases} [x_i, x_U] & \text{if } (f(x_i) \times f(x_L)) > 0 \\ [x_L, x_i] & \text{if } (f(x_i) \times f(x_U)) > 0 \end{cases} \tag{8.7}$$

Step 3: The iterative process continues until a halting condition is reached. One halting condition mentioned in the previous step, albeit an unlikely one, is that a middle point x_i is found to be exactly the root. More usually, one of the two halting conditions presented in Chap. 7 and Chap. 3 will apply: the algorithm will reach a preset maximum number of iterations (failure condition) or some error metric, such as the interval between the two brackets or the relative error on the point x_i, will become lower than some preset error value (success condition).

It is clear to see that the absolute error on the approximation of the root x_i is the current interval $[x_{i-2}, x_{i-1}]$. Moreover, since this interval is reduced at each iteration, it follows that the error is also reduced at each iteration. More formally, define h_0 as the width of the initial interval, and the initial absolute error value:

$$h_0 = |x_L - x_U| \tag{8.8}$$

At each iteration, the interval, and therefore the error, is reduced by a factor of 2 compared to its previous iteration value. If we assume a total number of n iterations performed, the final interval and error value is:

$$h_n = \frac{h_{n-1}}{2} = \frac{h_0}{2^n} \tag{8.9}$$

Equation (8.9) is the convergence rate of the algorithm to the solution, or the rate the error decreases over the iterations: it is a linear algorithm with $O(h)$. But the equation also makes it possible to predict the number of iterations that will be needed to converge to an acceptable solution. For example, if an initial interval on a root is [0.7, 1.5] and a solution is required with an error of no more than 10^{-5}, then the algorithm will need to perform $\lceil \log_2(0.8/10^{-5}) \rceil = 17$ iterations.

Example 8.1
Suppose a circuit represented by a modified version of Eq. (8.4) as follows:

$$-0.5 + 1.4\ln(I + 1) + 0.1I = 0$$

Perform six iterations of the bisection method to determine the value of the current running through this circuit, knowing initially that it is somewhere between 0 and 1 A.

Solution
First, evaluate the function at the given bounds. At $I = 0$ A the total voltage in the system is -0.5 V, and at $I = 1$ A it is 0.5704 A.

The first middle point between the initial bounds is $x_1 = 0.5$ A. Evaluating the function at that value gives 0.1177 V. This is a positive evaluation, just like at 1 A; the middle value thus replaces this bound, and the new interval is [0, 0.5].

(continued)

Example 8.1 (continued)

The second middle point is at $x_2 = 0.25$ A. The function evaluation is -0.1626 V. This is a negative evaluation, on the same side of the zero-crossing as 0 A. Consequently, it replaces that bound and the new interval at this iteration is [0.25,0.5].

The next four iterations are summarized in this table:

Initial interval (A)	Middle point (A)	Function evaluation (V)	Final interval (A)
[0.25, 0.5]	$x_3 = 0.3750$	-0.0167	[0.375, 0.5]
[0.375, 0.5]	$x_4 = 0.4375$	0.0518	[0.375, 0.4375]
[0.375, 0.4375]	$x_5 = 0.4063$	0.0179	[0.375, 0.4063]
[0.375, 0.4063]	$x_6 = 0.3906$	0.0007	[0.375, 0.3906]

At the end of six iterations, the root is known to be in the interval [0.375, 0.3906]. It could be approximated as 0.3828 ± 0.0078 A. For reference, the real root of the equation is 0.389977 A, so the approximation has a relative error of only 1.84 %. To further illustrate this method, the function is plotted in red in the figure below, with the five intervals of the root marked in increasingly light color on the horizontal axis. It can be seen that the method tries to keep the root as close as possible to the middle of each shrinking interval.

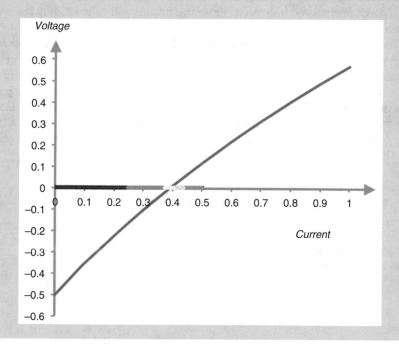

8.3   False Position Method

The false position method is an improvement of the bisection method. Like the bisection method, it is a bracketing algorithm. However, it does not simply use the middle point between the two bounds as a new bound, but instead interpolates a degree-1 polynomial between the two bounds and uses the root of that straight line as the new bound. This usually gives a better approximation of the root then blindly using the middle point, especially when the function being modeled can accurately be represented by a straight line, which will be the case as the interval around the root gets smaller and smaller, as discussed in Chap. 5. To understand how this works, consider the function plotted in blue in Fig. 8.3. Its brackets go from 6 to 8, and its root is at 6.3. The bisection method picks the middle point at each iteration: the middle point this iteration is at 7, which reduces the interval to [6, 7] and yields 6.5 ± 0.5 as an approximation of the root. On the other hand, the false position method interpolates a straight line between the two bounds, shown in red in Fig. 8.3, and uses the root of that interpolated polynomial at 6.45 to reduce the interval to [6, 6.45]. This interval is already a lot better than the one obtained after one iteration (or even two iterations) of the bisection method. After just one step, the root is approximated to be 6.225 ± 0.225, a much better approximation than the one obtained by the bisection method. Moreover, as can be seen in Fig. 8.3, the function in that new interval is practically linear, which means that the root of the polynomial interpolated in-between those two bounds will be practically the same as the root of the real function.

The algorithm for the false position method is a three-step iterative process that is very similar to the bisection method. In particular, the first step to setup the initial bounds is exactly the same as before, to get two points on either side of the zero-crossing. In the second step, the method iteratively tightens the bounds by using the root of the straight-line polynomial interpolated in-between the two bounds. Interpolating a straight line between two known points and then finding the zero of that line is trivially easy, and in fact both operations can be done at once using Eq. (8.10).

$$x_i = x_U - \frac{f(x_U)(x_L - x_U)}{f(x_L) - f(x_U)} \tag{8.10}$$

Fig. 8.3   The interpolating linear polynomial and its root

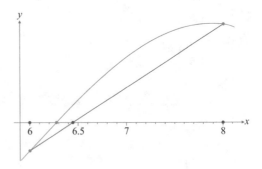

The method then evaluates the new function at the new point, $f(x_i)$, and substitutes x_i for the root on the same side of the zero-crossing. One distinctive feature of the false position method is that it will usually focus on updating only one bound of the interval. For a concave-down function such as the one in Fig. 8.3, the root of the interpolated polynomial will usually be on the right-hand side of the real root and only the right-hand side bound will be the one updated. And conversely, in the case of a concave-up function, the interpolated root will usually be on the left-hand side of the function and only the left-hand bound will be updated.

Finally, the iterative process terminates when one of three termination conditions are reached. Two of these conditions are exactly the same as for the bisection method: the algorithm might generate a point x_i that is exactly the root of the function (success condition), or reach a preset maximum number of iterations (failure condition). The third condition is that the root of the function is approximated to an acceptable preset error rate (success condition). However, this acceptable approximation is defined differently in this algorithm than it was in the bisection algorithm or in most bracketing algorithms. Since usually only one bound is updated, the relative error between two successive update values can be used to measure the error, using the usual relative error value introduced in Chap. 1:

$$E_i = \left| \frac{x_{i-1} - x_i}{x_i} \right| \tag{8.11}$$

The pseudocode for this method, in Fig. 8.4, will clearly be very similar to the one for the bisection method, which was presented in Fig. 8.2, with only a different formula to evaluate the new point.

Example 8.2

Suppose a circuit represented by a modified version of Eq. (8.4) as follows:

$$-0.5 + 1.4\ln(I + 1) + 0.1I = 0$$

Perform six iterations of the bisection method to determine the value of the current running through this circuit, knowing initially that it is somewhere between 0 and 1 A.

Solution

First, evaluate the function at the given bounds. At $I = 0$ A the total voltage in the system is -0.5 V, and at $I = 1$ A, it is 0.5704 A.

Using Eq. (8.10), the first interpolated root is found to be at:

$$x_1 = 1 - \frac{0.5704(0 - 1)}{-0.5 - 0.5704} = 0.4671 \text{ A}$$

and evaluating the function using that current value gives a voltage of 0.0833 V. This is a positive evaluation, just like at 1 V; the middle value thus replaces this bound, and the new interval is [0, 0.4671].

(continued)

Example 8.2 (continued)

The next interpolated root is at $x_2 = 0.4004$ A. The function evaluation gives 0.0115 V. This is again a positive evaluation, on the same side of the zero-crossing as the previous point. Consequently, it replaces that bound and the new interval at this iteration is [0 0.4004].

The next four iterations are summarized in this table:

Initial interval (A)	Interpolated root (A)	Evaluation (V)	Final interval (A)
[0,0.4004]	$x_3 = 0.3914$	0.0016	[0,0.3914]
[0,0.3914]	$x_4 = 0.3902$	0.0002	[0,0.3902]
[0,0.3902]	$x_5 = 0.3900$	0.000029	[0,0.3900]
[0,0.3900]	$x_6 = 0.38998$	0.000004	[0,0.38998]

At the end of six iterations, the root is known to be in the vicinity of 0.38998 A. This is a relative error of 0.000974 % compared to the real root at 0.389977 A, a major improvement compared to the bisection method, which only achieved 1.84 % relative error after the same number of iterations. Notice as well that, in all the iterations, only one bound was ever updated.

To further illustrate this method, the function is plotted in red in the figure below, with the five intervals of the root marked in increasingly light color on the horizontal axis. It can be seen that the method tries to keep the root as close as possible to the upper bound of the interval. As a result, by contrast to Example 6.1, the interval size decreases much more slowly and the final interval is a lot larger than it was with the bisection method, but the final approximation of the root is much more accurate.

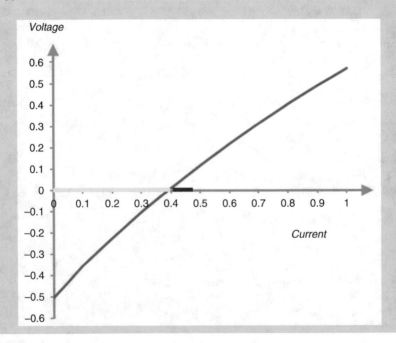

```
XL ← Input lower bound
XU ← Input upper bound
IterationMaximum ← Input maximum number of iterations
ErrorMinimum ← Input minimum relative error
x ← XL

IterationCounter ← 0
WHILE (TRUE)

    PreviousValue ← x
    x ← XU - (CALL F(XU)) × (XL - XU) / [(CALL F(XL)) - (CALL F(XU))]
    Evaluation ← CALL F(x)

    IF ( Evaluation = 0 )
        RETURN Success, x
    ELSE IF ( Evaluation is of the same sign as CALL F(XL) )
        XL ← x
    ELSE IF ( Evaluation is of the same sign as CALL F(XU) )
        XU ← x
    END IF

    CurrentError ← absolute value of [ (x - PreviousValue) / x ]
    IterationCounter ← IterationCounter + 1

    IF (CurrentError <= ErrorMinimum)
        RETURN Success, x
    ELSE IF (IterationCounter = IterationMaximum)
        RETURN Failure
    END IF

END WHILE


FUNCTION F(x)
    RETURN evaluation of the target function at point x
END FUNCTION
```

Fig. 8.4 Pseudocode of the false position method

8.3.1   Error Analysis

To simplify the error analysis, assume that one of the bounds is fixed. As explained previously, that assumption will be correct for all except occasionally the first few iterations of the false position method.

Let x_r be the root, bracketed between a lower bound a_0 and an upper bound b, and assume that the bound b is fixed. Then, the change in the moving bound will be proportional to the difference between the slope from $(x_r, 0)$ to $(b, f(b))$ and the derivative $f^{(1)}(x_r)$. To visualize this, define the error between the root and the moving bound $h_0 = |a_0 - r|$ and assume that the bound is sufficiently close to the root that the first-order Taylor series approximation $f(a_0) \approx f^{(1)}(r)h_0$ holds. In that case, the slope of the interpolated polynomial from $(a_0, f(a_0))$ to $(b, f(b))$ is approximately equal to $f(b)/(b - x_r)$. This is shown in Fig. 8.5.

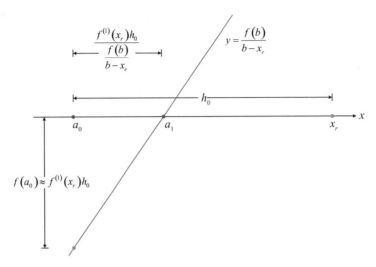

Fig. 8.5 One iteration of the false position method

After one iteration, the bound will move to a_1, the root of the interpolated line (in red), and the error will be reduced by the distance between a_0 and a_1, which can be computed by simple trigonometry as indicated in Fig. 8.5. The resulting error is:

$$h_1 = h_0 - \frac{f^{(1)}(x_r)h_0}{f(b)/(b-x_r)} = h_0\left(1 - \frac{f^{(1)}(x_r)(b-x_r)}{f(b)}\right) \tag{8.12}$$

and more generally at iteration n:

$$h_n = h_0\left(1 - \frac{f^{(1)}(x_r)(b-x_r)}{f(b)}\right)^n \tag{8.13}$$

The error rate thus decreases linearly as a function of h, or $O(h)$, just like for the bisection method. This result seems to clash with the observation earlier, both in the explanations and in Example 8.2 that the error of the false position method decreases faster than that of the bisection method. In fact, Eq. (8.13) shows that this method will not always converge faster than the bisection method, but only in the specific case where the factor multiplying h_0 is smaller, that is to say:

$$\frac{1}{2} < \frac{f^{(1)}(r)(b-x_r)}{f(b)} \tag{8.14}$$

Since the root and the derivative are beyond one's ability to change, the only ways to improve the convergence rate of the false position algorithm is to increase the difference between b and x_r, or to decrease the value of $f(b)$. It can be seen from Fig. 8.5 that the effect of these actions will be, respectively, to move the point b further right on the x-axis or down nearer to the x-axis. Either option will cause the root of the interpolated line between a_0 and b to be nearer to the root.

Example 8.3

Compare and contrast the error of the bisection method and the false position method from Examples 8.1 and 8.2.

Solution

The points computed in each of the six iterations are listed in the table below, along with the relative error of each one compared to the real root at 0.389977 A.

Iteration	Bisection		False position	
	Point (A)	Relative error (%)	Point (A)	Relative error (%)
1	$x_1 = 0.5$	28.21	$x_1 = 0.4671$	19.78
2	$x_2 = 0.25$	35.89	$x_2 = 0.4004$	2.67
3	$x_3 = 0.375$	3.84	$x_3 = 0.3914$	0.36
4	$x_4 = 0.4375$	12.19	$x_4 = 0.3902$	0.05
5	$x_5 = 0.40625$	4.17	$x_5 = 0.3900$	0.0068
6	$x_6 = 0.390625$	0.17	$x_6 = 0.38998$	0.000974

It is clear to see that the false position method reduces the error a lot more quickly than the bisection method: after three iterations the error achieved by the false position is comparable to that from six iterations of the bisection method, and after the fifth iteration, the false position method has surpassed by orders of magnitude the best error achieved by the bisection method.

But it is even more interesting to observe the progression of the error. The relative error of points generated by the false position method always decreases after each generation, and does so almost perfectly linearly. By contrast, the relative error of the points computed by the bisection method zigzags, it decreases but then increases again between iterations 3 and 4, and between iterations 5 and 6. This zigzag is another consequence of blindly selecting the middle point of the interval at each iteration. When the real root is near the center of the interval, the middle point selected by the bisection method will have a low relative error, but when the root is nearer the edge of the interval the middle point will have a high relative error. By contrast, the false position method selects points intelligently by interpolating an approximation of the function and using the approximation's root, and thus is not subject to these fluctuations. Moreover, as the interval becomes smaller after each iteration, the approximation becomes more accurate and the interpolated root is guaranteed to become closer to the real one.

8.3.2   Nonlinear Functions

The false position method introduced an additional assumption that was not present in the bisection method, namely that the function can be approximated within the interval by a linear polynomial. This is of course not always the case, and it is important to be mindful of it: when this assumption does not hold, the function cannot be approximated well by a straight line, and consequently the root of the straight line is a very poor approximation of the root and cannot be used to effectively tighten the bounds.

To illustrate, an example of a highly nonlinear function is given in Fig. 8.6: within the interval [0,1], the function looks like a straight horizontal line with a sudden sharp vertical turn near the root at 0.96. The function is concave-up, and as explained before the root of the interpolated polynomial falls on the left side of the real root, and the left-hand bound is the one updated. And indeed, it can be seen in that Figure that a straight line interpolated from 0 to 1 will have a root at a point before 0.96, and therefore the left bound will be updated. However, interpolated line's root will actually be at 0.04, very far from the real root at 0.96! This is a result of the fact that the straight line from 0 to 1 is not at all a good approximation of the concave function it is supposed to represent. Worse, the bounds will be updated to [0.04,1] and the function in that interval will still be nonlinear, so in the next iteration the false position method will again interpolate a poor approximation and poorly update the bounds. In fact, it will take over 20 iterations for the false position to generate a approximation of the root of this function within the interval [0.9,1]. By contrast, the bisection method, by blindly cutting out half the interval at each iteration, gets within that interval in four iterations.

Referring back to Eq. (8.13), it can be seen that the inequality necessary for the false position method to outperform the bisection method does not hold in this example. The difference between the root and the fixed bound is only 0.04, while

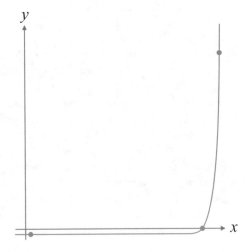

Fig. 8.6 A highly nonlinear function

the first derivative at the root is almost 1 because of the discontinuity and the value of $f(b)$ is exactly 1, so Eq. (8.13) evaluates to a result much smaller than 0.5.

Clearly, it is important to determine whether the function can be approximated by a straight line interpolated between the bounds before starting the false position method, otherwise many iterations will be wasted computing poor approximations of the root. The solution to that problem is also hinted at the end of the example: to switch to the bisection method, which will work the same regardless of whether the function is linear or nonlinear within the interval, for a few iterations, until the interval has been reduced to a region where the function is closer to linear.

8.4 Closed and Open Methods

The bisection and false position methods are both bracketing root-finding methods. They are called *closed* methods, because they enclose the root between bounds. These bounds constitute both an advantage and a limitation. On the one hand, they guarantee that the methods will converge, that they will succeed in finding the root. Indeed, these methods cannot possibly fail, since they begin by locking in the root between bounds and never lose sight of it. However, iteratively updating bounds is a slow process, and the two methods seen so far only have $O(h)$ convergence rates.

The alternatives to closed methods are *open* methods. As the name implies, these methods do not enclose the root between bounds. They do use initial points, but these points could all be on the same side of the root. The methods then use some mathematical formula to iteratively refine the value of the root. Since these algorithms can update their estimates without worrying about maintaining bounds, they typically converge a lot more efficiently than closed methods. However, for the same reason that they do not keep the root bracketed, they can sometimes diverge and fail to find the root altogether if they use a bad point or combination of points in their computations.

8.5 Simple Fixed-Point Iteration

The simple fixed-point iteration (SFPI) method is the simplest open root-finding method available. As will be seen, it is also the open method with the worst convergence rate in general and it diverges in many common situations, so it is far from the best. However, it will be useful to use as the first open numerical method in this book, to introduce fundamental notions that will be applied to all other methods.

As mentioned in the previous section, an open method is one that iteratively improves an (unbounded) estimate of the solution point x_i. The first necessary step to any open method is thus to write the system being studied in an iterative form of $x_{i+1} = f(x_i)$. In the case of root-finding methods, however, this is a special problem, since the solution point is the one where $f(x_i) = 0$. It is necessary to modify the equation of the system's model somehow. Each open root-finding method that will be presented in the next sections will be based on a different intuition to rewrite the equation into iterative form. For SPFI, it is done simply by isolating a single instance of x in the equation $f(x) = 0$ to get $g(x) = x$. In other words:

$$f(x) = g(x) - x = 0 \Rightarrow g(x) = x \tag{8.15}$$

This transformation is nothing more than a bit of algebra. If an instance of x is available in the equation $f(x)$, isolate it. If not, then simply add x to both side of $f(x) = 0$. Equation (8.16) gives examples of both situations.

$$f(x) = -5x^3 + 4x^2 + 3x - 2 = 0 \Rightarrow g(x) = \frac{-5x^3 + 4x^2 - 2}{-3} = x \tag{8.16}$$
$$f(x) = \ln(x) = 0 \Rightarrow g(x) = \ln(x) + x = x$$

The fixed-point equation of (8.15) is an iterative equation. Given an initial estimate of the root x_0, the method computes

$$x_{i+1} = g(x_i) \tag{8.17}$$

until the equation converges; and given Eq. (8.15), the value of x it converges to is the root of $f(x)$. As with any iterative method, the standard halting conditions apply. The method will be said to have converged and succeeded if the relative error between two successive iterations is less than a predefined threshold ε:

$$E_i = \left| \frac{x_{i+1} - x_i}{x_{i+1}} \right| < \varepsilon \tag{8.18}$$

And it will be said to have failed if the number of iterations reaches a preset maximum. The pseudocode for this method is presented in Fig. 8.7. Notice that, contrary to the pseudocode of the bisection and false position methods, this one does not maintain two bounds, only one current value x. Consequently, the iterative update is a lot simpler; while before it was necessary to check the evaluation of the current value against that of each bound to determine which one to replace, now the value is updated unconditionally.

Example 8.4
The power of a signal being measured by a tower is decreasing over time according to this model:

$$P(t) = e^{-t} \cos(t)$$

Starting at $t_0 = 0$ s, find the time that the signal will have lost all power to a relative error of less than 0.5 %.

(continued)

Example 8.4 (continued)
Solution
The iterative SFPI equation corresponding to this problem is:

$$e^{-t_i} \cos(t_i) + t_i = t_{i+1} = g(t_i)$$

At the initial value of $t_0 = 0$, $t_1 = e^0 \cos(0) + 0 = 1$, and the relative error computed by Eq. (8.18) is 100 %. This is usual for a first iteration when using 0 as an initial estimate. The following iterations, until the target relative error is achieved, are given in the table below. The figure next to the table illustrates the power function and the approximation of the root at each iteration.

Iteration	t_i (s)	E_i (%)	
0	0		Power
1	1	100.00	
2	1.20	16.58	
3	1.31	8.38	
4	1.38	5.09	
5	1.43	3.38	
6	1.46	2.36	
7	1.49	1.71	
8	1.51	1.26	
9	1.52	0.95	
10	1.53	0.72	
11	1.54	0.56	
12	1.55	0.43	

For reference, the real root of the equation is at $t_r = 1.5707963267949$ s, so the relative error of the final result compared to the real root is 1.56 %.

It is interesting to note as well that, from iteration 4 onward, the relative error of each iteration is approximately 0.75 of the previous one. This shows practically that the algorithm has a linear convergence rate comparable to that of the bisection method.

The SFPI method works by computing a sequence $x_i \to g(x_i) \to x_{i+1} \to g(x_{i+1})$ $\to \cdots$, and converges on the point where $x_{i+1} = g(x_i)$. Example 8.4 can be visualized graphically by making a graph of t_i against t_{i+1}, and plotting both sides of the equation, namely $t_{i+1} = g(t_i)$ and $t_{i+1} = t_i$, and seeing where both lines intersect. This is shown in Fig. 8.8, with $t_{i+1} = g(t_i)$ in blue and $t_{i+1} = t_i$ in red. The root, the intersection of both functions, is clearly found at $t_{i+1} = t_i = 1.5707963267949$ s. Figure 8.8 also shows the first four iterations of the example as arrows starting at $t_i = 0$, going to the corresponding value of $g(t_i) = t_{i+1} = 1$, then setting $t_i = 1$,

```
x ← Input initial approximation of the root
IterationMaximum ← Input maximum number of iterations
ErrorMinimum ← Input minimum relative error

IterationCounter ← 0
WHILE (TRUE)

    PreviousValue ← x
    x ← CALL G(x)

    IF ( (CALL G(x)) - x = 0 )
        RETURN Success, x
    END IF

    CurrentError ← absolute value of [ (x - PreviousValue) / x ]
    IterationCounter ← IterationCounter + 1

    IF (CurrentError <= ErrorMinimum)
        RETURN Success, x
    ELSE IF (IterationCounter = IterationMaximum)
        RETURN Failure
    END IF

END WHILE


FUNCTION G(x)
    RETURN evaluation of the transformed version of the function at point x
END FUNCTION
```

Fig. 8.7 Pseudocode of the SPFI method

Fig. 8.8 The convergence of Example 8.4

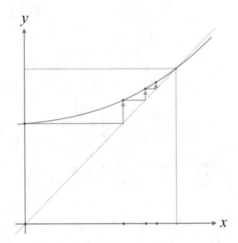

pointing to $g(t_i) = t_{i+1} = 1.20$, and so on. It can be seen that the method converges towards the root.

But, as explained in Sect. 8.4, convergence is not guaranteed for open methods like the SFPI. In fact, it can be shown that the SFPI method only converges if the

Fig. 8.9 The divergence
of Eq. (8.18)

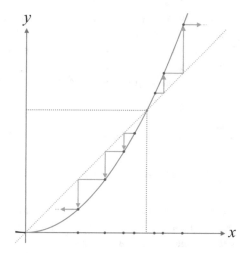

slope of $g(x)$ at the root point x_r is less than the slope of x at that point, or in other
words if the absolute value of the derivative of $g(x_r)$ is less than 1. That was not a
problem for the function of Example 8.8; its derivative is $g^{(1)}(x) = -e^{-x}(\cos(x) +
\sin(x)) + 1$, and evaluated at the root it gives 0.792120424, below the threshold
value of 1. However, consider for example the function $f(x) = x^2 - x$, which has two
roots at $x_{r0} = 0$ and $x_{r1} = 1$. In the SFPI method, this equation becomes:

$$x_{i+1} = g(x_i) = x_i{}^2 \tag{8.19}$$

The absolute value of the derivative is $|g^{(1)}(x)| = |2x_i|$, and evaluated at the roots it
gives $|g^{(1)}(x_{r0})| = 0$ and $|g^{(1)}(x_{r1})| = 2$. This means the SPFI will converge on the first
root, but cannot converge on the second root. The practical impact of this problem
is illustrated in Fig. 8.9. As this figure shows, picking an initial value x_0 that is less
than 1 will lead to the SPFI converging on the root at 0. However, picking an initial
value that is greater than 1 doesn't allow the SPFI to converge on the root at 1, but
instead it causes the method to diverge quickly away from both roots and towards
infinity. The root at 1 is simply unreachable by the SPFI, unless it is picked as the
initial value (in which case the method converges on it without iterating). This
discussion also illustrates two other problems with the SPFI method. First, one
cannot predict whether the method will converge or diverge ahead of time unless
one already knows the value of the root in order to evaluate the derivative at that
point, which is of course not a piece of information available ahead of time in a
root-finding problem. And second, the form of $g(x)$ that the equation $f(x)$ is
rewritten into actually matters. Multiple different forms of $g(x)$ are possible for
one equation $f(x)$, and not all of them will converge on the same values, or at all.

The convergence rate for the SPFI method is derived from a Taylor series
expansion of the function $g(x)$. Recall from Chap. 5 that this means that the error
rate will be proportional to the order of the first non-null term in the series. In other

words, much like the convergence test, the convergence rate will depend on evaluations of the derivative at the root. If $g^{(1)}(x_r)=0$, then the error rate will be $O(h^2)$, if in addition $g^{(2)}(x_r)=0$ then the error rate will be $O(h^3)$, and if in addition to those two $g^{(3)}(x_r)=0$ then the error rate will be $O(h^4)$, and so on. In the general case though, the assumption is that $g^{(1)}(x_r)\neq0$ and the error rate is $O(h)$.

8.6   Newton's Method

8.6.1   One-Dimensional Newton's Method

Newton's method, also called the Newton-Raphson method, is possibly the most popular root-finding method available. It has a number of advantages: it converges very efficiently (in fact it has the highest convergence rate of any root-finding methods covered in this book), it is simple to implement, and it only requires to maintain one past estimate of the root, like the SFPI but unlike any of the other root-finding methods available. Its main downside is that it requires knowing or estimating the derivative of the function being modelled.

The basic assumption behind Newton's method is that, for a small enough neighborhood around a point, a function can be approximated by its first derivative. Since this first derivative is a straight line, its root is straightforward to find. The derivative's root is used as an approximation of the original function's root, and as a new point to evaluate the derivative at to iteratively improve the approximation. As the approximation point gets closer to the root and the neighborhood approximated by the first derivative gets smaller, the first derivative becomes a more accurate approximation of the function and its root becomes a more accurate approximation of the function's root. To illustrate, a single iteration of Newton's method is represented graphically in Fig. 8.10.

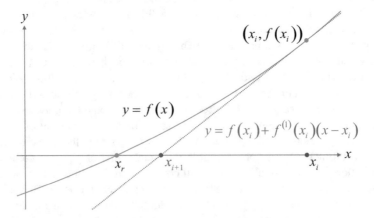

Fig. 8.10 One iteration of Newton's method

The two underlying ideas of Newton's method should be familiar: the approximation of a function by its first derivative is the first-order Taylor series approximation explained back in Chap. 4, and modelling the root of a function by the root of a straight-line approximation is the idea behind the false position method. In fact, Newton's method is derived directly from the first-order Taylor series approximation:

$$f(x_{i+1}) = f(x_i) + f^{(1)}(x_i)(x_{i+1} - x_i) \qquad (8.20)$$

Since the function is converging to a root, then the value $f(x_{i+1}) = 0$. With that value set, Eq. (8.20) can be rewritten as an iterative formula of x:

$$x_{i+1} = x_i - \frac{f(x_i)}{f^{(1)}(x_i)} \qquad (8.21)$$

And that is Newton's method.

There are three halting conditions for Newton's method. Two are the standard conditions: the success condition if the function converges and the relative error between two successive approximations is less than a predefined threshold ε, as defined in Eq. (8.18), and the failure condition if the method reaches a preset maximum number of iterations. To this, it is necessary to add a third failure condition, if the first derivative $f^{(1)}(x_i) = 0$. In that case, the point generated is in a discontinuity, which causes a division by zero in Eq. (8.21), and the method cannot continue.

The pseudocode for Newton's method is given in Fig. 8.11. Notice the introduction of a new function to evaluate the derivative of a target function, which is necessary to compute Eq. (8.21), as well as the additional failure test if the first derivative is zero.

Since the equation for Newton's method is the first-order Taylor series approximation, the error of the method is the same as that of the approximation. Recall from Chap. 4 that the error of a Taylor series approximation is the next non-zero term after the cut-off point of the series. In this case, it is the second-order term, and so the error is $O(h^2)$. Newton's method thus has a quadratic convergence rate, much better than the linear methods seen so far. To further understand this, consider the second-order Taylor series approximation of the root x_r:

$$f(x_r) = f(x_i) + f^{(1)}(x_i)(x_r - x_i) + \frac{f^{(2)}(x_i)}{2}(x_r - x_i)^2 \qquad (8.22)$$

Following the same steps used to get from Eq. (8.20) to (8.21) but putting the entire Newton's method equation on one side of the equation gives:

$$x_r - x_i + \frac{f(x_i)}{f^{(1)}(x_i)} = \frac{f^{(2)}(x_i)}{2f'(x_i)}(x_r - x_i)^2 \qquad (8.23)$$

```
x ← Input initial approximation of the root
IterationMaximum ← Input maximum number of iterations
ErrorMinimum ← Input minimum relative error

IterationCounter ← 0
WHILE (TRUE)

    PreviousValue ← x
    x ← x - (CALL F(x)) / (CALL Derivative(F(x)))

    IF ( CALL F(x) = 0 )
        RETURN Success, x
    ELSE IF ( CALL Derivative(F(x)) = 0 )
        RETURN Failure
    END IF

    CurrentError ← absolute value of [ (x - PreviousValue) / x ]
    IterationCounter ← IterationCounter + 1

    IF (CurrentError <= ErrorMinimum)
        RETURN Success, x
    ELSE IF (IterationCounter = IterationMaximum)
        RETURN Failure
    END IF

END WHILE


FUNCTION F(x)
    RETURN evaluation of the target function at point x
END FUNCTION

FUCNTION Derive(F(x))
    RETURN evaluation of the derivative of the function at point x
END FUNCTION
```

Fig. 8.11 Pseudocode of Newton's method

Note that Newton's method formula for x_{i+1} is on the left of the equation. Both sides of the equation then have a subtraction of x_r to an approximation, which is the error h at that point. The equation thus simplifies to:

$$h_{i+1} = \frac{f^{(2)}(x_i)}{2f^{(1)}(x_i)} h_i^2 \qquad (8.24)$$

Which is indeed an $O(h^2)$ error rate.

Example 8.5

The power of a signal being measured by a tower is decreasing over time according to this model:

$$P(t) = e^{-t} \cos(t)$$

Starting at $t_0 = 0$ s, find the time that the signal will have lost all power to a relative error of less than 0.5 %.

Solution

The derivative of this function is:

$$P'(t) = -e^{-t}(\sin(t) + \cos(t))$$

And the iterative formula for Newton's method, from Eq. (8.21), is thus:

$$t_{i+1} = t_i - \frac{e^{-t_i} \cos(t_i)}{-e^{-t_i}(\sin(t_i) + \cos(t_i))}$$

Starting at $t_0 = 0$, the iterations computed are:

Iteration	t_i (s)	E_i (%)
0	0	
1	1	100
2	1.39	28.1
3	1.54	10.0
4	1.570	1.6
5	1.571	0.04

For reference, the real root of the equation is at $t_r = 1.5707963267949$ s, so the relative error of the final result compared to the real root is less than 3e-5.

The benefits of quadratic convergence are clear to see: after three iterations the approximation obtained by Newton's method was the same as that obtained by the SFPI method after 11 iterations, and the final result of this method is orders of magnitude better than that of the SFPI method despite requiring less than half the total number of iterations.

As with any open method, there is a risk that Newton's method will diverge and fail to find a root altogether. In fact, Eq. (8.24) suggests that there are three conditions when the method can diverge. The first condition is if x_i is very far from the root x_r. In that case, the error h_i will be a large value and squared, which will cause the method to fail. This highlights the need to pick an initial value that is

Fig. 8.12 Divergence
because the first derivative
is near zero

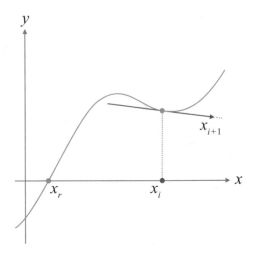

somewhat in the vicinity of the root, and not any random value anywhere in the
x-axis. The second condition that can cause the method to diverge is if the first
derivative at the current point is near zero. This has already been included as a
failure halting condition in the algorithm. Conceptually, this means the point x_i is at
an optimum of the function, and the first derivative is horizontal. In such a case, the
next point computed by Eq. (8.21) will shoot out very far from the current point and
the root. This situation is illustrated in Fig. 8.12. The third and final condition that
can cause the method to diverge is if the second derivative is very large. Again, it is
clear to see from Eq. (8.24) that this will cause the error to increase between
iterations. Conceptually, a high second derivative means that the current point is
near a saddle of the function. In that case, Eq. (8.21) will generate points that
oscillate around this saddle and get progressively further and further away. This
situation is illustrated in Fig. 8.13.

8.6.2   Multidimensional Newton's Method

The examples used so far in this chapter have all been one-dimensional $y = f(x)$
root-finding problems. However, one of the main advantages of Newton's method,
especially when contrasted to the bracketing methods, is that it can easily be
adapted to more complex multidimensional problems. In such a problem, an
engineer must deal with n independent variables constrained by n separate model
equations. The root of the system is the simultaneous root of all n equations.

Begin by defining $\mathbf{x} = [x_0, x_1, \ldots, x_{n-1}]^T$, the vector of n independent variables of
the system, and $\mathbf{f}(\mathbf{x}) = [f_0(\mathbf{x}), f_1(\mathbf{x}), \ldots, f_{n-1}(\mathbf{x})]^T$, the vector of n n-dimensional
functions that model the system. Since this is now a vector problem, Newton's
method equation (8.21) needs to be rewritten to eliminate the division as:

Fig. 8.13 Divergence because the second derivative is high

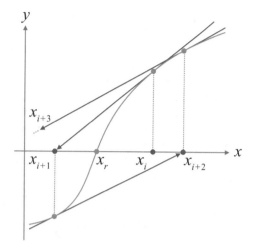

$$f^{(1)}(x_i)(x_{i+1} - x_i) = -f(x_i) \tag{8.25}$$

And then, substituting the newly defined vectors:

$$\mathbf{f}^{(1)}(\mathbf{x}_i)(\mathbf{x}_{i+1} - \mathbf{x}_i) = -\mathbf{f}(\mathbf{x}_i) \tag{8.26}$$

The derivative of the vector of functions is the *Jacobian matrix* $\mathbf{Jf(x)}$, or the $n \times n$ matrix of partial derivatives of each of the n functions with respect to each of the n variables, arranged as shown in Eq. (8.27):

$$\mathbf{Jf(x)} = \begin{bmatrix} \dfrac{\partial f_0(\mathbf{x})}{\delta x_0} & \dfrac{\partial f_0(\mathbf{x})}{\partial x_1} & \cdots & \dfrac{\partial f_0(\mathbf{x})}{\partial x_{n-1}} \\ \dfrac{\partial f_1(\mathbf{x})}{\partial x_0} & \dfrac{\partial f_1(\mathbf{x})}{\partial x_1} & \cdots & \dfrac{\partial f_1(\mathbf{x})}{\partial x_{n-1}} \\ \vdots & & \ddots & \vdots \\ \dfrac{\partial f_{n-1}(\mathbf{x})}{\partial x_0} & \dfrac{\partial f_{n-1}(\mathbf{x})}{\partial x_1} & \cdots & \dfrac{\partial f_{n-1}(\mathbf{x})}{\partial x_{n-1}} \end{bmatrix} \tag{8.27}$$

With this new definition, Eq. (8.26) can be rewritten as:

$$\mathbf{Jf(x}_i) \times \Delta \mathbf{x}_i = -\mathbf{f}(\mathbf{x}_i) \tag{8.28}$$

At each iteration, the Jacobian function can be evaluated, and only the step size $\Delta \mathbf{x}_i$ is unknown. The problem has thus become an $\mathbf{Mx = b}$ linear algebra equation to solve, which can be done using any of the methods learned in Chap. 4. Finally, the next vector is obtained simply with:

$$\mathbf{x}_{i+1} = \mathbf{x}_i + \Delta \mathbf{x}_i \tag{8.29}$$

The halting conditions for the iterative algorithm are the same as for the one-dimensional Newton's method, but adapted to matrices and vectors. The success condition is that the relative error between two successive approximations of the root is less than a preset threshold value, defined now as the Euclidean distance between the two vectors $\mathbf{x}_i$ and $\mathbf{x}_{i+1}$ introduced in Chap. 3. There is a failure condition if the derivative is zero, as there was with the one-dimensional Newton's method. This is defined here as the case where the determinant of the Jacobian matrix is zero. Finally, as always, the algorithm fails if it reaches a preset maximum number of iterations. The pseudocode for Newton's method, updated to handle multidimensional problems, has been updated from Fig. 8.11 and is presented in Fig. 8.14.

Since the multidimensional Newton's method equation of (8.28) is derived from the first-order Taylor series approximation, just like the one-dimensional case, it will also have $O(h^2)$ convergence rate.

```
x ← Input initial approximation of the root as vector of length n
IterationMaximum ← Input maximum number of iterations
ErrorMinimum ← Input minimum relative error

IterationCounter ← 0
WHILE (TRUE)

    PreviousValue ← x
    Delta ← solution of system [CALL Jacobian(F(x))] × Delta
              = -1 × [CALL F(x)]
    x ← x + Delta

    IF ( CALL F(x) = 0 )
        RETURN Success, x
    ELSE IF ( Determinant of [CALL Jacobian(F(x))] = 0 )
        RETURN Failure
    END IF

    CurrentError ← Euclidean distance between x and PreviousValue
    IterationCounter ← IterationCounter + 1

    IF (CurrentError <= ErrorMinimum)
        RETURN Success, x
    ELSE IF (IterationCounter = IterationMaximum)
        RETURN Failure
    END IF

END WHILE


FUNCTION F(x)
    RETURN vector of length n of evaluations of the n target functions at
           point x
END FUNCTION

FUCNTION Jacobian(F(x))
    RETURN nxn matrix of evaluation of the partial derivatives of the n
           functions with respect to the n variables   at point x
END FUNCTION
```

Fig. 8.14 Pseudocode of the multidimensional Newton's method

Example 8.6

The shape of the hull of a sunken ship is modelled by this equation:

$$z = f_0(x, y) = x^2 + 2y^2 - xy - x - 1$$

where z is the height of the sunken hull above the seabed. An automated submarine is scanning the hull to find the damaged point where the ship has hit the seabed. It has been programmed to explore it in a 2D grid pattern, starting at coordinates (0,0) and following this program:

$$z = f_1(x, y) = 3x^2 + 2y^2 + xy - 3y - 2$$

Determine if the probe will find the damage it is looking for with a relative error of 0.001.

Solution

This underwater exploration can be modelled by the following system of equations:

$$\mathbf{f}(\mathbf{x}) = \begin{bmatrix} x^2 + 2y^2 - xy - x - 1 \\ 3x^2 + 2y^2 + xy - 3y - 2 \end{bmatrix}$$

$$\mathbf{x} = \begin{bmatrix} x \\ y \end{bmatrix}$$

The point where the hull has hit the sea bed is at $z = 0$, and that is the damaged point the probe is looking for. It is therefore a root-finding problem. To use Newton's method, first compute the Jacobian following Eq. (8.27):

$$\mathbf{Jf}(\mathbf{x}) = \begin{bmatrix} 2x - y - 1 & 4y - x \\ 6x + y & 4y + x - 3 \end{bmatrix}$$

Then, starting at $\mathbf{x}_0 = [0, 0]^T$, the first iteration will compute

$$\mathbf{Jf}(\mathbf{x}_0) \times \mathbf{\Delta x}_0 = -\mathbf{f}(\mathbf{x}_0)$$

$$\begin{bmatrix} -1 & 0 \\ 0 & -3 \end{bmatrix} \mathbf{\Delta x}_0 = \begin{bmatrix} 1 \\ 2 \end{bmatrix}$$

$$\mathbf{\Delta x}_0 = \begin{bmatrix} -1.0000 \\ -0.6667 \end{bmatrix}$$

$$\mathbf{x}_1 = \mathbf{x}_0 + \mathbf{\Delta x}_0$$

$$\mathbf{x}_1 = \begin{bmatrix} -1.0000 \\ -0.6667 \end{bmatrix}$$

(continued)

Example 8.6 (continued)

The relative error is computed using the Euclidean distance between $\mathbf{x}_0$ and $\mathbf{x}_1$:

$$E_1 = \sqrt{(0-(-1))^2 + (0-(-0.6667))^2} = 1.2019$$

which is well above the required threshold of 0.001. The next iterations are given in the following table:

i	$\mathbf{x}_i$	$\Delta\mathbf{x}_i$	E_i
0	$[0, 0]^{\mathrm{T}}$	$[-1.0000, -0.66667]^{\mathrm{T}}$	
1	$[-1.0000, -0.66667]^{\mathrm{T}}$	$[0.12483, 0.55851]^{\mathrm{T}}$	1.2019
2	$[-0.87517, -0.10816]^{\mathrm{T}}$	$[0.20229, -0.079792]^{\mathrm{T}}$	0.5723
3	$[-0.67288, -0.18795]^{\mathrm{T}}$	$[0.032496, -0.0040448]^{\mathrm{T}}$	0.2175
4	$[-0.64038, -0.19199]^{\mathrm{T}}$	$[0.00056451, 0.00016391]^{\mathrm{T}}$	0.0328
5	$[-0.63982, -0.19183]^{\mathrm{T}}$		0.0009

For reference, the real root is found at $[-0.6397934171, -0.1918694996]^{\mathrm{T}}$, so the approximation found by Newton's method after five iterations has a real relative error of only 4.8e-5. To further illustrate, the curves of the ship's hull and of the probe's exploration pattern are represented in the figure below, with the common root of both equations marked.

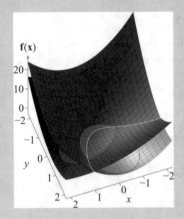

8.7   Secant Method

One limitation of Newton's method is that it requires computing the derivative of the function being studied at multiple points. There are many cases where that can be a problem: situations where the derivative is unknown and cannot be estimated to a good accuracy, for example, or situations where the derivative is too difficult or computationally expensive to evaluate. In these cases, Newton's method cannot be used. One alternative is to approximate the derivative using the secant line of the curve, a line passing through (or interpolating) two points on the function. As these two points iteratively become closer to the root and to each other, the secant line will become an approximation of the tangent near the root and this secant method will approximate Newton's method.

From the first-order Taylor series approximation, the approximation of the first derivative at a point x_i computed near a previous point x_{i-1} is given as:

$$f^{(1)}(x_i) \approx \frac{f(x_{i-1}) - f(x_i)}{x_{i-1} - x_i} \tag{8.30}$$

This immediately adds a new requirement into the method: instead of keeping only one current point with Newton's method, it is necessary to keep two points at each iteration. This is one of the costs of eliminating the derivative from the method. Next, the derivative approximation formula is used to replace the actual derivative in Eq. (8.21):

$$x_{i+1} = x_i - \frac{f(x_i)}{f(x_{i-1}) - f(x_i)/_{x_{i-1} - x_i}} = x_i - \frac{f(x_i)(x_{i-1} - x_i)}{f(x_{i-1}) - f(x_i)} \tag{8.31}$$

And Newton's method is now the secant method. The halting conditions for the iterations are the same as for Newton's method: the method will fail if it reaches a preset maximum number of iterations or if the denominator becomes zero, which will be the case if two points are generated too close to each other (this situation will also introduce the risk of subtractive cancellation explained in Chap. 2), and it will succeed if the relative error between two iterations is less than a preset threshold. The pseudocode for Newton's method in Fig. 8.11 can be updated for the secant method, and is presented in Fig. 8.15.

Since the secant method approximates Newton's method and replaces the derivative with an approximation of the derivative, it should be no surprise that its convergence rate is not as good as Newton's method. In fact, while the proof is outside the scope of this book, the convergence rate of the secant method is $O(h^{1.618})$, less than the quadratic rate Newton's method boasted but better than the linear rate of the other methods presented so far in this chapter.

Equation (8.31) should be immediately recognizable: it is the same as the false position method's equation (8.10). In fact, both methods work in the same way: they both estimate the root by modelling the function with a straight line interpolated from two function points, and use the root of that line as an approximation of the root. The difference between the two methods is in the update process once a new approximation of the root is available. As explained back in Sect. 8.3, the false position method will update the one boundary point on the same side of the zero-crossing as the new point. Moreover, the method will usually generate points only on one side of the zero-crossing, which means that only one of the two bounds is updated, while the other keeps its original value in most of the computations. This will insure that the root stays within the brackets, and guarantee that the method will

```
PreviousValue, x ← Input two initial approximations of the root
IterationMaximum ← Input maximum number of iterations
ErrorMinimum ← Input minimum relative error

IterationCounter ← 0
WHILE (TRUE)

    TemporaryValue ← x
    x ← x - (CALL F(x)) × (PreviousValue - x) /
        [ (CALL F(PreviousValue)) - (CALL F(x)) ]
    PreviousValue ← TemporaryValue


    IF ( CALL F(x) = 0 )
        RETURN Success, x
    END IF

    CurrentError ← absolute value of [ (x - PreviousValue) / x ]
    IterationCounter ← IterationCounter + 1

    IF (CurrentError <= ErrorMinimum)
        RETURN Success, x
    ELSE IF (IterationCounter = IterationMaximum)
        RETURN Failure
    END IF

END WHILE


FUNCTION F(x)
    RETURN evaluation of the target function at point x
END FUNCTION
```

Fig. 8.15 Pseudocode of the secant method

converge, albeit slowly. By contrast, the secant method will update the points in the order they are generated: the new approximation and the previous one are kept and used to compute the next one, and the approximation from two iterations back is discarded. This is done without checking whether the points are on the same side of the zero-crossing or on opposite sides. This allows faster convergence, since the two newest and best estimates of the root are always used in the computations. However, it also introduces the risk that the function will diverge, which was impossible for the false position method.

To understand the problem of divergence with the secant method, consider the example in Fig. 8.16. On the top side, a secant line (in blue) is interpolated between two points x_{i-1} and x_i of the function (in red) and an approximation of the root x_{i+1} is obtained. This approximation is then used along with x_i to interpolate a new secant line, which is a very good approximation of the function. It can clearly be seen that the next approximation x_{i+2} will be very close to the real root of the function. But what if the exact same points had been considered in the opposite order? The result is shown on the bottom side of Fig. 8.16. Initially the same secant

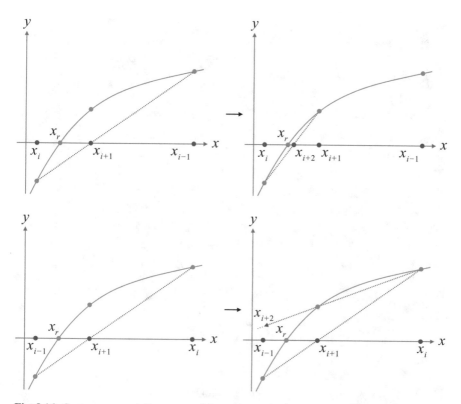

Fig. 8.16 Convergence and divergence of the secant method

line is interpolated and the same approximation x_{i+1} is obtained. However, now the next secant line interpolated between x_i and x_{i+1} diverges and the next point x_{i+2} will be very distant from the root. The problem is that, in this new situation, the points x_i and x_{i+1} are interpolating a section of the function that is very dissimilar to the section that includes the root. As a result, while the interpolation is a good approximation of that section of the function, it is not at all useful for the purpose of root-finding. Meanwhile, because the false position method only updates the point on the same side of the zero-crossing, it can only generate the situation on the top side of Fig. 8.16 regardless of the order the points are fed into the algorithm, and can never diverge in the way shown on the bottom side. Note however that this constraint is not necessary to avoid divergence: it is only necessary for the secant method to use points that interpolate a section of the function similar to the section that has the zero-crossing. For example, using two points both on the negative side of the function would allow the secant method to generate a very good approximation of the root.

Example 8.7
The power of a signal being measured by a tower is decreasing over time according to this model:

$$P(t) = e^{-t} \cos(t)$$

Starting at $t_{-1} = 0$ s and $t_0 = 1$ s, find the time that the signal will have lost all power to a relative error of less than 0.5 %.

Solution
The secant method equation, using Eq. (8.31), is:

$$t_{i+1} = t_i - \frac{(e^{-t_i} \cos(t_i))(t_{i-1} - t_i)}{(e^{-t_{i-1}} \cos(t_{i-1})) - (e^{-t_i} \cos(t_i))}$$

Given the two initial points given, the iterations computed are:

Iteration	t_i (s)	E_i (%)	
-1	0		
0	1	100	
1	1.25	19.9	
2	1.46	14.4	
3	1.54	5.51	
4	1.568	1.61	
5	1.571	0.2	

For reference, the real root of the equation is at 1.5707963267949 s, and the relative error of the final result compared to the real root is 0.005 %. This is comparable to the error of the estimate obtained by Newton's method after 4 iterations in Example 8.5. Conversely, to get the same error as the final result of Newton's method would require six iterations of the secant method. This example shows that the method is very efficient, but not quite as efficient as Newton's method: it requires one more iteration to reach the same relative error.

8.8   Muller's Method

The secant method and the false position method both approximate the function using a straight line interpolated from two points, and use the root of that line as the approximation of the root of the function. The weakness of these methods, as illustrated in Figs. 8.6 and 8.16, is that a straight line is not always a good approximation of a function. To address that problem, a simple solution is available: to use more points and compute a higher-degree interpolation, which would be

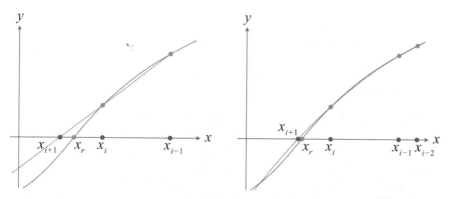

Fig. 8.17 Approximating the root using a degree 1 (*left*) and degree 2 (*right*) interpolation of a function

a better approximation of the function and would make it possible to get a closer approximation of the root. Of course, there is a limit to this approach; the higher the degree of the interpolating polynomial, the more roots it will have and the more difficult it will be to find them all easily and efficiently. Muller's method offers a good compromise. It uses three points to compute a degree-2 interpolation (a parabola) of the function to model. The degree-2 polynomial offers a better approximation of the function than a straight line, as illustrated in Fig. 8.17, while still being easy enough to handle to find the roots.

The first step of Muller's method is thus to approximate the function $f(x)$ using a parabola interpolated from three points x_{i-2}, x_{i-1}, and x_i. The equation for a parabola is well known to be $p(x) = ax^2 + bx + c$. Given three points on the function to model, it can be computed using the Vandermonde method from Chap. 4 as the solution to the system:

$$\begin{bmatrix} x_{i-2}^2 & x_{i-2} & 1 \\ x_{i-1}^2 & x_{i-1} & 1 \\ x_i^2 & x_i & 1 \end{bmatrix} \begin{bmatrix} a \\ b \\ c \end{bmatrix} = \begin{bmatrix} f(x_{i-2}) \\ f(x_{i-1}) \\ f(x_i) \end{bmatrix} \tag{8.32}$$

In order to use these equations in an iterative formula of the form $x_{i+1} = x_i + \Delta x_i$, substitute x for $x - x_i$. This changes the parabola equation to $p(x) = a(x - x_i)^2 + b(x - x_i) + c$ and the Vandermonde system to:

$$\begin{bmatrix} (x_{i-2} - x_i)^2 & (x_{i-2} - x_i) & 1 \\ (x_{i-1} - x_i)^2 & (x_{i-1} - x_i) & 1 \\ 0 & 0 & 1 \end{bmatrix} \begin{bmatrix} a \\ b \\ c \end{bmatrix} = \begin{bmatrix} f(x_{i-2}) \\ f(x_{i-1}) \\ f(x_i) \end{bmatrix} \tag{8.33}$$

Written in that form, the Vandermonde system is trivial to solve. In fact, a solution can be obtained immediately as:

$$a = \frac{\left(\big(f(x_i) - f(x_{i-1})\big)\big/x_i - x_{i-1}\right) - \left(\big(f(x_{i-1}) - f(x_{i-2})\big)\big/x_{i-1} - x_{i-2}\right)}{x_i - x_{i-2}}$$

$$b = a(x_i - x_{i-1}) + \left(\big(f(x_i) - f(x_{i-1})\big)\big/x_i - x_{i-1}\right)$$

$$c = f(x_i)$$

(8.34)

Note that in both the parabola equation and the Vandermonde system, the solution remains unchanged. This is because the subtraction represents only a horizontal shift of the function. All values of the function are moved along the x-axis by a factor of x_i, but they remain unchanged along the y-axis. This is akin to a time-shifting operation in signal processing, and is illustrated in Fig. 8.18 for clarity.

Once the coefficients a, b, and c for the parabola equation are known, the next step is to find the roots of the parabola, which will serve as approximations of the root of $f(x)$. The standard quadratic equation to find the roots of a polynomial is:

$$\frac{r_0}{r_1} = \frac{-b \pm \sqrt{b^2 - 4ac}}{2a}$$

(8.35)

This equation will yield both roots of the polynomial. The one that is useful for the iterative system is the one obtained by setting the $\pm$ sign to the same sign as b. Note however that this will introduce the risk of that the problem of subtractive cancellation described in Chap. 2 will occur in cases where $b^2 \gg 4\,ac$. To avoid this, an alternative form of Eq. (8.35) exists that avoids this issue:

$$\frac{r_0}{r_1} = \frac{-2c}{b \pm \sqrt{b^2 - 4ac}}$$

(8.36)

Finally, the iterative algorithm of Muller's method can be written as:

$$x_{i+1} = x_i + \frac{-2c}{b \pm \sqrt{b^2 - 4ac}}$$

(8.37)

Where the $\pm$ is set to the same sign as b and the values a, b, and c are computed by solving the $\mathbf{Mx} = \mathbf{b}$ system of Eq. (8.33). The algorithm has only two halting conditions: a success conditions if the relative error between two successive values

Fig. 8.18 Horizontal shift of the parabola $f(x)$ to $f(x + 3)$

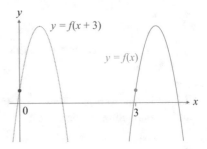

x_i and x_{i+1} is less than a preset threshold, and a failure condition if a preset maximum number of iterations is reached. The pseudocode for this algorithm, using the solution of Eq. (8.34) and a simple test to set the $\pm$ sign, is presented in Fig. 8.19.

The convergence rate of Muller's method is $O(h^{1.839})$, slower than Newton's method but better than the secant method. Intuitively, the fact that Muller's method performs better than the secant method should not be a surprise, since it follow the same idea of interpolating a model of the function and using the model's root as an approximation, but does so with more information (one more point) to get a better model. On the other hand, Newton's method uses information from the function itself, namely its derivative, instead of an approximation, so it should naturally perform better than any approximation-based method.

```
PreviousValue2, PreviousValue, x ← Input three initial approximations of
    the root
IterationMaximum ← Input maximum number of iterations
ErrorMinimum ← Input minimum relative error

IterationCounter ← 0
WHILE (TRUE)

    A ← {[(CALL F(x) - CALL F(PreviousValue)] / [x - PreviousValue] -
        [(CALL F(PreviousValue) - CALL F(PreviousValue2)] /
        [PreviousValue - PreviousValue2]} / (x - PreviousValue2)
    B ← A × (x - PreviousValue) +
        [(CALL F(x) - CALL F(PreviousValue)]/[x - PreviousValue]
    C ← CALL F(x)

    IF (B < 0)
        Sign ← -1
    ELSE
        Sign ← 1
    END IF

    PreviousValue2 ← PreviousValue
    PreviousValue ← x
    x ← x + (-2 × C) / [ B + Sign × square root of (B × B - 4 × A × C) ]

    IF ( CALL F(x) = 0 )
        RETURN Success, x
    END IF

    CurrentError ← absolute value of [ (x - PreviousValue) / x ]
    IterationCounter ← IterationCounter + 1

    IF ( CurrentError <= ErrorMinimum )
        RETURN Success, x
    ELSE IF (IterationCounter = IterationMaximum)
        RETURN Failure
    END IF

END WHILE


FUNCTION F(x)
    RETURN evaluation of the target function at point x
END FUNCTION
```

Fig. 8.19 Pseudocode of Muller's method

Example 8.8

The power of a signal being measured by a tower is decreasing over time according to this model:

$$P(t) = e^{-t} \cos(t)$$

Starting at $t_{-2} = 0$ s, $t_{-1} = 0.1$ s, $t_0 = 0.2$ s, find the time that the signal will have lost all power to a relative error of less than 0.5%.

Solution

Note to begin that this example starts with some initial bad values. Indeed, $p(0) = 1$, $p(0.1) = 0.900$, and $p(0.2) = 0.802$. Nonetheless, use these values to compute the first iteration of Muller's method. Applying Eq. (8.34) finds the coefficients of the interpolated parabola to be:

$$a = \frac{\left(0.802-0.900/0.2-0.1\right) - \left(0.900-0/0.1-0\right)}{0.2 - 0} = 0.089$$

$$b = 0.089(0.2 - 0.1) + \left(0.802+0.900/0.2-0.1\right) = -0.970$$

$$c = 0.802$$

Using these values in Eq. (8.37) finds the relevant root of the interpolated parabola, and the first approximation of the root:

$$t_1 = 0.2 + \frac{-2 \times 0.802}{0.970 - \sqrt{(-0.970)^2 - 4 \times 0.089 \times 0.802}} = 1.101$$

Note that the actual root of this equation is at $t_r = 1.5707963267949$ s. Given the initial points used for the computations, a first approximation at 1.101 is already a huge step forward. The relative error compared to the previous approximation of 0.2 is still too high, and the iterations continue. The next steps are listed in the table below.

Iteration	t_i (s)	E_i (%)	
-2	0		
-1	0.1		
0	0.2		
1	1.101	81.8	
2	1.481	25.6	
3	1.583	6.5	
4	1.5706	0.82	
5	1.5708	0.01	

(continued)

Example 8.8 (continued)

Notice how, as the bad initial values are replaced by better approximations in each successive iteration, the approximation of the root improves a lot. This is a feature of every iterative algorithm: the better the approximations used in its computation are, the better the result will be. However, since Muller's method uses three values in the computation of each iteration, as opposed to only one for Newton's method or two for the secant method, this means that bad initial guesses will negatively affect the result of the iterative algorithm for up to the first three iterations! This is what is observed in this example. Iterations 1 and 2 yield rather poor approximations of the root because they use three and two of the bad initial guesses in their computations, respectively. Iteration 3 gives a much better result because it uses only one bad initial guess and two computed approximations near the real root in its computations. And iteration 4, computed from the three approximations of the first three iterations and not influenced by the bad initial guesses at all, is almost spot-on.

8.9   Engineering Applications

Root-finding problems arise often in engineering design, namely when a system has been modelled by an equation or set of equations with known parameter and property values, and the value of a dependent (controllable) variable of the system must be discovered. If the model used only a linear equation, it would be a simple matter to isolate the variable to be controlled and to compute its optimal value. Unfortunately, most real engineering situations are modelled by more complex equations where the variable is part of exponentials, logarithms, or trigonometric terms, and cannot be isolated. This was the case of the value of the current I in Eq. (8.4); it is simply impossible to isolate it in that equation. Many other such situations can also occur in engineering practice. Some examples are listed here.

- The van der Waals equation relates the pressure p, volume V, number of moles N, and temperature T of a fluid or gas in a container. The equation is:

$$\left(p + a\frac{N^2}{V^2}\right)\left(\frac{V}{N} - b\right) = RT \tag{8.38}$$

where a and b are substance-specific constants and R is the universal gas constant. In that case, suppose a system is designed to handle a known maximum pressure and temperature. It becomes necessary to know what is the maximum volume (or molal volume V/N) of each type of substance that can be safely contained in the system, in order for example to properly document the system's safe operating parameters or to insure that it can be used in an specific application.

- A column supporting an off-center load will suffer from bending stress and compression stress. The maximum stress σ_{max} that the column can sustain is given by the secant formula:

$$\sigma_{max} = \frac{P}{A}\left[1 + \frac{ec}{r^2}\sec\left(\frac{L}{2r}\sqrt{\frac{P}{EA}}\right)\right] \qquad (8.39)$$

where P is the axial load, A is the cross-section area of the column, ec/r^2 is the eccentricity ratio, L/r is the slenderness ratio, and E is Young's modulus for the material of the column. Structural design often requires using this equation to determine the area of a column that will support a given load.

- It is well-known that an object thrown will follow a parabolic trajectory. More specifically, the (x, y) coordinates of the object following this trajectory after being thrown at an angle θ with initial velocity v will obey the equation:

$$y = x\tan(\theta) - \frac{gx^2}{2v^2\cos^2(\theta)} \qquad (8.40)$$

where g is the Earth's gravitational acceleration and the object's starting position is assumed to be the origin $(0, 0)$. While the initial speed will often be determined by the nature of the object's propulsion mechanism, a common challenge is to determine the initial angle θ to use in order to reach a specific target destination or to intercept a point in mid-trajectory.

In all these situations, the exact value of a specific variable must be known to properly design the system, but that variable cannot be isolated from the system's equation in order for the value to be computed. However, simply by subtracting one side of the equation from the other, the equation becomes equal to zero and the needed value becomes the root of the equation, and can thus be approximated to a known error rate by any of the methods seen in this chapter.

8.10   Summary

Many engineering situations can be modelled and solved by finding the value of some parameters of the system for which the system balances out to zero. These are root-finding problems, and this chapter introduced several numerical methods to solve them. The two closed methods, the bisection and false position methods, setup bounds around the root and either blindly pick the middle point between these bounds or interpolate a line through the function to get closer to the root. Because they bracket the root, these two methods are guaranteed to converge on the root

Table 8.1 Summary of root-finding methods

Method	Requires	Error
Bisection method	2 bounds	$O(h)$
False position method	2 bounds	$O(h)$
Simple fixed-point iteration	1 point	$O(h)$
Newton's method	1 point + derivative	$O(h^2)$
Secant method	2 points	$O(h^{1.618})$
Muller's method	3 points	$O(h^{1.839})$

eventually, albeit slowly. Next, three open methods were introduced, namely Newton's method, the secant method, and Muller's method. These methods all work by approximating the function, either using its derivative at one point, a straight line interpolated through two points, or a parabola interpolated through three points. Since none of them are burdened by maintaining possibly inaccurate brackets, they all perform faster than the closed methods. However, they all have a risk of diverging and failing to find the root in certain conditions. Of these three open methods, Newton's method was the most efficient and the most versatile since it could easily be expanded to multidimensional and multivariate problems. Table 8.1 summarizes the methods covered in this chapter.

8.11 Exercises

1. Approximate the root of the following equations in the respective intervals using the bisection method to a relative error of 0.1.

    (a) $f(x) = x^3 - 3$; interval $[1, 2]$
    (b) $f(x) = x^2 - 10$; interval $[3, 4]$
    (c) $f(x) = e^{-x}(3.2 \sin(x) - 0.5 \cos(x))$; interval $[3, 4]$

2. Write an algorithm to use the bisection method to find a root of $f(x) = \sin(x)$ starting with the interval $[1, 99]$ with a relative error of 0.00001. Comment on the result.

3. Approximate the root of the following equations in the respective intervals using the false position method to a relative error of 0.1.

    (a) $f(x) = x^3 - 3$; interval $[1, 2]$
    (b) $f(x) = x^2 - 10$; interval $[3, 4]$
    (c) $f(x) = e^{-x}(3.2 \sin(x) - 0.5 \cos(x))$; interval $[3, 4]$

4. Use Newton's method to find a root of the function $f(x) = e^{-x} \cos(x)$ starting with $x_0 = 1.3$ to a relative error of 10^{-5}.

5. Use Newton's method to find a root of the function $f(x) = x^2 - 7x + 3$ starting with $x_0 = 0$ and with an accuracy of 0.1.

6. Perform three steps of Newton's method for the function $f(x) = x^2 - 2$ starting with $x_0 = 1$.

7. Perform three iterations of Newton's method to approximate a root of the following multivariate systems given their starting points:

(a) $\mathbf{f}(\mathbf{x}) = \begin{bmatrix} x^2 + y^2 - 3 \\ -2x^2 - 0.5y^2 + 2 \end{bmatrix}$, $\mathbf{x}_0 = [1, \, 1]^T$.

(b) $\mathbf{f}(\mathbf{x}) = \begin{bmatrix} x^2 - xy + y^2 - 3 \\ x + y - xy \end{bmatrix}$, $\mathbf{x}_0 = [-1.5, \, 0.5]^T$.

8. Perform three steps of the secant method for the function $f(x) = x^2 - 2$ starting with
$x_{-1} = 0$ and $x_0 = 1$.

9. Perform four steps of the secant method for the function $f(x) = \cos(x) + 2 \sin(x) + x^2$ starting with $x_{-1} = 0.0$ and $x_0 = -0.1$.

10. Use the secant method to find a root of the function $f(x) = x^2 - 7x + 3$ starting with
$x_{-1} = -1$ and $x_0 = 0$ and with an accuracy of 0.1.

11. Perform six iterations of Muller's method on the function $f(x) = x^7 + 3x^6 + 7x^5$ $+ x^4 + 5x^3 + 2x^2 + 5x + 5$ starting with the three initial values $x_{-2} = 0$, $x_{-1} = -0.1$, and $x_0 = -0.2$.

Chapter 9
Optimization

9.1 Introduction

One major challenge in engineering practice is often the need to design systems that must perform as well as possible given certain constraints. Working without constraints would be easy: when a system can be designed with no restrictions on cost, size, or components used, imagination is the only limit on what can be built. But when constraints are in place, as they always will be in practice, then not only must engineering designs respect them, but the difference between a good and a bad design will be which one can get the most done within the stated constraints.

Take for example the design of a fuel tank. If the only design requirement is "hold a certain amount of fuel," then there are no constraints and the tank could be of any shape at all, provided the shape's volume is greater than the amount of fuel it must contain. However, when the cost of the materials the tank is made up of is taken into account, the design requirement becomes "hold at least a certain amount of fuel at the least cost possible," and this new constraint means the problem becomes about designing a fuel tank while minimizing its surface area, a very different one from before. A clever engineer would design the fuel tank to be a sphere, the shape with the lowest surface to volume ratio, in order to achieve the optimal result within the constraints. This design will be superior to the one using, say, a cube-shaped fuel tank, that would have a higher surface area and higher cost to hold the same volume of fuel.

To make the example more interesting, suppose the shape of the fuel tank is also constrained by the design of the entire system: it must necessarily be a cylinder closed at the top and made of a metal that costs 300\$/m^2, while the bottom of the tank is attached to a nozzle shaped as a cone with height equal to its radius and made of a plastic that costs 500\$/m^2. The entire assembly must hold at least 2000 m^3 of fuel. How to determine the optimal dimensions of the tank and the connected nozzle? First, model the components. For a given radius r and height h of the cylinder tank, the surface of the side and top of the cylinder will be:

© Springer International Publishing Switzerland 2016
R. Khoury, D.W. Harder, *Numerical Methods and Modelling for Engineering*,
DOI 10.1007/978-3-319-21176-3_9

$$A_1 = 2\pi r h + \pi r^2 \tag{9.1}$$

while the area of the nozzle will be:

$$A_2 = \pi r^2 \left(1 + \sqrt{2}\right) \tag{9.2}$$

Likewise, the volume of the cylinder of radius r and height h will be:

$$V_1 = \pi r^2 h \tag{9.3}$$

And the volume of the nozzle will be:

$$V_2 = \frac{\pi r^3}{3} \tag{9.4}$$

By looking at the cost (area) and volume of the entire assembly, this model becomes two equations with two unknown parameters that can be controlled, r and h:

$$\pi r^2 h + \frac{\pi r^3}{3} = 2000 \text{m}^3$$
$$(2\pi r h + \pi r^2)300 + \pi r^2 \left(1 + \sqrt{2}\right)500 = ?\$ \tag{9.5}$$

Normally a system of two equations and two unknowns would be easy to solve. The problem in the system of (9.5) is that one of the equations does not have a known result. The area and cost of the tank is not specified in the problem, the only requirement is that they must be as low as possible.

The problem could be further simplified by writing the parameter h as a function of r in the volume equation, and inserting that function of r into the price equation, to get:

$$\frac{2000}{\pi r^2} - \frac{r}{3} = h$$
$$\left(\frac{4000}{r} - \frac{2\pi r^2}{3} + \pi r^2\right)300 + \pi r^2 \left(1 + \sqrt{2}\right)500 = ?\$ \tag{9.6}$$

Now the price in Eq. (9.6) is only dependent on the radius; the height will be automatically adjusted to generate a container of 2000 m³ of fuel. The cost of a fuel tank with a radius from 1 to 10 m can be computed, and will give the values illustrated on the graph of Fig. 9.1. The ideal tank with minimal cost can also be found to have a radius of 5.27 m, a height of 17,679 m, and a cost of \$341,750.

This type of problem is called *optimization*, since it is seeking the optimal value of a function. This optimum can be the minimum value of the function, as it was for the cost function in the preceding example, or its maximum, for example if one was trying to design a fuel tank that can hold the greatest volume given a fixed budget.

Fig. 9.1 Radius and cost
of fuel tanks

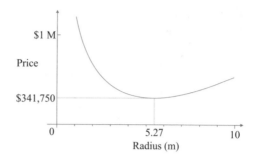

In the former case the problem can be called *minimization*, and in the latter *maximization*. It is important to note that these are not different types of problems though, but simply sign differences on the same optimization problem.

9.2   Golden-Mean Search

The *golden-mean search*, sometimes also called the *golden-section search*, is a simple and straightforward bracketing optimization method. A basic outline of its algorithm would be very similar to the other bracketing algorithms in Chap. 7 and Chap. 8: set an upper and a lower bound that bracket one (and only one) optimum of the function, then iteratively reduce the interval between these bounds to get a better and better approximation of the optimum. One important difference however concerns the iterative update of the bounds. In other algorithms, such as the bisection and false position methods of Chap. 8, one point was generated in the interval and used to replace one of the bounds. This cannot work in an optimization problem, because one point alone is not enough information to determine where the optimum might be. To visualize this problem, consider the example of a function with a minimum between the bounds of $x = 1$ and $x = 3$. At $x = 1$ the function evaluates at 4, at $x = 3$ it evaluates at 3, and in the middle point of $x = 2$ the function evaluates to 1. Should the new bounds be [1, 2] or [2, 3]? In fact there is not enough information provided to decide: the function might have reached its minimum between $x = 1$ and $x = 2$ and be on an upward slope from $x = 2$ to $x = 3$, or it might be on a downward slope from $x = 1$ to $x = 2$ to reach a minimum somewhere between $x = 2$ and $x = 3$ before increasing again. Both of these scenarios are illustrated in Fig. 9.2. Note that the issue is not with the selection of the point in the middle of the interval as opposed to somewhere else in the interval. One measurement in the interval, no matter where in the interval it is taken, will never be enough information to determine the location of the optimum.

If one point does not provide sufficient information to make a decision, then more points must be considered. In fact, two points dividing the bracketed section into three intervals are enough to determine which two intervals of the optimum

Fig. 9.2 Two functions with points $(1, 4), (2, 1)$, and $(3, 3)$, with a minimum in the $[1, 2]$ interval (*left*) and in the $[2, 3]$ interval (*right*)

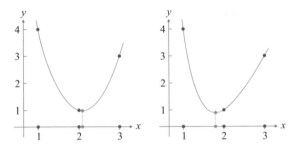

Fig. 9.3 Two functions with points $(1,4)$, $(1.66,0.95)$, $(2.33,1.6)$, and $(3,3)$, with a minimum in the $[1, 1.66]$ interval (*left*) and in the $[1.66, 2.33]$ interval (*right*)

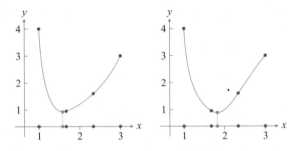

must be in and which one can be safely discarded. Consider once again the example of the function with a minimum between the bounds of $x = 1$ and $x = 3$ and which evaluates to $(1,4)$ and $(3,3)$. Points evaluated at 1.66 and 2.33 divide the function neatly into three equal intervals. Suppose the function evaluates to $(1.66, 0.95)$ and $(2.33, 1.6)$. The fact that the function has a lower value at the one-third point than at the two-third point means that a minimum must have been reached somewhere within those two intervals, to allow the function to turn around and increase again. In fact, two cases are possible: either the minimum is in the interval $[1, 1.66]$ and the function is on an upward slope through 1.66 and 2.33 to 3, or the function is decreasing from 1 through 1.66 to reach a minimum in the $[1.66, 2.33]$ interval, and is increasing again through 2.33 to 3. The only impossible case is for the minimum to be in the $[2.33, 3]$ interval, as that would require the function two have two minimums, one in the $[1, 2.33]$ interval to allow the decrease from 1 to 1.66 and increase from 1.66 to 2.33, and the second one in the $[2.33, 3]$ interval, and it has already been stated that the function has only one minimum within the bounds. Consequently, the interval $[2.33, 3]$ can safely be discarded, and the new bounds can be reduced to $[1, 2.33]$. This situation is illustrated in Fig. 9.3.

To formalize the bound update rule demonstrated above, assume that, at iteration i, the algorithm has a lower bound x_{iL} and an upper bound x_{iU} bracketing an optimum of the function $f(x)$. Two points are generated for the iteration within the bounds, x_{i0} and x_{i1} where $x_{i0} < x_{i1}$, and they are evaluated. Then, in the case of a minimization problem, the bounds are updated according to the following rule:

$$[x_{iL}, x_{iU}] \rightarrow \begin{cases} [x_{iL}, x_{i1}] & \text{if } (f(x_{i0}) < f(x_{i1})) \\ [x_{i0}, x_{iU}] & \text{if } (f(x_{i0}) > f(x_{i1})) \end{cases} \quad (9.7)$$

In the case of a maximization problem, the update rule is simply inverted:

$$[x_{iL}, x_{iU}] \rightarrow \begin{cases} [x_{iL}, x_{i1}] & \text{if } (f(x_{i0}) > f(x_{i1})) \\ [x_{i0}, x_{iU}] & \text{if } (f(x_{i0}) < f(x_{i1})) \end{cases} \quad (9.8)$$

Note that these rules are independent of the step between x_{iL}, x_{i0}, x_{i1}, and x_{iU}. The decision to use the one-third and two-third points in the previous example was made only for the sake of simplicity. More generally, the step can be represented by a value λ, and the two points are computed as:

$$\begin{aligned} x_{i0} &= \lambda x_{iL} + (1 - \lambda)x_{iU} \\ x_{i1} &= (1 - \lambda)x_{iL} + \lambda x_{iU} \end{aligned} \quad (9.9)$$

In these definitions, at each iteration, the bracketed interval is reduced by a factor of λ, to a size of $\lambda(x_{iL} - x_{iU})$. In the previous example using the one-third and two-third points, the value was $\lambda = 2/3$. While that is a natural value to use when the question is to split an interval into three parts, it is also a suboptimal value for the computations in the algorithm. Indeed, consider how these two points and the update rules of Eqs. (9.7) and (9.8) will interact with each other. An example case is detailed in Table 9.1, with the initial bounds being 0 and 1 and the first rule of Eq. (9.7) being used every time.

Notice that both values x_{i0} and x_{i1} are new at each step, which means that both $f(x_{i0})$ and $f(x_{i1})$ need to be recomputed each time. This is twice the amount that would be needed if one point could be reused, and in cases where evaluating the function is time-consuming, it can become a major drawback for the algorithm. Now consider what would happen in the case where $\lambda = \varphi^{-1} = 0.6180$, where φ is the golden ratio. Table 9.2 runs through the example again, using this new value of λ.

Table 9.1 Sample iterations using $\lambda = 2/3$

Iteration	x_{iL}	x_{i0}	x_{i1}	x_{iU}
0	0	0.33333	0.66666	1
1	0	0.22222	0.44444	0.66666
2	0	0.14815	0.29629	0.44444
3	0	0.09876	0.19753	0.29629

Table 9.2 Sample iterations using $\lambda = 0.6180$

Iteration	x_{iL}	x_{i0}	x_{i1}	x_{iU}
0	0	0.3820	0.6180	1
1	0	0.2361	0.3820	0.6180
2	0	0.1459	0.2361	0.3820
3	0	0.0902	0.1459	0.2361

This time, when one inner value becomes the new bound, the interval is reduced in such a way that the other inner value becomes the new opposite inner value. In Table 9.2, whenever x_{i1} becomes the new bound, x_{i0} becomes x_{i1}. This is a natural result of using the golden ratio: the ratio of the distance between x_{iL} and x_{iU} to the distance between x_{iL} and x_{i1} is the same as the ratio of the distance between x_{iL} and x_{i1} to the distance between x_{i1} and x_{iU} and the same as the ratio of the distance between x_{iL} and x_{i1} to the distance between x_{iL} and $x_{i0,}$. Consequently, when the interval between x_{i1} and x_{iU} is taken out and the new complete interval is x_{iL} to x_{i1}, x_{i0} is at the correct distance from x_{iL} to become the new inner point x_{i1}. Moreover, with this value of λ, the interval is reduced at each iteration to 0.6180 of its previous size, which is smaller than the reduction to 0.6666 of its previous size when $\lambda = 2/3$. In other words, using $\lambda = 0.6180$ leads to an algorithm that both requires only half the computations in each iteration and that converges faster. There are no downsides.

As with any iterative algorithm, it is important to define termination conditions. There are two conditions for the golden-mean search, the two usual conditions apply. If the absolute error between the bounds after the update is less than a predefined threshold, then an accurate enough approximation of the optimum has been found and the algorithm terminates in success. If however the algorithm first reaches a predefined maximum number of iterations, it ends in failure. The pseudocode of the complete golden-mean search method is given in Fig. 9.4.

The convergence rate of this algorithm has already been hinted to previously, when it was mentioned that each iteration reduces the interval by a factor of λ. When the value of λ is set to the golden ratio and the initial interval between the bounds is $h_0 = |x_{0L} - x_{0U}|$, then after the first iteration it will be $h_1 = \varphi h_0$, and after the second iteration it will be:

$$h_2 = \varphi h_1 = \varphi^2 h_0 \tag{9.10}$$

and more generally, after iteration n it will be:

$$h_n = \varphi^n h_0 = 0.6180^n h_0 \tag{9.11}$$

This is clearly a linear $O(h)$ convergence rate. One advantage of Eq. (9.11) is that it makes it possible to predict an upper bound on the number of iterations the golden-mean algorithm will reach the desired error threshold. For example, if the initial search interval was $h_0 = 1$ and an absolute error of 0.0001 is required, the algorithm will need to perform at most $\log_{0.6180}(0.0001) \approx 19$ iterations.

```
XL ← Input lower bound
XU ← Input upper bound
ProblemType ← Input minimization or maximization
IterationMaximum ← Input maximum number of iterations
ErrorMinimum ← Input minimum relative error

X0 ← 0.6180 × XL + (1 - 0.6180) × XU
X1 ← (1 - 0.6180) × XL + 0.6180 × XU
IterationCounter ← 0
WHILE (TRUE)

    IF (ProblemType = minimization)
        IF ( [CALL F(X0)] < [CALL F(X1)] )
            XU ← X1
            X1 ← X0
            X0 ← 0.6180 × XL + (1 - 0.6180) × XU
        ELSE
            XL ← X0
            X0 ← X1
            X1 ← (1 - 0.6180) × XL + 0.6180 × XU
        END IF
    ELSE IF (ProblemType = maximization)
        IF ( [CALL F(X0)] > [CALL F(X1)] )
            XU ← X1
            X1 ← X0
            X0 ← 0.6180 × XL + (1 - 0.6180) × XU
        ELSE
            XL ← X0
            X0 ← X1
            X1 ← (1 - 0.6180) × XL + 0.6180 × XU
        END IF
    END IF

    CurrentError ← absolute value of (XU - XL)
    IterationCounter ← IterationCounter + 1

    IF (CurrentError <= ErrorMinimum)
        RETURN Success, x
    ELSE IF (IterationCounter = IterationMaximum)
        RETURN Failure
    END IF

END WHILE

FUNCTION F(x)
    RETURN evaluation of the target function at point x
END FUNCTION
```

Fig. 9.4 Pseudocode of the golden-mean method

Example 9.1

A solar panel is connected to a house, connected also to the city's power grid. When the house consumes more power than can be generated by the solar panel it draws from the city, and when it consumes less the extra power is fed into the city's power grid. The power consumption of the house over time has been modelled as $P(t) = t(t-1)$, where a positive value is extra power generated by the house and a negative value is power drain from the city. Find the maximum amount of power the house will need from the city over the time interval [0, 2] to an absolute error of less than 0.01 kW.

Solution

Begin by noting that Eq. (9.11) gives:

$$0.01 = 2 \times 0.6180^n$$
$$n \approx 8$$

In other words, the solution should be found at the eighth iteration of the golden-mean method.

The first two middle points computed from Eq. (9.9) are:

$$x_{00} = 0.6180 \times 0 + 0.3820 \times 2 = 0.76393$$
$$x_{01} = 0.3820 \times 0 + 0.6180 \times 2 = 1.2361$$

The power consumption can then be evaluated from the model at those two points:

$$P(x_{00}) = 0.76393(0.76393 - 1) = -0.18034$$
$$P(x_{01}) = 1.2361(1.2361 - 1) = 0.29180$$

Since this is a minimization problem, the rule of Eq. (9.7) applies, and the upper bound is replaced by x_{01}. The absolute error after this first iteration is $|0 - 1.2361| = 1.2361$. At the second iteration, the new middle points are:

$$x_{10} = 0.6180 \times 0 + 0.3820 \times 1.2361 = 0.47214$$
$$x_{11} = 0.3820 \times 0 + 0.6180 \times 1.2361 = 0.76393$$

Notice that x_{11} is exactly the same as x_{00}; this was expected from the earlier explanations, and as a result that middle point does not need to be re-evaluated, its value can simply be carried over from the previous iteration. The other middle point does need to be evaluated:

$$P(x_{10}) = 0.47214(0.47214 - 1) = -0.24922$$

(continued)

Example 9.1 (continued)

Once again, using the update rule of Eq. (9.7), the upper bound is the one that is updated. The absolute error is now 0.76393. The table below summarizes all the iterations needed for this example, and the following figure illustrates the function and the decreasing size of the interval around the optimum as the iterations increase, from darker to lighter shade.

i	x_{iL}	x_{iU}	x_{i0}	x_{i1}	$P(x_{i0})$	$P(x_{i1})$	E_i
0	0	2	0.76393	1.2361	−0.18034	0.29180	1.2361
1	0	1.2361	0.47214	0.76393	−0.24922	−0.18034	0.76393
2	0	0.76393	0.29180	0.47214	−0.20665	−0.24922	0.47213
3	0.29180	0.76393	0.47214	0.58359	−0.24924	−0.24301	0.29179
4	0.29180	0.58359	0.40325	0.47214	−0.24064	−0.24922	0.18034
5	0.40325	0.58359	0.47214	0.51471	−0.24922	−0.24978	0.11145
6	0.47214	0.58359	0.51471	0.54102	−0.24978	−0.24832	0.06888

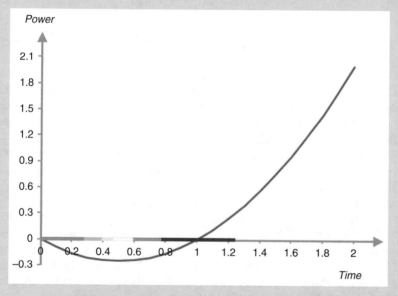

At the beginning of iteration 7 (the eighth one, as predicted), the minimum is known to be in [0.47214, 0.54102] with an absolute error of 0.06888, and one middle point at 0.51471 carries over from iteration 6. That middle point has the lowest evaluated value, at $P(0.51471) = -0.24978$ kW, and can be used as the minimum without any further function evaluations.

9.3 Newton's Method

It is well-known that the optimum of a function $f(x)$ is an inflection point where its derivative $f^{(1)}(x) = 0$. This means that an optimization method for $f(x)$ is the same as a root-finding method for $f^{(1)}(x)$, and any of the root-finding methods learned in Chap. 8 could be used. Most interestingly, if the first and second derivatives of the function are known, it is possible to use Newton's method, the most efficient of the root-finding methods learned. Recall from Sect. 8.6 that the equation for Newton's method to find a root of $f(x)$ is:

$$x_{i+1} = x_i - \frac{f(x_i)}{f^{(1)}(x_i)} \tag{9.12}$$

Then the equation to find a root of $f^{(1)}(x)$, an optimum of $f(x)$, is simply:

$$x_{i+1} = x_i - \frac{f^{(1)}(x_i)}{f^{(2)}(x_i)} \tag{9.13}$$

As was proven in Chap. 8 using Taylor series, this method will iteratively converge towards the nearest root of $f^{(1)}(x)$ at a quadratic rate $O(h^2)$.

Once issue is that there is no indication in Eq. (9.13) as to whether this root will be a maximum or a minimum of $f(x)$; the root of the derivative only indicates that it is an optimum. Real-world functions will usually have both maxima and minima, and a problem will require finding a specific one of the two, not just the nearest optimum regardless of whether it is a maximum or a minimum. One way of checking if the function is converging on a maximum or a minimum is of course to evaluate $f(x_i)$ and see if the values are increasing or decreasing. However, this will require additional function evaluations, since evaluating $f(x)$ is not needed for Newton's method in Eq. (9.13), as well as a memory of one past value $f(x_{i-1})$ to compare $f(x_i)$ to. To avoid these added costs in the algorithm, another way of checking using only information available in Eq. (9.13) is to consider the sign of the second derivative at the final value x_i. If $f^{(2)}(x_i) < 0$ the optimum is a maximum, and if $f^{(2)}(x_i) > 0$ then the optimum is a minimum. If the method is found to have converged to the wrong type of optimum, then the only solution is to start over from another, more carefully chosen initial point.

The same three halting conditions seen for Newton's method in Chap. 8 still apply. To review, if the relative error between two successive approximations is less than a predefined threshold ε, then the iterative algorithm has converged successfully. If however a preset maximum number of iterations is reached first, then the method has failed to converge. Likewise, if the evaluation of the second derivative $f^{(2)}(x_i) = 0$, then the point generated is in a discontinuity of $f^{(1)}(x)$ and the method cannot continue. The pseudocode for Newton's optimization method is given in Fig. 9.5; it can be seen that it is only a minor modification of the code of Newton's root-finding method presented in the previous chapter, to replace the

```
x ← Input initial approximation of the optimum
IterationMaximum ← Input maximum number of iterations
ErrorMinimum ← Input minimum relative error

IterationCounter ← 0
WHILE (TRUE)

    PreviousValue ← x
    x ← x - [CALL Derivative(F(x))] / [CALL Derivative(Derivative(F(x)))]

    IF ( CALL Derivative(F(x)) = 0 )
        RETURN Success, x
    ELSE IF ( CALL Derivative(Derivative(F(x))) = 0 )
        RETURN Failure
    END IF

    CurrentError ← absolute value of [ (x - PreviousValue) / x ]
    IterationCounter ← IterationCounter + 1

    IF (CurrentError <= ErrorMinimum)
        RETURN Success, x
    ELSE IF (IterationCounter = IterationMaximum)
        RETURN Failure
    END IF

END WHILE


FUNCTION F(x)
    RETURN evaluation of the target function at point x
END FUNCTION

FUCNTION Derive(F(x))
    RETURN evaluation of the derivative of the function at point x
END FUNCTION
```

Fig. 9.5 Pseudocode of Newton's method

function calls F(x) and Derivative(F(x)) with Derivative(F(x)) and Derivative(Derivative(F(x))) respectively. On that point, notice that the second derivative of a function is computed simply by calling the Derive function twice in a row. In essence, Newton's optimization method could be implemented using exactly the code of Newton's root-finding method, but by calling it with the derivative of the function $f(x)$ instead of the function itself.

Example 9.2
A sudden electrical surge is known to cause a nearly one-second-long power spike in an electrical system. The behavior of the system during the spike has been studied, and during that event the power (in kW) is modelled as:

$$P(t) = \sin(t) - t^5$$

(continued)

Example 9.2 (continued)

Determine the maximum power that the system must be designed, in order to handle the electrical surge, to a relative error of 0.01 %.

Solution

The first and second derivatives of the model are:

$$P'(t) = \cos(t) - 5t^4$$
$$P''(t) = -\sin(t) - 20t^3$$

For an initial value, since none are provided but the spike is said to be one-second-long, the search could start in the middle of the interval, at $t_0 = 0.5$. In that case, Eq. (9.13), using the model's derivatives gives:

$$t_1 = 0.5 - \frac{0.565}{-2.979} = 0.6900\text{s}$$

$$E_1 = \left| \frac{0.5 - 0.6900}{0.6900} \right| = 27.501\%$$

The next iterations are given in the table below, and illustrated in the accompanying figure along with the power function:

i	t_i (s)	$P^{(1)}(t_i)$	$P^{(2)}(t_i)$	t_{i+1} (s)	E_i (%)
0	0.5	0.565	-2.979	0.6900	27.501
1	0.690	-0.360	-7.197	0.6397	7.813
2	0.640	-0.035	-5.832	0.6337	0.945
3	0.634	-0.0005	-5.682	0.6336	0.013
4	0.634	-8×10^{-8}	-5.680	0.6336	0.000

The method converges in five iterations on $t = 0.6336$ s, at which point the power spike is 0.4899 V. For reference the real optimum is at $t = 0.63361673$ s with a power spike of 0.489938055 kW. The relative error of the method compared to the real result is thus less than 2.3×10^{-6} % on the time of the spike peak and 5.1×10^{-8} % on the power at the maximum the system must be able to handle. Note as well that the second derivative of the function is negative at the final result, and in fact throughout the iterations, confirming that the method is converging on a maximum and not on a minimum.

9.4   Quadratic Optimization

It has been observed several times already that an optimum in a function $f(x)$ is an inflection point where the function turns around. Locally, the inflection point region could be approximated as a degree-2 polynomial, a parabola $p(x)$. As was learned in Chap. 5, all that is required for this is to be able to evaluate three points on the function to interpolate the polynomial from. The situation is illustrated in Fig. 9.6. The equation for a degree-2 polynomial is:

$$p(x) = c_0 + c_1 x + c_2 x^2 \tag{9.14}$$

where $p(x) = f(x)$ at three points $x = x_{i-2}$, x_{i-1}, and x_i. Given this information, Chap. 5 has covered several methods to discover the values of the coefficients, such as solving the matrix–vector system of the Vandermonde method:

$$\begin{bmatrix} 1 & x_{i-2} & x_{i-2}^2 \\ 1 & x_{i-1} & x_{i-1}^2 \\ 1 & x_i & x_i^2 \end{bmatrix} \begin{bmatrix} c_0 \\ c_1 \\ c_2 \end{bmatrix} = \begin{bmatrix} f(x_{i-2}) \\ f(x_{i-1}) \\ f(x_i) \end{bmatrix} \tag{9.15}$$

If the interpolated parabola serves as a local approximation of the inflection point of the function, then the optimum of the parabola can serve as an approximation of the optimum of the function. This is of great advantage, since the optimum of the parabola is also a lot easier to find. Once the equation of the parabola has been interpolated, its optimum is simply the point where its derivative is zero:

$$p^{(1)}(x) = c_1 + 2c_2 x = 0$$
$$x = -\frac{c_1}{2c_2} \tag{9.16}$$

The *quadratic optimization method* is an iterative version of this approach. As a new approximation of the optimum is computed at each iteration, it replaces the

Fig. 9.6 The optimum of a function (*solid blue line*) and an interpolated parabola (*dashed red line*)

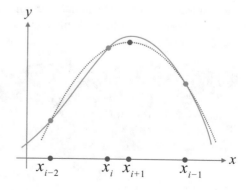

oldest of the three points used in the interpolation. The region covered by the interpolation becomes iteratively a smaller section of the inflection point, the interpolated polynomial thus becomes a better approximation of the function in that region, and the optimum of the parabola becomes closer to the function's optimum.

For the iterative version of this approach, given three past approximations of the optimum x_{i-2}, x_{i-1}, and x_i, it is possible to interpolate this iteration's polynomial $p_i(x)$. Then, the new approximation of the optimum is computed as:

$$x_{i+1} = -\frac{c_{1,i}}{2c_{2,i}} \tag{9.17}$$

or, integrating Eq. (9.15) in, as:

$$x_{i+1} = \frac{f(x_{i-2})\left(x_{i-1}^2 - x_i^2\right) + f(x_{i-1})\left(x_i^2 - x_{i-2}^2\right) + f(x_i)\left(x_{i-2}^2 - x_{i-1}^2\right)}{2f(x_{i-2})(x_{i-1} - x_i) + 2f(x_{i-1})(x_i - x_{i-2}) + 2f(x_i)(x_{i-2} - x_{i-1})} \tag{9.18}$$

This new point replaces x_{i-2} to compute $p_{i+1}(x)$ in the next iteration. There are three halting conditions to the iterative algorithm. If the relative error between two successive approximations of the optimum is less than a preset threshold ε, then the algorithm has successfully converged. On the other hand, if the algorithm first reaches a preset maximum number of iterations, it has failed. There is a second failure condition to watch out for: if the interpolated polynomial becomes a degree-1 polynomial, a straight line, then the algorithm has diverged and is no longer in the region of the inflection point at all. From Eq. (9.17), it can be seen that in that case the equation would have a division by zero, a sure sign of divergence. The pseudocode of a version of the quadratic optimization method using the matrix–vector system of Eq. (9.15) and including the additional failure condition check is presented in Fig. 9.7.

The convergence rate for this method is $O(h^{1.497})$, although the proof is outside the scope of this book. This method thus converges more efficiently than the golden-mean method, which is normal when comparing an open method like this one to a closed method that must maintain brackets. On the other hand it converges more slowly than the Newton's method. Again, this was to be expected: Newton's method uses actual features of the function, namely its first and second derivatives, to find the optimum, while this method uses an interpolated approximation of the function to do it, and therefore cannot get as close at each iteration.

Example 9.3
A sudden electrical surge is known to cause a one-second-long power spike in an electrical system. The behavior of the system during the spike has been studied, and during that event the power (in kW) is modelled as:

(continued)

Example 9.3 (continued)

$$P(t) = \sin(t) - t^5$$

Determine the maximum power the system must be designed, in order to handle the electrical surge, to a relative error of 0.01 %. Use the samples at $t = 0, 0.5$, and 1 s as initial values.

Solution
Begin by evaluating the power at the three initial points:

$$t_{-2} = 0 \text{ s} \qquad p(t_{-2}) = 0 \text{ kW}$$
$$t_{-1} = 0.5 \text{ s} \qquad p(t_{-1}) = 0.448 \text{ kW}$$
$$t_0 = 1 \text{ s} \qquad p(t_0) = -0.159 \text{ kW}$$

From there, the Vandermonde method can be used to interpolate the parabola:

$$\begin{bmatrix} 1 & 0 & 0 \\ 1 & 0.5 & 0.25 \\ 1 & 1 & 1 \end{bmatrix} \begin{bmatrix} c_{00} \\ c_{10} \\ c_{20} \end{bmatrix} = \begin{bmatrix} 0 \\ 0.488 \\ -0.159 \end{bmatrix}$$

$$p_0(x) = 0 + 1.951x - 2.110x^2$$

The optimum of this parabola, the first approximation of the optimum computed by the method, is then obtained from the derivative of the parabola, as given in Eq. (9.17):

$$t_1 = \frac{-1.951}{2 \times (-2.110)} = 0.4624\text{s}$$

Alternatively, Eq. (9.18) could be used to compute the approximation directly without solving the Vandermonde system:

$$t_1 = \frac{0(0.25 - 1) + 0.448(1 - 0) - 0.159(0 - 0.25)}{0(0.5 - 1) + 0.896(1 - 0) - 0.318(0 - 0.5)} = 0.4624\text{s}$$

Either way, the relative error on this approximation is:

$$E_1 = \left| \frac{1 - 0.462}{0.462} \right| = 116.249\%$$

(continued)

Example 9.3 (continued)

This is far higher than the required error, and the algorithm continues. The next iterations are given in the table below, and illustrated on the accompanying figure along with the power function:

i	t_{i-2} (s)	t_{i-1} (s)	t_i (s)	t_{i+1} (s)	E_i (%)
0	0.0000	0.5000	1.0000	0.4624	116.249
1	0.5000	1.0000	0.4624	0.5719	19.136
2	1.0000	0.4624	0.5719	0.5849	2.237
3	0.4624	0.5719	0.5849	0.6538	10.527
4	0.5719	0.5849	0.6538	0.6342	3.088
5	0.5849	0.6538	0.6342	0.6329	0.207
6	0.6538	0.6342	0.6329	0.6336	0.117
7	0.6342	0.6329	0.6336	0.6336	0.000

The method converges in eight iterations on $t = 0.6336$ s, the same optimum that Newton's method found in five iterations in Example 9.2. Having to carry three bad initial guesses for several iterations slows down the initial convergence; in iteration 3, the first one computed using only approximations computed in previous iterations, the result shows a large jump in accuracy, from a relative error of 7.8 % compared to the real optimum to 3.1 %.

9.5   Gradient Descent

The *gradient descent* optimization method is also known by its more figurative name of *hill climbing*. It has been described as "what you would do if you needed to find the top of Mount Everest with amnesia in a fog." What would this unfortunate climber, unable to remember where they've been or to see where they are going, do? Simply feel around the ground in one step in every direction to find the one that goes up the fastest, and proceed along that way step by step. Once the climber has reached a point where the ground only goes down in all directions, they can assume they have reached the top of the mountain. In mathematical terms, the direction of the step that gives the greatest change (be it increase or decrease) in the value of a function is called its *gradient*, and the basic idea of the gradient descent method is simply to take step after step along the gradient until a point is reached where no step can be taken to improve the value.

The gradient descent is different from the other methods covered in this chapter so far by the fact that it is a multidimensional optimization method, instead of a one-dimensional one. In fact, as will be shown, it makes it possible to reduce the multidimensional optimization problem into a one-dimensional problem of optimizing the step size along the gradient that optimizes the multidimensional function.

```
PreviousValue2, PreviousValue, x ← Input three initial approximations of
    the optimum
IterationMaximum ← Input maximum number of iterations
ErrorMinimum ← Input minimum relative error

IterationCounter ← 0
WHILE (TRUE)

    V ← 3×3 Vandermonde matrix with the following rows:
        Row 1: 1, PreviousValue2, PreviousValue2 × PreviousValue2
        Row 2: 1, PreviousValue, PreviousValue × PreviousValue
        Row 3: 1, x, x × x
    Y ← vector of length 3 with the following values:
        CALL F(PreviousValue2), CALL F(PreviousValue), CALL F(x)
    C ← vector of length 3 that is the solution of the system V × C = Y

    IF [ (Third value of C) = 0 ]
        RETURN Failure
    END

    PreviousValue2 ← PreviousValue
    PreviousValue ← x
    x ← [-1 × (Second value of C)] / [2 × (Third value of C)]

    CurrentError ← absolute value of [ (x - PreviousValue) / x ]
    IterationCounter ← IterationCounter + 1

    IF (CurrentError <= ErrorMinimum)
        RETURN Success, x
    ELSE IF (IterationCounter = IterationMaximum)
        RETURN Failure
    END IF

END WHILE


FUNCTION F(x)
    RETURN evaluation of the target function at point x
END FUNCTION
```

Fig. 9.7 Pseudocode of the quadratic optimization method

Assume a function $\mathbf{y} = f(\mathbf{x})$ where $\mathbf{x}$ and $\mathbf{y}$ are n-dimensional vectors, an initial point $\mathbf{x}_0$, and a step size h. At each iteration i, the gradient descent algorithm evaluates the function at steps of h around the current point $\mathbf{x}_i$, and takes one step in the orientation that evaluates closest to the optimum (the maximum for maximization and the minimum for minimization). The next point $\mathbf{x}_{i+1}$ is $\mathbf{x}_i$ plus the appropriate step. This leaves only two questions: how to determine the orientation of the best step, and how to determine the size of the step.

As indicated earlier, the orientation of the step is the gradient, the direction of maximum change of the function. This is the orientation that will bring the iterations close to the optimum fastest. The gradient is the vector of partial derivatives of $f(\mathbf{x})$ with respect to each dimension of $\mathbf{x}$:

$$\nabla \mathbf{f}(\mathbf{x}) = \begin{bmatrix} \dfrac{\partial f(\mathbf{x})}{\partial x_0} \\[2ex] \dfrac{\partial f(\mathbf{x})}{\partial x_1} \\[2ex] \vdots \\[2ex] \dfrac{\partial f(\mathbf{x})}{\partial x_{n-1}} \end{bmatrix} \qquad (9.19)$$

Knowing the direction of maximum change limits the options for the orientation of the step from $360°$ around the current point to only two directions, either positively or negatively along the line of the gradient. In fact this is the distinction between maximization and minimization in this method: for a maximization problem the algorithm should positively follow the gradient to increase as quickly as possible, and for a minimization problem the algorithm should go in the negative direction to decrease as quickly as possible. Thus, the next point $\mathbf{x}_{i+1}$ will be the current iteration's $\mathbf{x}_i$ plus or minus one step along the gradient at that point, as so:

$$\mathbf{x}_{i+1} = \mathbf{x}_i \pm h\nabla \mathbf{f}(\mathbf{x}_i) \qquad (9.20)$$

Example 9.4
Compute the gradient of this function:

$$f(\mathbf{x}) = x_0^2 + 3x_0x_1 - x_1^3 + x_2 + 4x_3^2 - x_1x_2x_3$$

Solution
Following Eq. (9.19), the gradient is:

$$\nabla \mathbf{f}(\mathbf{x}) = \begin{bmatrix} \dfrac{\partial f(\mathbf{x})}{\partial x_0} \\[1.5ex] \dfrac{\partial f(\mathbf{x})}{\partial x_1} \\[1.5ex] \dfrac{\partial f(\mathbf{x})}{\partial x_2} \\[1.5ex] \dfrac{\partial f(\mathbf{x})}{\partial x_3} \end{bmatrix} = \begin{bmatrix} 2x_0 + 3x_1 \\ 3x_1 - 3x_1^2 - x_2x_3 \\ 1 - x_2x_3 \\ 8x_3 - x_1x_2 \end{bmatrix}$$

The second question is how to pick the step size. After all, a value of h too large will give a poor approximation, as the method will step over the optimum. On the other hand, a small step will make the algorithm converge slowly. One solution is to use an iteratively decreasing step size h_i, that begins with a large value to take large steps towards the optimum and then decreases it in order to pinpoint the optimum.

A better solution though would be to actually compute the length of the step h_i to take at each iteration to get as close as possible to the optimum; in other words, to optimize the step size at that iteration! The optimal step size is of course the one that will give the value of $\mathbf{x}_{i+1}$, as computed in Eq. (9.20), which will in turn allow $f(\mathbf{x}_{i+1})$ to evaluate to its optimal value. This leads to a simple but important realization: the only unknown value in Eq. (9.20) missing to compute the value of $\mathbf{x}_{i+1}$ is the value of h_i, and as a result the value of the function evaluation $f(\mathbf{x}_{i+1})$ will only vary based on h_i:

$$f(\mathbf{x}_{i+1}) = f(\mathbf{x}_i \pm h_i \nabla \mathbf{f}(\mathbf{x}_i)) = g(h_i) \qquad (9.21)$$

In other words, optimizing the value of the multidimensional function $f(\mathbf{x}_{i+1})$ is the same as optimizing the single-variable function $g(h_i)$ obtained by the simple variable substitution of Eq. (9.20). And a single-variable, single-dimension optimization problem is one that can easily be done using the golden-mean method, Newton's method, or the quadratic optimization method. The optimal value of h_i that is found is the optimal step size to use in iteration i to compute $\mathbf{x}_{i+1}$; it corresponds to the step to take to get to the local optimum of the gradient line $\nabla \mathbf{f}(\mathbf{x}_i)$.

To summarize, at each iteration i, the gradient descent algorithm will perform these steps:

1. Evaluate the gradient at the current point, $\nabla \mathbf{f}(\mathbf{x}_i)$, using Eq. (9.19).
2. Rewrite the function $\mathbf{x}_{i+1}$ with the variable substitution of Eq. (9.21) as $f(\mathbf{x}_i \pm h_i \nabla \mathbf{f}(\mathbf{x}_i))$ to get a function of h_i.
3. Use the method of your choice to find the value of h_i that is the optimum of $f(\mathbf{x}_i \pm h_i \nabla \mathbf{f}(\mathbf{x}_i))$.
4. Compute the next value of $\mathbf{x}_{i+1}$ using Eq. (9.20) as $\mathbf{x}_i \pm h_i \nabla \mathbf{f}(\mathbf{x}_i)$.
5. Evaluate termination conditions, either to terminate or continue the iterations.

The pseudocode of an algorithm implementing all these steps is presented in Fig. 9.8. The success termination condition for this algorithm is that the Euclidean distance between $\mathbf{x}_i$ and $\mathbf{x}_{i+1}$ becomes less than a preset error threshold ε, at which point the method has converged on the optimum of the function with sufficient accuracy. There are two failure termination conditions. The first is, as always, if the algorithm reaches a preset maximum number of iterations without converging. The second condition is if the gradient $\nabla \mathbf{f}(\mathbf{x}_i)$ evaluates to a vector of zeros. From Eq. (9.20), it can be seen that in that case, the method is stuck in place and $\mathbf{x}_{i+1} = \mathbf{x}_i$. Mathematically, a point where the gradient is null is a plateau in the function, a point where there is no change to the evaluation of the function in any direction. The method cannot continue anymore since all orientations from that point are equivalent and none improve the function at all. Note however that, since $\mathbf{x}_{i+1} = \mathbf{x}_i$ in that situation, the Euclidean distance between $\mathbf{x}_i$ and $\mathbf{x}_{i+1}$ will be zero, which corresponds to the success condition despite it being actually a failure of the

```
x ← Input vector of length n; the initial approximation of the optimum
h ← Input the initial step size
ProblemType ← Input minimization or maximization
IterationMaximum ← Input maximum number of iterations
ErrorMinimum ← Input minimum step size

IterationCounter ← 0
WHILE (TRUE)

    IF (ProblemType = minimization)
        x(h) ← new function of variable h as:
                [CALL F(x)] - h × [CALL Gradient(F(x))]
    ELSE IF (ProblemType = maximization)
        x(h) ← new function of variable h as:
                [CALL F(x)] + h × [CALL Gradient(F(x))]
    END IF

    h ← optimum of [CALL F(x(h))]
    PreviousValue ← x
    x ← x(h)

    IF [ (CALL Gradient(F(X))) = zero vector ]
        RETURN Failure
    END IF

    CurrentError ← Euclidean distance between x and PreviousValue
    IterationCounter ← IterationCounter + 1

    IF (CurrentError <= ErrorMinimum)
        RETURN Success, x
    ELSE IF (IterationCounter = IterationMaximum)
        RETURN Failure
    END IF

END WHILE


FUNCTION F(x)
    RETURN vector of length n of evaluation of the n-dimensional target
            function at point x
END FUNCTION

FUNCTION Gradient(F(x))
    RETURN vector of length n of evaluation of the partial derivatives of
            the function with respect to the n variables at point x
END FUNCTION
```

Fig. 9.8 Pseudocode of the gradient descent method

method. It is thus important to evaluate this condition first, before the Euclidean distance of the success condition, to avoid potentially disastrous mistakes in the interpretation of the results.

Example 9.5

The strength of the magnetic field of a sheet of magnetic metal has been modelled by the 2D function:

$$H(\mathbf{x}) = x^2 - 4x + 2xy + 2y^2 + 2y + 14$$

which is in amperes per meter. The origin (0,0) corresponds to the center of the sheet of metal. Determine the point with the weakest magnetic field, to the nearest centimeter. Use the corner of the sheet at $(4, -4)$ meters as a starting point.

Solution

The first thing to do is to compute the gradient of the function using equation (9 19). This gives the vector of functions:

$$\nabla \mathbf{H}(\mathbf{x}) = \begin{bmatrix} 2x + 2y - 4 \\ 2x + 4y + 2 \end{bmatrix}$$

For the first iteration, the value of $\mathbf{x}_0$ is given as $[4, -4]^T$. Since this is a minimization problem, Eq. (9.20) becomes:

$$\mathbf{x}_{i+1} = \mathbf{x}_i - h_i \nabla \mathbf{H}(\mathbf{x}_i)$$

And for the first iteration, it evaluates to:

$$\mathbf{x}_1 = \begin{bmatrix} 4 \\ -4 \end{bmatrix} - h_0 \begin{bmatrix} 2 \times 4 + 2 \times (-4) - 4 \\ 2 - 4 + 4 \times (-4) + 2 \end{bmatrix} = \begin{bmatrix} 4 \\ -4 \end{bmatrix} - h_0 \begin{bmatrix} -4 \\ -6 \end{bmatrix}$$

The next step of the iteration is the variable substitution of $\mathbf{x}_1$ to make $H(\mathbf{x}_1)$ into a function of h_0 that can be easily optimized. To do this, write down the original $H(\mathbf{x})$ replacing x with $(4 + 4h_0)$ and y with $(-4 + 6h_0)$. The new equation, which will be labelled $g(h_0)$ for convenience, is:

$$g(h_0) = (4 + 4h_0)^2 - 4(4 + 4h_0) + 2(4 + 4h_0)(-4 + 6h_0) + 2(-4 + 6h_0)^2$$
$$+ 2(-4 + 6h_0) + 14 = 136h_0^2 - 52h_0 + 6$$

The step h_0 is the optimum of $g(h_0)$, which can be easily found using any single-variable optimization method, or by setting the derivative of $g(h_0)$ to zero (which is the most straightforward way to get the optimum of a degree-2 polynomial). The value is $h_0 = 0.191$. Using this step value in Eq. (9.20) gives the next point $\mathbf{x}_1 = [4.764, -2.854]$. The magnetic field model and the first

(continued)

Example 9.5 (continued)

iteration are represented in the figure below. The gradient line is represented as the solid red line on the figure, starting at the initial $(4, -4)$ point; it can be seen visually that it is indeed a parabola and indeed follows the direction of maximum change from $(4, -4)$. The new point $\mathbf{x}_1$ is found at the minimum of the gradient parabola, and in fact very near to the minimum of the magnetic field.

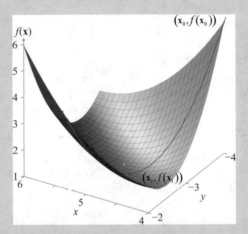

The Euclidean distance between $\mathbf{x}_1$ and the initial $\mathbf{x}_0$ is 1.378 m. This is much more than the centimeter precision required, so the iterations continue. The table below summarizes the next steps.

i	$\mathbf{x}_i$	$\nabla\mathbf{H}(\mathbf{x}_i)$	h_i	$\mathbf{x}_{i+1}$	E_i
0	$\begin{bmatrix} 4 \\ -4 \end{bmatrix}$	$\begin{bmatrix} -4 \\ -6 \end{bmatrix}$	0.191	$\begin{bmatrix} 4.764 \\ -2.854 \end{bmatrix}$	1.378
1	$\begin{bmatrix} 4.764 \\ -2.854 \end{bmatrix}$	$\begin{bmatrix} -0.176 \\ 0.112 \end{bmatrix}$	1.300	$\begin{bmatrix} 4.994 \\ -3.006 \end{bmatrix}$	0.276
2	$\begin{bmatrix} 4.994 \\ -3.006 \end{bmatrix}$	$\begin{bmatrix} -0.024 \\ -0.035 \end{bmatrix}$	0.191	$\begin{bmatrix} 4.999 \\ -2.999 \end{bmatrix}$	0.008

The method converges to 8 mm of precision by the third iteration. The optimum at that point is at $(4.999, -2.999)$ meters, almost exactly on the real optimum at $(5, -3)$ meters.

9.6   Stochastic Optimization

Engineering models of complex real-world systems will usually have no one optimum but several. Oscillating signals, repeating features, periodic events, and the interactions of multiple independent variables moving in opposite directions, all lead to systems that peak, drop, and peak again, repeatedly. A single one of these peaks or drops, the optimum within a limited region of the system, is called a *local optimum*. This is to differentiate it from the single greatest optimum of the entire function, which is the *global optimum*. To take a simple example, every single mountain in the Himalayas is a geographic local optimum, but Mount Everest is the single global optimum of the mountain range. For a mathematical example, consider the fluctuating function illustrated in Fig. 9.9. It has one global maximum and one global minimum indicated, as well as four additional local maxima, a plateau, and five local minima.

In such situations, a good optimization method should be able to converge on the global optimum, not just on a local optimum. It should discover the best solution possible. This is a problem with the methods presented so far in this chapter: the golden-means, quadratic, and gradient descent methods all converge on the nearest maximum or minimum without any checks to make sure it is the global one, and Newton's method is even worse, converging on the nearest optimum without even checking if it is a maximum or minimum.

What would be needed for a method to discover the global optimum? The method cannot somehow "know" that it has only converged on a local optimum and to continue searching, since that would imply that it already knows the value of the global optimum to compare its current result to. Instead, it has to "decide" somehow to diverge away from the optimum it has found and continue searching the function in case there is a better one elsewhere, and then to converge on that better optimum. Such a change in behavior of the method over time, especially to include a behavior that involves deliberately diverging away from an apparent solution, is very different from the behavior of the methods studied so far. In fact this is impossible for any of the methods seen so far, as they have all been designed

Fig. 9.9 Local and global maxima and minima

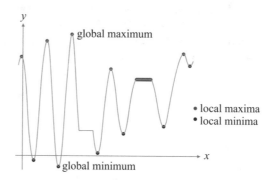

to behave in only one way, to always take the best step they can find, the one that makes them converge on the nearest optimum in the fastest way possible. They are *deterministic* methods: their behavior is entirely known and predictable. Given the same function and initial values, they will always compute the same steps and converge on the same optimum.

The alternative to a deterministic method is a *stochastic* method: an algorithm that includes an element of randomness. This random element is what will allow the method to have two different behaviors, by giving it a chance to escape local optima but not the global optimum. This can be implemented in practice in a number of ways, such as for example by including a random variable in an equation or by having a decision step that includes an element of chance. It should be noted that a random element is not necessarily one whose value is a result of complete and unbiased chance, such as a lottery draw. Rather, it is simply a term whose result is not known for sure in advance. It is completely acceptable to skew the probabilities towards a preferred outcome, or to change the probabilities as the algorithm progresses to reduce the impact of randomness over the iterations. A flip of a coin with a 99 % probability of landing on heads is still a stochastic event: even though one outcome is more likely than the other, it is still a result of chance.

Using randomness in optimization methods eliminates some of the certainty offered by deterministic algorithms. As mentioned already, a stochastic method is no longer guaranteed to converge on the nearest local optimum, which can be a desirable feature. However, this should not be mistaken for a certainty to converge on the global optimum; stochastic methods can make no such guarantee. Given the use of randomness in their algorithms and decision-making, no outcome can be certain or guaranteed. Another important point to keep in mind is that running the same stochastic method twice on the same function with the same initial values can lead to two very different final results, unlike with stochastic methods which guaranteed the exact same result both times. The reason for this is of course the inclusion of a random element in the algorithm, which can take very different values in successive runs.

Stochastic optimization algorithms are an intense area of ongoing research. Dozens of algorithms already exist, and new algorithms, variations, and enhancements are being proposed every year. Popular algorithms include genetic algorithms, ant colony algorithms, particle swarm algorithms, and many others. A complete review of these methods is beyond the scope of this book. The next two sections will introduce two stochastic optimization methods, to give an overview of this class of optimization methods.

9.7 Random Brute-Force Optimization

The *random brute-force search* is the simplest stochastic search method available. However, despite its inefficiency, it remains a functional and useful tool. Moreover, its simplicity makes it a good method to study as an initiation to stochastic optimization.

A *brute-force search* refers to any search algorithm that systematically tries possible solutions one after the other until it finds one that is acceptable or until a preset maximum number of attempts. A brute-force optimization algorithm would thus simply evaluate value after value for a given time, and return the value with the optimal result as its solution at the end. And a random brute-force search is one that selects the values to evaluate stochastically.

While the random brute-force search may seem unsophisticated, it does have the advantage of being able to search any function, even one that has a complex and irregular behavior, multiple local optima, and even discontinuities. By trying points at random and always keeping the optimal one, it is likely to get close to the global optimum and certain not to get stuck in a local optimum. The random brute-force search can be useful to deal with "black box" problems, when no information is available on the behavior of the function being optimized. The method makes no assumptions on the function and does not require a starting point, an interpolation or a derivative; it only needs an interval to search in. It can therefore perform its search and generate a good result in a situation of complete ignorance.

However, when this method does get a point close to the global optimum, it does not improve on it except by possibly randomly generating an even closer point. In other words, while the random brute-force approach is likely to find a point close to the global optimum, it is very unlikely to actually find the global optimum itself. For that reason, the algorithm is often followed by a few iterations of a deterministic algorithm such as Newton's method, which can easily and quickly converge to the global optimum from the starting point found by the brute-force search.

It should be instinctively clear that testing more points increases the algorithm's odds of getting closer to the optimum. However, even that rule of thumb is not a certainty given the stochastic nature of the algorithm. It could easily be the case that in one short run, the algorithm fortuitously generates a point very close to the optimum while in another much longer run, the algorithm is a lot unluckier and does not get as close. This is one of the risks of working with stochastic algorithms.

The iterative algorithm of the random brute-force search is straightforward: at each iteration, generate a random value within the search interval and evaluate it. Compare the result to the best value discovered so far in previous iterations. If the new result is better, keep it as the new best value; otherwise, discard it. This continues until the one and only termination condition, that a maximum number of iterations is reached.

Example 9.6

A two-dimensional periodic signal $s(x,y)$ is generated by the combination of two sinusoids, modelled by this equation:

$$s(x,y) = e^{-y^2} \cos(3x) + e^{-x^2} \cos(3y)$$

The emitter is hidden somewhere in a 12 km $\times$ 12 km field, marked from -6 to 6 km in each direction. Determine where the emitter is.

Solution

The emitter will be at the position where the signal is strongest, so this is a maximization problem. However, a combination of periodic signals such as this one will have multiple optima, making the search difficult. A visual plotting of the function in the interval, shown below, confirms that there are multiple local optima surrounding the global optimum where a deterministic optimization method could get stuck, as well as four large plateaux where the evaluation of the function is constant and deterministic optimization methods cannot work. This visual inspection also shows that the optimum is found at $(0, 0)$ and evaluates to 2.0.

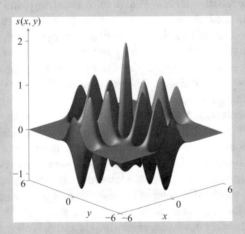

A stochastic optimization method can work well in this case, however. To illustrate, ten separate runs of the random brute-force search algorithm were performed, by increasing each time the number of random points generated and evaluated by 1000. The optimal point discovered in each run of the algorithm is listed in the following table. From the results shown, it can be seen that the algorithm never returned one of the local maxima and always got close to the global maximum. Although it never actually found the maximum, in some runs it got extremely close.

(continued)

Example 9.6 (continued)

Number of points	Optimum found	Optimum value
1000	(0.0284, −0.1005)	1.6886
2000	(0.0219, −0.0730)	1.9453
3000	(−0.0090, −0.0459)	1.9581
4000	(−0.0317, −0.0011)	1.9410
5000	(0.0067, −0.0161)	1.8775
6000	(−0.0137, −0.0010)	1.9999
7000	(−0.0064, 0.0005)	1.9922
8000	(−0.0098, 0.0149)	1.9909
9000	(−0.0108, 0.0065)	1.9816
10,000	(0.0245, −0.0159)	1.9981

The table also shows that, on average, checking more points leads to a better result: the five runs of 6000 points or more all returned better maxima than the five runs with 5000 points or less. However, this relationship is not perfect. The maximum found in the run with 6000 points is the single best one of all ten runs, better even than the maximum found in the run with 10,000 points, while the one found in the run at 5000 points is second-worst, better only than the one found in 1000 points. This illustrates nicely one of the important differences between stochastic and deterministic optimization mentioned earlier. In a deterministic algorithm, more iterations will always lead to a better result (unless the method diverges), while in a stochastic search, more iterations will on average, but not necessarily improve the result.

9.8 Simulated Annealing

Annealing is a metallurgical process used to temper metals through a heating and cooling treatment. The weaknesses in the metal that are eliminated by annealing are the result of atomic irregularities in the crystalline structure of the metal. These irregularities are due to atoms being stuck in the wrong place of the structure. In the process of annealing, the metal is heated up and then allowed to cool down slowly. Heating up gives the atoms the energy they need to get unstuck, and the slow cool-down period allows them to move to their correct location in the structure. Annealing can be seen as a multiple-optima optimization problem. A weakness in the metal is due to an atom having converged on a local optimum in the metal's crystalline structure. Heating the metal gives that atom the ability to escape the local optimum, and the slow cool-down period allows it to converge on its global optimum.

Simulated annealing is a stochastic optimization method based on the annealing process and on the gradient descent method studied previously. In fact, may stochastic optimization methods take their inspiration on natural phenomena; genetic algorithms, and ant colony algorithms are two more popular examples.

The parallel between real-world annealing and simulated annealing is straightforward: in one case an atom moves towards an optimal position in the crystal while avoiding getting stuck in attractive but suboptimal positions, and in the other steps are taken on a function to find the global optimum while avoiding getting trapped in a local optimum. But the real insight comes by studying how to escape the local optimum. In annealing, this is done by heating the metal to give it energy. When the metal is hot and the atoms are energized, they are more likely to move out of the local optimum (a high-energy movement), and as the metal cools down over time the atoms are more less energized and more likely to simply converge on the nearest (hopefully global) optimum. This can be simulated by using a "temperature" parameter that starts off at a high value and decreases iteratively. This temperature is directly related to the probability of a bad step (one that causes the value of the function to become less optimal) being accepted. At the higher initial value, bad moves are accepted more often and steps are taken away from the local optimum, and at the lower temperature of later iterations, bad steps are unlikely to be accepted and the method converges.

An iteration of the simulated annealing algorithm is thus:

1. Select a step h_i a random orientation around $\mathbf{x}_i$ to generate $\mathbf{x}_{i+1}$. Compute Δf_i, the difference in function evaluation between $f(\mathbf{x}_i)$ and $f(\mathbf{x}_{i+1})$, defined as:

$$\Delta f_i = f(\mathbf{x}_i) - f(\mathbf{x}_{i+1}) \tag{9.22}$$

   for maximization problems and:

$$\Delta f_i = f(\mathbf{x}_{i+1}) - f(\mathbf{x}_i) \tag{9.23}$$

   for minimization problems. Either way, the value of Δf_i will be negative if the step brings the method closer to an optimum and positive if it moves away from it, and the magnitude of the value will be proportional to the significance of the change.
2. If the value of Δf_i is negative, accept the step to $\mathbf{x}_{i+1}$.
3. If the value of Δf_i is positive, compute a probability of accepting the step based on the current temperature parameter value T_i:

$$P = e^{\frac{-\Delta f_i}{T_i}} \tag{9.24}$$

4. Reduce the temperature and step size for the next iteration.
5. Terminate the search if the termination condition is reached, which is that $T_{i+1} \leq 0$.

   These steps are implemented in the pseudocode of Fig. 9.10.

   The stochastic behavior of the method thus comes from step 3, where a step that worsens the value of the function might or might not be accepted based on a probability P. Equation (9.24) shows that P depends on two variables: the change in value of the step Δf_i, so that steps with a weak negative impact are more likely to be accepted than steps that massively worsen results, and the temperature T_i, so that

```
Optimum ← Input initial approximation of the optimum
Temperature ← Input initial temperature
DeltaTemperature ← Input iterative temperature decrease value
h ← Input initial neighbourhood size
DeltaH ← Input iterative neighbourhood size decrease value
ProblemType ← Input minimization or maximization

WHILE (TRUE)

    x ← random neighbour of Optimum at distance h

    IF (ProblemType = minimization)
        DeltaF ← [CALL F(x)] - [CALL F(Optimum)]
    ELSE IF (ProblemType = maximization)
        DeltaF ← [CALL F(Optimum)] - [CALL F(x)]
    END IF

    IF (DeltaF < 0)
        Optimum ← x
    ELSE
        Threshold ← Random value between 0 and 1
        P ← Exponential of ( -1 × DeltaF / T )
        IF (Threshold < P)
            Optimum ← x
        END IF
    END IF

    Temperature ← Temperature - DeltaTemperature
    h ← h - DeltaH

    IF (Temperature <= 0)
        RETURN Success, x
    END IF

END WHILE


FUNCTION F(x)
    RETURN vector of length n of evaluation of the n-dimensional target
            function at point x
END FUNCTION
```

Fig. 9.10 Pseudocode of the simulated annealing method

a bad move is more likely to be accepted at the beginning of the method when the temperature is high than at the end when the temperature is low.

Simulated annealing has the advantage of being able to explore a complex solution space and to escape local optima. The method used to explore the solution space in step 1 is simple and only requires that it be possible to numerically evaluate and compare two possible solutions. For that reason, simulated annealing is also very good at optimizing complex problems, including problems where the optimum depends on multiple interdependent variables and where the optimum is found by maximizing certain variables while minimizing others.

Example 9.7

A two-dimensional periodic signal $s(x,y)$ is generated by the combination of two sinusoids, modelled by this equation:

$$s(x,y) = e^{-y^2} \cos(3x) + e^{-x^2} \cos(3y)$$

The emitter is hidden somewhere in a 12 km × 12 km field, marked from −6 to 6 km in each direction. Determine where the emitter is.

Solution

Example 9.6 has already explored this function, and demonstrated how the solution space is too irregular for a deterministic optimization method. The random brute-force optimization method found points very near the optimum at (0, 0), but by the pure random chance of generating a point to try near the optimum, without any method to the search. A simulated annealing method could be used instead, to actually search the space step by step. The following figures show in white the path followed by this method—which, given the stochastic nature of the method, is only one of the countless possible paths it could randomly take.

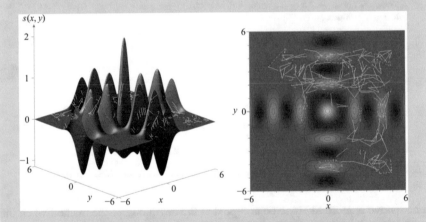

It can be seen that the method moves randomly through the space, and visits six of the local optima and three of the plateaux, before finally finding the global optimum. For each of the local optima, the method eventually steps away, quickly earlier in the iterations or after a longer exploration later in the iterations. Being able to explore and eventually leave the plateaux is another advantage of simulated annealing over other methods; Newton's method, quadratic optimization, and the gradient descent would all fail and terminate if they reached a constant region of the function they were optimizing.

9.9   Engineering Applications

Engineering design is constrained by reality, both physical (the laws of nature that dictate the performance limits of their systems) and economic (the need to keep costs and resource consumption down). In that sense, engineers are constantly confronted by optimization problems, to get the most out of systems within the limits of their constraints. The fuel tank design problem of Sect. 9.1 was a simple illustration of that common problem: the design had to minimize the tank's surface area and cost while respecting the system's requirements in terms of shape and volume. It is nonetheless representative of real-world problems; a similar optimization challenge led to the selection of the cylindrical 330 mL soft-drink can as the most cost-efficient design. Optimization problems are also encountered elsewhere in engineering practice, whenever conflicting requirements and constraints will arise.

- The design of many electrical components can be reduced to finding optimal points in equations. Indeed, the equations representing individual resistors, capacitors, inductances, and voltage sources are well-known, as are the formulae to combine them in parallel and serial connections. An entire circuit can thus be modelled in that manner, and once this model is available, the values of specific components can be optimized. For example, the impedance value Z for a resistor is known to be R, for an inductance it is ωL, and for a capacitor it is $(\omega C)^{-1}$, where ω is the frequency of the power supply. In turn, the impedance of a serial RLC circuit is given by:

$$Z(\omega) = \sqrt{R^2 + \left(\omega L - \frac{1}{\omega C}\right)^2} \tag{9.25}$$

From this model, it is possible to select a power supply with a frequency appropriate to maximize or minimize the circuit's impedance.

- The growth rate of yeast can be modelled by an exponential equation in function of the environment's temperature t:

$$G(t) = at^b e^{-ct} \tag{9.26}$$

where a, b, and c are constants dependent on the specific type of yeast studied. From this equation, it is possible to determine the temperature that will maximize growth.

- Scheduling problems are among the most popular optimization problems encountered in practice. Suppose for example a production line that can manufacture two different products, each with an associated unit production cost C_k and unit profit P_k. The aim is to manufacture a number of units of each product N_k in order to maximize profits P; however, the production line must operate within its allocated budget B. In other words:

$$N_0 C_0 + N_1 C_1 = B$$
$$N_0 P_0 + N_1 P_1 = P$$
$$N_0 P_0 + \frac{B - N_0 C_0}{C_1} P_1 = P \qquad (9.27)$$

Equation (9.27) can be used to maximize the number of units N_0 to manufacture, and from that value the number of units N_2 can be determined easily.

9.10   Summary

One of the most common challenges in engineering is to try to determine the value of some parameter of a system being designed to either maximize or minimize its output value. This value of the parameter is the optimum, and this challenge is optimization. This chapter has introduced several methods designed to solve an optimization model. The golden-mean method is a closed method, which sets up bounds around the optimum and uses the golden ratio property to get closer to the value. As with the closed methods seen for root-finding, this closed method is the least efficient one available but also the only one guaranteed not to diverge, because of its requirement to keep the optimum bracketed between the bounds. Two more open methods were examined, Newton's method and the quadratic optimization method. Both are more efficient than the golden-mean method, but both require more information, namely the first and second derivative for Newton's method and three points to interpolate a parabola with for the quadratic optimization, and both have a risk of diverging and failing in certain conditions. All three of these methods are also designed for two-dimensional problems; the next method learned was the gradient method, and it is a more general method designed to deal with multidimensional optimization problems. Finally, the topic of stochastic optimization was discussed. This topic is huge, worthy of an entire textbook to itself, and highly active in the scientific literature, so the discussion in this chapter is meant as nothing more than an introduction. Nonetheless, two stochastic methods were presented, the random brute-force search and simulated annealing. Table 9.3 summarizes the methods covered in this chapter.

Table 9.3 Summary of optimization methods

Method	Requires	Error
Golden-mean search	2 bounds	$O(h)$
Newton's method	1 point + first and second derivatives	$O(h^2)$
Quadratic optimization	3 points	$O(h^{1.497})$
Gradient descent	1 point + derivatives	$O(h^2)$
Random Brute-Force search	1 point + thousands of tries	Unbounded
Simulated annealing	1 point	Unbounded

9.11 Exercises

1. Use the golden-mean search to find a minimum of:

   (a) $f(x) = x^2$ starting with the interval $[-1, 2]$ to an absolute error of 0.1.
   (b) $f(x) = x^4$ starting with the interval $[0, 1]$ to an absolute error of 0.1.
   (c) $f(x) = (x - 1) \, x \, (x + 1)$ starting with the interval $[0, 2]$ to an absolute error of 0.1.
   (d) $f(x) = \sin(x)$ starting with the interval $[4, 5]$ to an absolute error of 0.1, working in radians.
   (e) $f(x) = x^2(x - 2)$ starting with the interval $[1, 2]$ to an absolute error of 0.1.
   (f) $f(x) = e^{-x} \sin(x)$ starting with the interval $[3, 5]$ to an absolute error of 0.1.

2. Using the golden-mean search with an initial interval of width h, how many iterations would be required to get to a width of less than ε?

3. Use Newton's method to find a minimum of:

   (a) $f(x) = x^2$ starting with the point $x_0 = 1$ to a relative error of 0.1.
   (b) $f(x) = x^4$ starting with the point $x_0 = 2$ to a relative error of 0.1.
   (c) $f(x) = (x - 1) \, x \, (x + 1)$ starting with the point $x_0 = 1$ to a relative error of 0.1.
   (d) $f(x) = \sin(x)$ starting with the point $x_0 = 4$ radians to a relative error of 0.01.
   (e) $f(x) = x^2(x - 2)$ starting with the point $x_0 = 1.5$ to a relative error of 0.01.
   (f) $f(x) = e^{-x} \sin(x)$ starting with $x_0 = 4$ to a relative error of 0.001.

4. Use the quadratic optimization method to find a minimum of:

   (a) $f(x) = x^2$ starting with the points $x_{-2} = 1$, $x_{-1} = 0.9$, and $x_0 = 0.8$ to a relative error of 0.1.
   (b) $f(x) = x^4$ starting with the points $x_{-2} = 1$, $x_{-1} = 0.9$, and $x_0 = 0.8$ and iterating for four steps.
   (c) $f(x) = (x - 1) \, x \, (x + 1)$ starting with the point $x_{-2} = 1.5$, $x_{-1} = 1$, and $x_0 = 0.5$ to a relative error of 0.1.
   (d) $f(x) = \sin(x)$ starting with the points $x_{-2} = 4$, $x_{-1} = 4.1$, $x_0 = 4.2$ to a relative error of 0.00001.
   (e) $f(x) = x^2(x - 2)$ starting with the points $x_{-2} = 2$, $x_{-1} = 1$, and $x_0 = 1.5$ to a relative error of 0.01.
   (f) $f(x) = e^{-x} \sin(x)$ with the points $x_{-2} = 3$, $x_{-1} = 4$, and $x_0 = 5$ to a relative error of 0.001.

5. Perform two iterations of gradient descent to find a minimum of the function $f(\mathbf{x}) = x_0^2 + x_1^2 - x_0 x_1 + x_0 - 2x_1$ starting with $\mathbf{x} = [1, 1]^T$.

Chapter 10
Differentiation

10.1 Introduction

Differentiation and its complement operation, *integration*, allow engineers to measure and quantify change. Measuring change is essential in engineering practice, which often deals with systems that are changing in some way, by moving, growing, filling up, decaying, discharging, or otherwise increasing or decreasing in some way.

To visualize the relationship, consider a simple example: a robot that was initially at rest begins moving at time 0 s in a straight line with a constant acceleration of 5 m/s^2. From this information, it is possible to model the entire situation and to track the robot's speed and position. Since the robot was at rest before 0 s and has a constant acceleration after 0 s, at time 0 s exactly it has an impulse jerk (variation in acceleration) of 5 m/s^3 that goes back to zero after that point. Its speed was also 0 m/s before time 0 s, but with a constant acceleration it will increase at a constant rate after that, being 5 m/s after 1 s, 10 m/s after 2 s, 15 m/s after 3 s, and so on. And the position will also be increasing, but faster than the speed, as the robot moves further per second at higher speeds. It will be 2.5 m from the starting position after 1 s, at 10 m after 2 s, at 22.5 m after 3 s, and so on. These four measures are illustrated in Fig. 10.1. And to further formalize this example, the equations for each of the metrics are as follows: The jerk, as indicated, is a single impulse of 5 m/s^3 at time 0 s:

$$\text{Jerk} = \begin{cases} 5 \text{ m/s}^3 & t = 0 \\ 0 \text{ m/s}^3 & t > 0 \end{cases} \tag{10.1}$$

The acceleration is constant:

$$\text{Acceleration} = 5 \text{ m/s}^2 \tag{10.2}$$

© Springer International Publishing Switzerland 2016
R. Khoury, D.W. Harder, *Numerical Methods and Modelling for Engineering*,
DOI 10.1007/978-3-319-21176-3_10

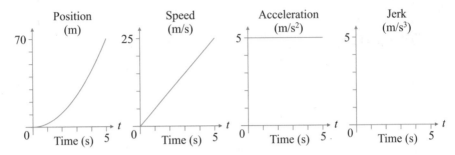

Fig. 10.1 Jerk (*top-right*), acceleration (*top-left*), speed (*bottom-left*), and position (*bottom-right*) with respect to time, for a robot at constant acceleration of 5 m/s²

And the speed and position both increase with respect to time:

$$\text{Speed} = 5t \text{ m/s} \tag{10.3}$$

$$\text{Position} = 2.5t^2 \text{ m} \tag{10.4}$$

Looking at the relationship, the graphs of Fig. 10.1 in clockwise order from the bottom-right shows the differentiation operation, while looking at their relationship in counter-clockwise order from the top-right shows the integration operation. Indeed, the counter-clockwise relationship demonstrates the rate of change of the previous curve. The position curve is increasing exponentially, as Eq. (10.4) shows, and the speed curve is thus one whose value is constantly increasing. But since it is increasing at a constant rate, the acceleration curve is a constant line, aside from the initial jump from 0 to 5 m/s² when movement started. And since a constant line is not changing, the jerk curve is zero, save for the impulse of 5 m/s³ at time 0 s when the acceleration changes from 0 to 5 m/s². On the other hand, integration, which will be covered in Chap. 11, is the area under the curve of the previous graph. The jerk is a single impulse of 5 m/s³ at time 0 s, which has an area of 5 m/s², followed by a constant zero line with null area. Consequently, the acceleration value jumps to 5 m/s² at time 0 s and remains constant there since no additional area is added. The area under this acceleration graph will be increasing constantly, by a value of 5 m/s. Consequently, the speed value that reflects its area increases constantly by that rate. And the area under the linearly increasing speed graph is actually increasing exponentially: it covers 2.5 m from 0 to 1 s, 10 m from 0 to 2 s, 22.5 m from 0 to 3 s, and so on, and as a result the position value is an exponentially increasing curve over time.

Being able to measure the differentiation and integral of systems is thus critical if the system being modelled is changing; if the system is not in a steady-state or if the model is not meant to capture a snapshot of the system at a specific moment in time. If the equation of the system is known or can be determined, as is the case with Eq. (10.2) in the previous example, then its derivates and integrals can be computed exactly using notions learned in calculus courses. This chapter and the next one, however, will deal with the case where the equation of the system is unknown and cannot be determined precisely, and the only information available is discrete measurements of the system. As these two chapters will show, it is still possible to compute the integral and derivative of the attributes of a system and thus model its changing nature with as little as two discrete measurements of it.

10.2   Centered Divided-Difference Formulae

Recall from Chap. 5 the first-order Taylor series approximation formula:

$$f(x_i + h) = f(x_i) + f^{(1)}(x_i)h \tag{10.5}$$

This formula makes it possible to approximate the value of a function at a point a step h after a point x_i where the value is known, using the known value of the first derivative at that point. Alternatively, if there was a need to approximate the value of the function at a point a step h before the known point x_i, there would only be a sign difference in the formula:

$$f(x_i - h) = f(x_i) - f^{(1)}(x_i)h \tag{10.6}$$

Now assume that the measurements of the function are known at all three points, but the derivative is unknown. It is immediately clear that either (10.5) or (10.6) could be solved to find the value of the derivative, since each is an equation with only one unknown. But, for reasons that will become clear soon, it is possible and preferable to do even better than this, by taking the difference of both equations:

$$f(x_i + h) - f(x_i - h) = f(x_i) + f^{(1)}(x_i)h - f(x_i) + f^{(1)}(x_i)h \tag{10.7}$$

And solving that equation for $f^{(1)}(x_i)$:

$$f^{(1)}(x_i) = \frac{f(x_i + h) - f(x_i - h)}{2h} \tag{10.8}$$

The formula of Eq. (10.8) is called the *second-order centered divided-difference formula*, and it gives a good approximation of the derivative of a function at point x_i given only two measurements of the function at equally spaced intervals before and after x_i. How good is the approximation? Since it is derived from the Taylor series, recall that the error is proportional to the next non-zero term. In this case, it will be the third-order term, since the second-order terms will cancel out. This can be verified by expanding the Taylor series approximations of Eqs. (10.5) and (10.6) and solving Eq. (10.7) again:

$$f(x_i + h) = f(x_i) + f^{(1)}(x_i)h + \frac{f^{(2)}(x_i)h^2}{2!} + \frac{f^{(3)}(x_i)h^3}{3!}$$

$$f(x_i - h) = f(x_i) - f^{(1)}(x_i)h + \frac{f^{(2)}(x_i)(-h)^2}{2!} + \frac{f^{(3)}(x_i)(-h)^3}{3!}$$

$$f(x_i + h) - f(x_i - h) = f(x_i) + f^{(1)}(x_i)h + \frac{f^{(2)}(x_i)h^2}{2!} + \frac{f^{(3)}(x_i)h^3}{3!} - f(x_i)$$

$$+ f^{(1)}(x_i)h - \frac{f^{(2)}(x_i)h^2}{2!} + \frac{f^{(3)}(x_i)h^3}{3!} \tag{10.9}$$

$$= 2f^{(1)}(x_i)h + \frac{2f^{(3)}(x_i)h^3}{3!}$$

$$f^{(1)}(x_i) = \frac{f(x_i + h) - f(x_i - h)}{2h} - \frac{f^{(3)}(x_i)h^2}{3!}$$

$$= \frac{f(x_i + h) - f(x_i - h)}{2h} + O(h^2)$$

Since the error is quadratic, or second-order, this gives the formula its name. Moreover, the development of Eq. (10.9) demonstrates why it was preferable to take the difference between Eqs. (10.5) and (10.6) to approximate the derivative, rather than simply solving either one of these equations. With only one equation, the second-order term of the Taylor series would have nothing to cancel out with, and the final formula would be $O(h)$, a less-accurate first-order formula.

Example 10.1
A 1-L reservoir is getting filled with water. It was initially empty, but reached a quarter-full after 1.16 s, half-full after 2.39 s, three-quarter-full after 3.45 s, and completely full after 4 s. Estimate the rate it was getting filled up by the time it was half-full using the centered divided-difference formula and steps of 0.5 and 0.25 L.

Solution
The fill-up rate is the volume filled per unit of time. The information given in the problem statement is instead the time needed to fill certain units of volume. The derivative of these values will be the time per volume, and the inverse will be the rate asked for. The derivative can be obtained by a straightforward application of Eq. (10.8):

$$f^{(1)}_{h=0.5}(0.5) = \frac{f(1) - f(0)}{2h} = \frac{4 - 0}{2 \times 0.5} = 4.00 \text{ s/L} \Rightarrow 0.25 \text{ L/s}$$

$$f^{(1)}_{h=0.25}(0.5) = \frac{f(0.75) - f(0.25)}{2h} = \frac{3.45 - 1.16}{2 \times 0.25} = 4.57 \text{ s/L} \Rightarrow 0.22 \text{ L/s}$$

From Eq. (10.9), it has been demonstrated that the error on the approximation is proportional to the square of the step size. Reducing the step size by half, from h to $h/2$, should thus reduce the error by a factor of 4, from $O(h^2)$

(continued)

Example 10.1 (continued)

to $O((h/2)^2) = O(h^2/4)$. To verify this, note that the equation modelling the time needed to fill the reservoir is:

$$f(V) = -3V^6 + 4V^5 + V^4 - 6V^3 + 4V^2 + 4V$$

Evaluating the derivative of this equation at $V = 0.5$ gives a rate of 4.69 s/L, or a fill-up rate of 0.21 L/s. The relative error of each approximation of the derivative is:

$$E_{h=0.5} = \left| \frac{4.00 - 4.69}{4.69} \right| = 14.7\%$$

$$E_{h=0.25} = \left| \frac{4.57 - 4.69}{4.69} \right| = 2.4\%$$

The error has in fact been reduced to less than a quarter of its original value, though it is still within the quadratic reduction range that was expected.

Finally, it is interesting to visualize the situation. The function describing the time needed to fill a given volume is plotted below. The derivative value is the slope of the tangent to the function at point $V = 0.5$. The approximation of that tangent with $h = 0.5$ is plotted in purple, and it can be seen to be rather off, and in fact clearly intersects the function. The approximation with $h = 0.25$, in red, is clearly a lot better, and is in fact almost overlaps with the real tangent, plotted in green.

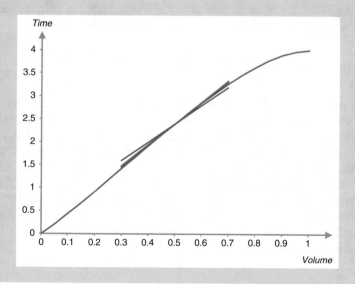

The previous example used two pairs of two points to compute two second-order divided-difference approximations of the derivative. But if four points are available, an intuitive decision would be rather to use all four of them to compute a single approximation with a higher accuracy. Such a formula can be derived from the Taylor series approximations, as before. Begin by writing out the four Taylor series approximations, to a sufficiently high order to model the error of the approximation as well. To determine which order to go up to, note that for with the divided-difference formula with two points in Eq. (10.9), the even second-order term cancelled out and the error was due to the third-order term. Consequently, it can be intuitively expected that, with the more accurate formula with four points, the next even-order term will also cancel out and the error will be due to the fifth-order term. The four fifth-order Taylor series approximations are:

$$f(x_i + 2h) = f(x_i) + 2f^{(1)}(x_i)h + \frac{4f^{(2)}(x_i)h^2}{2!} + \frac{8f^{(3)}(x_i)h^3}{3!} + \frac{16f^{(4)}(x_i)h^4}{4!} + \frac{32f^{(5)}(x_i)h^5}{5!}$$

$$f(x_i + h) = f(x_i) + f^{(1)}(x_i)h + \frac{f^{(2)}(x_i)h^2}{2!} + \frac{f^{(3)}(x_i)h^3}{3!} + \frac{f^{(4)}(x_i)h^4}{4!} + \frac{f^{(5)}(x_i)h^5}{5!}$$

$$f(x_i - h) = f(x_i) - f^{(1)}(x_i)h + \frac{f^{(2)}(x_i)h^2}{2!} - \frac{f^{(3)}(x_i)h^3}{3!} + \frac{f^{(4)}(x_i)h^4}{4!} - \frac{f^{(5)}(x_i)h^5}{5!}$$

$$f(x_i - 2h) = f(x_i) - 2f^{(1)}(x_i)h + \frac{4f^{(2)}(x_i)h^2}{2!} - \frac{8f^{(3)}(x_i)h^3}{3!} + \frac{16f^{(4)}(x_i)h^4}{4!} - \frac{32f^{(5)}(x_i)h^5}{5!}$$

$$\tag{10.10}$$

Taking the difference of the series one step before and after, as before, cancels out the second- and fourth-order terms, but leaves the third-order term:

$$f(x_i + h) - f(x_i - h) = 2f^{(1)}(x_i)h + \frac{2f^{(3)}(x_i)h^3}{3!} + \frac{2f^{(5)}(x_i)h^5}{5!} \tag{10.11}$$

That's a problem; the third-order term must get cancelled out as well, otherwise it will be the dominant term for the error. Fortunately, there are two more series to incorporate into the formula. Looking back at the set of equations (10.10), it can be seen that the third-order term in the series $f(x_i \pm h)$ are eight times less than the third-order term in the series $f(x_i \pm 2h)$. To get them to cancel out, the series $f(x_i \pm h)$ should thus be multiplied by 8 in Eq. (10.11) and the series $f(x_i \pm 2h)$ should be of opposite signs from their counterparts one step before or after. And by having the series $f(x_i \pm 2h)$ be of opposite signs from each other, the second- and fourth-order terms will cancel with each other as they did in Eq. (10.11). The resulting formula is:

$$-f(x_i + 2h) + 8f(x_i + h) - 8f(x_i - h) + f(x_i - 2h) = 12f^{(1)}(x_i)h - \frac{48f^{(5)}(x_i)h^5}{5!}$$

$$f^{(1)}(x_i) = \frac{-f(x_i + 2h) + 8f(x_i + h) - 8f(x_i - h) + f(x_i - 2h)}{12h} + O(h^4) \quad (10.12)$$

The third-order terms now cancel out, leaving only the fifth-order terms and a division by h for an error of $O(h^4)$. As expected, using two points before and after rather than just one has greatly increased the accuracy of the formula. This is now the *fourth-order centered divided-difference formula*.

Example 10.2
A 1-L reservoir is getting filled with water. It was initially empty, but reached a quarter-full after 1.16 s, half-full after 2.39 s, three-quarter-full after 3.45 s, and completely full after 4 s. Estimate the rate it was getting filled up by the time it was half-full using the fourth-ordered centered divided-difference formula.

Solution
The fill-up rate is the volume filled per unit of time. The information given in the problem statement is instead the time needed to fill certain units of volume. The derivative of these values will be the time per volume, and the inverse will be the rate asked for. The derivative can be obtained by a straightforward application of Eq. (10.12):

$$f^{(1)}(0.5) = \frac{-f(1) + 8f(0.75) - 8f(0.25) + f(0)}{12 \times h}$$

$$= \frac{-4 + 8 \times 3.45 - 8 \times 1.16 + 0}{12 \times 0.25} = 4.77 \text{ s/L} \Rightarrow 0.21 \text{ L/s}$$

Compared to the real rate of 4.69 s/L (fill-up rate of 0.21 L/s), the relative error of this approximation is:

$$E_{h=0.25} = \left| \frac{4.77 - 4.69}{4.69} \right| = 1.7\%$$

This result is clearly more accurate than either ones obtained with the same data using the second-order centered divided-difference formula in Example 10.1.

10.3   Forward and Backward Divided-Difference Formulae

The best approximation of the derivative is obtained by using measurements both before and after a target point. Unfortunately, oftentimes in engineering, it is not possible to get measurements on both sides of a target point. For example, it may be necessary to use the current rate of change of an incoming signal to adjust buffer capacity and processing resources in real-time, in which case only past measurements are available. Or it may be necessary to reconstruct how temperatures were changing on Earth long before climate records were kept for environmental and historical studies, in which case only future measurements (from the point of view of the values being estimated) are available. To deal with such cases, the *backward divided-difference formula* (using only past values) and the *forward divided-difference formula* (using only future values) can be derived from the Taylor series using only steps in the acceptable direction. The trade-off will be the need for more measurements to maintain the same accuracy as the centered divided-difference formulae.

In the previous section, the second-order centered divided-difference formula and its error term were developed from two steps of the third-order Taylor series approximation. Likewise, the *second-order backward divided-difference formula* and its error term can be derived from the same starting point. However, in this case, the two steps will be two previous steps, as such:

$$f(x_i - h) = f(x_i) - f^{(1)}(x_i)h + \frac{f^{(2)}(x_i)h^2}{2!} - \frac{f^{(3)}(x_i)h^3}{3!}$$

$$f(x_i - 2h) = f(x_i) - 2f^{(1)}(x_i)h + \frac{4f^{(2)}(x_i)h^2}{2!} - \frac{8f^{(3)}(x_i)h^3}{3!}$$

(10.13)

The second-order term of the Taylor series approximations will need to cancel out in order to keep the third-order term as the dominant error term. It can be seen from Eq. (10.13) that this second-order term is four times greater in $f(x_i - 2h)$ than it is in $f(x_i - h)$, and of the same sign. Consequently, the series $f(x_i - h)$ will need to be multiplied by -4 for the second-order terms to cancel out:

$$f(x_i - 2h) - 4f(x_i - h) = -3f(x_i) + 2f^{(1)}(x_i)h - \frac{4f^{(3)}(x_i)h^3}{3!}$$

$$f^{(1)}(x_i) = \frac{f(x_i - 2h) - 4f(x_i - h) + 3f(x_i)}{2h} + O(h^2)$$

(10.14)

The resulting equation is indeed $O(h^2)$. However, it can be noted that this accuracy required three measurements, at the current point and at one and two steps before, while the second-order centered divided-difference formula achieved it with only two measurements. As indicated before, this is because more measurements are needed to even out the loss of information that comes from using measurements all on one side of the target point, rather than centered around the target point.

The second-order forward divided-difference formula is computed using the same development as its backward counterpart. The final equation and error term are:

$$f^{(1)}(x_i) = \frac{-f(x_i + 2h) + 4f(x_i + h) - 3f(x_i)}{2h} + O(h^2) \tag{10.15}$$

Example 10.3
A robot is observed moving in a straight line. It starts off at the 8 m mark, and is measured every second:

Time t (s)	0	1	2	3	4
Position $f(t)$ (m)	8	16	34	62	100

Estimate its current speed, at time 4 s.

Solution
Speed is the first derivative of position. Since only past measurements and the current-time measurement are available, the second-order backward divided-difference formula can be used. Applying Eq. (10.14) gives:

$$f^{(1)}(4) = \frac{f(2) - 4f(3) + 3f(4)}{2 \times 1}$$

$$= \frac{34 - 4 \times 62 + 3 \times 100}{2}$$

$$= 43 \text{ m/s}$$

To verify this result, note that the equation modelling the robot's position is:

$$f(t) = 5t^2 + 3t + 8$$

The derivative of this equation is trivial to compute, and evaluated at $t = 4$ it does give 43 m/s.

10.4 Richardson Extrapolation

It has been demonstrated in the previous sections that the error on the approximation of the derivative is function of h, the step size between the measurements. In fact, Example 10.1 even demonstrated practically that a smaller step size leads to a better approximation. However, this also leads to a major problem with the divided-difference formulae: the risk of subtractive cancellation, introduced back in Chap. 2. Consider for example the second-order centered divided-difference formula of Eq. (10.9). As the value of h is reduced with the expectation of increasing accuracy, there will come a point where $f(x_i - h) \approx f(x_i + h)$ and the effects of subtractive cancellation will be felt. At that point, it would be wrong, and potentially dangerous, to continue to use smaller and smaller values of h and to advertise the results as "more accurate." And this issue of subtractive cancellation at smaller values of h will occur with every divided-difference formula available, as they are all based on taking the difference between measurements at regular intervals.

Yet the fundamental problem remains; smaller values of h are the only way to improve the accuracy of the approximation of the derivative. *Richardson extrapolation* offers a solution to this problem. Instead of decreasing the value of h and computing the divided-difference formula, this method makes it possible to compute the divided-difference formula with a large value of h then iteratively decrease it to increase accuracy.

To begin, note that in every divided-difference formula, the even-order terms from the Taylor series approximation cancel out, leaving the odd-order terms with even-valued exponents of h once the final division by h is performed. The first non-zero term after the formula becomes the error term, and all other terms are ignored. But if the first non-zero term is cancelled out, as it was when going from the second-order centered divided-difference formula to the fourth-order one, then the error drops by a factor of h^2 for that reason.

Now consider again the centered divided-difference formula of Eq. (10.9). If the equation is expanded to include all terms from the Taylor series, and since the even-order terms cancel out, the formula becomes:

$$f^{(1)}(x_i) = \frac{f(x_i + h) - f(x_i - h)}{2h} - \frac{f^{(3)}(x_i)h^2}{3!} - \frac{f^{(5)}(x_i)h^4}{5!} - \frac{f^{(7)}(x_i)h^6}{7!}$$
$$ - \cdots \tag{10.16}$$

Note again that since every other term cancels out, every term k actually appearing in the series represents an error of $O(h^{2k})$. The error is of course dominated by the largest term, which in this case is $O(h^2)$. But keeping that first error term written out explicitly, Eq. (10.16) can be rewritten equivalently as:

$$D_{exact} = D_1(h) + K_1 h^2 + O(h^4) \tag{10.17}$$

where D_{exact} represents the real exact value of the derivative and:

$$D_1(h) = \frac{f(x_i + h) - f(x_i - h)}{2h} \tag{10.18}$$

$$K_1 = -\frac{f^{(3)}(x_i)h^2}{3!} \tag{10.19}$$

As mentioned already, the way to improve the accuracy of this formula is to decrease the step size. Therefore, replace h with $h/2$ in Eq. (10.17) to improve its accuracy. This gives:

$$\begin{aligned}
D_{\text{exact}} &= D_1\left(\frac{h}{2}\right) + K_1\left(\frac{h}{2}\right)^2 + O\left(\left(\frac{h}{2}\right)^4\right) \\
&= D_1\left(\frac{h}{2}\right) + K_1\frac{h^2}{4} + O(h^4)
\end{aligned} \tag{10.20}$$

Note that the division by 4 in the big O term has disappeared; as explained in Chap. 1, big O notation is indifferent to constant values and only function of the variable, in this case h. As for the derivative approximation, nothing special has happened. The formula is the same with half the step size, and the risk of subtractive cancellation is still present if h becomes too small. But notice that Eqs. (10.17) and (10.20) each have a different parameter for D_1 but the same term h^2, with the only difference being that one is four times larger than the other. This should give an idea for cancelling out the h^2 term: taking four times Eq. (10.20) and subtracting Eq. (10.17):

$$4D_{\text{exact}} - D_{\text{exact}} = 4D_1\left(\frac{h}{2}\right) + 4K_1\frac{h^2}{4} + 4O(h^4) - D_1(h) - K_1h^2 - O(h^4)$$

$$3D_{\text{exact}} = 4D_1\left(\frac{h}{2}\right) - D_1(h) + O(h^4)$$

$$D_{\text{exact}} = \frac{4D_1\left(\frac{h}{2}\right) - D_1(h)}{3} + O(h^4) \tag{10.21}$$

The error is now $O(h^4)$, an important improvement, and more importantly this was done without risking subtractive cancellation! And this is after only one iteration; Richardson extrapolation is an iterative process, so it can be done again. Begin by rewriting Eq. (10.21) in the same form as Eq. (10.17):

$$D_{\text{exact}} = D_2\left(\frac{h}{2}\right) + K_2h^4 + O(h^6) \tag{10.22}$$

where:

$$D_2\left(\frac{h}{2}\right) = \frac{4D_1\left(\frac{h}{2}\right) - D_1(h)}{3} \tag{10.23}$$

Then compute again a more accurate version by dividing h into half:

$$D_{exact} = D_2\left(\frac{h}{4}\right) + K_2\frac{h^4}{16} + O(h^6) \tag{10.24}$$

This time, comparing Eqs. (10.22) and (10.24), the h^4 term is 16 times larger before. Therefore the second equation will need to be multiplied by 16 to cancel out this next error term.

$$16D_{exact} - D_{exact} = 16D_2\left(\frac{h}{4}\right) + 16K_2\frac{h^4}{16} + 16O(h^6) - D_2\left(\frac{h}{2}\right) - K_2h^4 - O(h^6)$$

$$15D_{exact} = 16D_2\left(\frac{h}{4}\right) - D_2\left(\frac{h}{2}\right) + O(h^6)$$

$$D_{exact} = \frac{16D_2\left(\frac{h}{4}\right) - D_2\left(\frac{h}{2}\right)}{15} + O(h^6) \tag{10.25}$$

This process can go on iteratively forever, or until one of the usual termination conditions applies: a threshold relative error between the approximation of the derivative of two iterations is achieved (success condition), or a preset maximum number of iterations is reached (failure condition). Richardson extrapolation can be summarized as follows:

$$D_{exact} = D_k\left(\frac{h}{2^{k-1}}\right) + O(h^{2k}) \quad \text{for } k \geq 1 \tag{10.26}$$

where:

$$D_k\left(\frac{h}{2^j}\right) = \frac{4^{k-1}D_{k-1}\left(\frac{h}{2^j}\right) - D_{k-1}\left(\frac{h}{2^{j-1}}\right)}{4^{k-1} - 1} \quad \text{if } k > 1 \tag{10.27}$$

$$D_k\left(\frac{h}{2^j}\right) = \frac{f\left(x_i + \frac{h}{2^j}\right) - f\left(x_i - \frac{h}{2^j}\right)}{2\frac{h}{2^j}} \quad \text{if } k = 1 \tag{10.28}$$

Then, for each value of j starting from 0 and going up to the termination condition, compute Eq. (10.27) for all values of k from 1 to $j+1$. The first iteration will thus only compute one instance of Eq. (10.28), and each subsequent iteration will add one more instance of Eq. (10.27). Moreover, each new instance of Eq. (10.27) will be computed from a lower-k-valued instance of it, down to Eq. (10.28). The value of Eq. (10.27) with the highest values of j and k at the final iteration will be the approximation of the derivative in Eq. (10.26).

Algorithmically, the Richardson extrapolation method can be implemented by filling up a table left to right using values computed and stored in the previous columns, much like the table of divided-differences in Newton's interpolation method. Each column in this table thus represents an increment of the value of k in Eqs. (10.26) to (10.28). The elements of the first, left-most column represent $k = 1$ and are computed using Eq. (10.28), while the following columns represent $k > 1$ and are computed using Eq. (10.27). Likewise, each row of the table represents a factor of 2 in the division, or an exponent of j in Eqs. (10.27) and (10.28). At each iteration, the algorithm adds a row at the bottom of the first column (one new division by 2), and then moves right using the new value to compute higher-order approximations. Since each step right in the table requires computing a new value using two previous values, the number of elements in each column is one less than the previous, until the right-most column has only one element and the iteration ends. This process then repeats until one of the termination conditions mentioned before is realized. The pseudocode for this method is presented in Fig. 10.2.

```
x ← Input target value to find the derivative at
h ← Input initial step size
IterationMaximum ← Input maximum number of iterations
ErrorMinimum ← Input minimum relative error

RichardsonTable ← empty IterationMaximum × IterationMaximum table
BestValue ← 0

RowIndex ← 0
WHILE (RowIndex < IterationMaximum)
     element at column 0, row RowIndex of RichardsonTable ←
             [F(x + h) - F(x - h)] / (2 × h)
     h ← h / 2

     ColumnIndex ← 1
     WHILE (ColumnIndex <= RowIndex)
          element at column ColumnIndex, row RowIndex of RichardsonTable ←
               [ (4 to the power ColumnIndex-1) × (element at column
               ColumnIndex-1, row RowIndex of RichardsonTable) - (element
               at column ColumnIndex-1, row RowIndex-1 of RichardsonTable)
               ] / [(4 to the power ColumnIndex-1) - 1]
          ColumnIndex ← ColumnIndex + 1
     END WHILE

     PreviousValue ← BestValue
     BestValue ← element at column RowIndex, row RowIndex of
               RichardsonTable
     CurrentError ← absolute value of [ (BestValue - PreviousValue) /
               BestValue ]
     IF (CurrentError <= ErrorMinimum)
         RETURN Success, BestValue
     END IF

     RowIndex ← RowIndex + 1
END WHILE
RETURN Failure

FUNCTION F(x)
     RETURN evaluation of the target function at point x
END FUNCTION
```

Fig. 10.2 Pseudocode of Richardson extrapolation

Example 10.4
A 1-L reservoir is getting filled with water. It was initially empty, but reached a quarter-full after 1.16 s, half-full after 2.39 s, three-quarter-full after 3.45 s, and completely full after 4 s. Estimate the rate it was getting filled up by the time it was half-full using the centered divided-difference formula and Richardson extrapolation.

Solution
The fill-up rate is the volume filled per unit of time. The information given in the problem statement is instead the time needed to fill certain units of volume. The derivative of these values will be the time per volume, and the inverse will be the rate asked for.

Richardson extrapolation starts with $j=1$. The only possible value of i from 1 to $j+1$ is thus $k=1$, and the only equation to compute is (10.28). Putting in the values naturally gives the second-order centered divided-difference formula as it was computed in Example 10.1:

$$D_1\left(\frac{h}{2^0}\right) = \frac{f\left(x+\frac{h}{2^0}\right) - f\left(x-\frac{h}{2^0}\right)}{2\frac{h}{2^0}}$$

$$= \frac{f(1) - f(0)}{2h}$$

$$= \frac{4 - 0}{2 \times 0.5} = 4.00 \text{ s/L}$$

At the next iteration, $j=1$ and $k=\{1, 2\}$. There are now two equations to compute, again one of which was already computed in Example 10.1:

$$D_1\left(\frac{h}{2^1}\right) = \frac{f\left(x+\frac{h}{2^1}\right) - f\left(x-\frac{h}{2^1}\right)}{2\frac{h}{2^1}}$$

$$= \frac{f(0.75) - f(0.25)}{2 \times 0.25}$$

$$= \frac{3.45 - 1.16}{2 \times 0.25} = 4.57 \text{ s/L}$$

$$D_2\left(\frac{h}{2^1}\right) = \frac{4^1 D_1\left(\frac{h}{2^1}\right) - D_1\left(\frac{h}{2^0}\right)}{4^1 - 1}$$

$$= \frac{4 \times 4.57 - 4}{3} = 4.77 \text{ s/L}$$

(continued)

Example 10.4 (continued)
The relative error between the derivative approximations of these two iterations is:

$$E = \left| \frac{4.77 - 4.00}{4.77} \right| = 16.1\%$$

As expected, the equation for D_2 in this new equation is computed from two
values for D_1, one from the previous iteration and one computed just this
iteration. If another iteration were computed, D_3 would likewise be evaluated
using this value of D_2 and from one computed in the third iteration.

However, given the information in the problem statement, this second
iteration is the maximum that can be computed. The final result is thus at $i = 2$
and:

$$D_{exact} = D_2\left(\frac{h}{2^1}\right) + O(h^4)$$

$$= 4.77 \text{ s/L} + O(h^4)$$

Note this same $O(h^4)$ approximation of the derivative that was computed with
the fourth-order centered divided-difference formula in Example 10.2, but
using only the second-order centered divided-difference results of Example
10.1 and an iterative process to combine and refine them.

10.5   Second Derivatives

So far, this chapter has focused on approximating the first derivative of a system
being modelled. This is reasonable, as the rate of change of the parameters of a
system over time is often critically important to include in a complete model.
However, these rates of change are themselves often not constant, and modelling
them as such will lead the model to diverge from reality over time. To remedy
that, it is important to include their rate of change over time as well; in other words,
to compute higher derivatives. In engineering practice, the second derivative of
the system (modelling the rate of change of the rate of change of parameters) is the
one most often included, and the one this chapter will focus on, but the same
technique described here could be used to develop equations for third derivatives
and higher.

The technique for finding the nth derivative is the same as that for finding the
first derivative. Given a set of measurements of the system at equally spaced
intervals, expand the Taylor series approximations for each measurement, then
combine them with multiplications and subtractions to eliminate all non-zeroth-

order terms except the one of the same order as the desired derivative and the highest-possible-order one for the error term.

Consider the case with one measurement before and after the target point. The two third-order Taylor series approximations were expanded in Eq. (10.9), and they were subtracted from each other to derive the second-order centered divided-difference formula to approximate the first derivative. But if the goal is to keep the second derivative and cancel out the first, then the two series should be summed together rather than subtracted. The result is:

$$
\begin{aligned}
f(x_i + h) + f(x_i - h) &= \left(f(x_i) + f^{(1)}(x_i)h + \frac{f^{(2)}(x_i)h^2}{2!} + \frac{f^{(3)}(x_i)h^3}{3!} + \frac{f^{(4)}(x_i)h^4}{4!} \right) + \\
&\quad \left(f(x_i) - f^{(1)}(x_i)h + \frac{f^{(2)}(x_i)h^2}{2!} - \frac{f^{(3)}(x_i)h^3}{3!} + \frac{f^{(4)}(x_i)h^4}{4!} \right) \\
&= 2f(x_i) + \frac{2f^{(2)}(x_i)h^2}{2!} + \frac{2f^{(4)}(x_i)h^4}{4!}
\end{aligned}
$$

$$
f^{(2)}(x_i) = \frac{f(x_i + h) + f(x_i - h) - 2f(x_i)}{h^2} + O(h^2) \tag{10.29}
$$

And this is the *second-derivative second-order divided-difference formula*. As before, the formula has an error of $O(h^2)$. However, by summing up, this time it is the odd-order terms of the series that cancel out in the centered formula, rather than the even-order ones. Note also that the measurement at the target point, $f(x_i)$, is needed in this formula. Mathematically, this corresponds to the zeroth-order term of the series, which like other even-order terms was cancelled out when computing the first derivative but is not anymore. Practically, it fits engineering intuition that getting more information (the second derivative rather than the first) requires more data (a third measurement). In general, using the centered divided-difference formula to get the nth derivative of a measurement with $O(h^2)$ error will require $n + 1$ points.

As with the first derivative, a fourth-order divided-difference formula can be computed for the second derivative by using one more point before and one more after the target point. Since the development of Eq. (10.29) has already shown that the final formula is divided by h^2, and since the final error term of a fourth-order formula after that division must be h^4, then the Taylor series approximations must be expanded to the sixth order. The fifth-order series were already given in Eq. (10.10), and the sixth-order term can be appended easily. Looking at Eq. (10.10), it can be seen that all the odd-order terms will cancel each other out provided that the two series one step away from the target point are summed together with the same sign and multiplier, and that the two series two steps away from the target point are also summed up together with the same sign and multiplier:

$$f(x_i + h) + f(x_i - h) = 2f(x_i) + \frac{2f^{(2)}(x_i)h^2}{2!} + \frac{2f^{(4)}(x_i)h^4}{4!} + \frac{2f^{(6)}(x_i)h^6}{6!}$$

$$f(x_i + 2h) + f(x_i - 2h) = 2f(x_i) + \frac{8f^{(2)}(x_i)h^2}{2!} + \frac{32f^{(4)}(x_i)h^4}{4!} + \frac{128f^{(6)}(x_i)h^6}{6!}$$

$$(10.30)$$

This leaves the zeroth-order term, which is the measurement at the target point and therefore available, the second-order term, which has the second derivative, the sixth-order term, which will be the error term, and the fourth-order term, which must be eliminated in order for the error to be the sixth-order term. The problem is that the fourth-order term is positive in all four series, so it cannot be cancelled out by adding them together, and it is 16 times greater in the two series two steps away. The solution is to multiply the two series one step away by 16 to make their fourth-order term of the correct magnitude, and the two series two steps away by -1 to insure that the terms cancel out with the series one step away without affecting the other terms eliminated by summation. The final result is:

$$f^{(2)}(x_i) = \frac{-f(x_i + 2h) + 16f(x_i + h) + 16f(x_i - h) - f(x_i - 2h) - 30f(x_i)}{12h^2} + O(h^4)$$

$$(10.31)$$

The same process can also be used to devise forward and backward formulae. Recall that the second-order backward divided-difference formula for the first derivative was estimated from two past measurements of the system. The discussion about the centered divided-difference formulae has already shown that an additional point is needed to estimate the second derivative to the same error value, as well as an additional order term in the Taylor series approximation to account for the division by h^2. Consequently, three past measurements will be needed for a second-derivative second-order backward divided-difference formula, and the corresponding Taylor series approximations will need to be expanded to the fourth order, as such:

$$f(x_i - h) = f(x_i) - f^{(1)}(x_i)h + \frac{f^{(2)}(x_i)h^2}{2!} - \frac{f^{(3)}(x_i)h^3}{3!} + \frac{f^{(4)}(x_i)h^4}{4!}$$

$$f(x_i - 2h) = f(x_i) - 2f^{(1)}(x_i)h + \frac{4f^{(2)}(x_i)h^2}{2!} - \frac{8f^{(3)}(x_i)h^3}{3!} + \frac{16f^{(4)}(x_i)h^4}{4!}$$

$$f(x_i - 3h) = f(x_i) - 3f^{(1)}(x_i)h + \frac{9f^{(2)}(x_i)h^2}{2!} - \frac{27f^{(3)}(x_i)h^3}{3!} + \frac{81f^{(4)}(x_i)h^4}{4!}$$

$$(10.32)$$

Clearly, cancelling out the first and third-order terms will require more than a simple addition as was the case with the centered divided-difference formulae.

However, this can be done in a simple methodical way, by starting with $f(x_i - 3h)$, the formula with the largest coefficients multiplying terms, and figuring out the multiple of $f(x_i - 2h)$ needed to eliminate the largest coefficients. It will not be exact, but it should be rounded up, and then $f(x_i - h)$ can be used to cancel out the remainders. In the case of Eq. (10.32), the third-order term of $f(x_i - 3h)$ is 3.4 times larger than that of $f(x_i - 2h)$, so rounding up the latter series will be multiplied by 4 and subtracted:

$$f(x_i - 3h) - 4f(x_i - 2h) = -3f(x_i) + 5f^{(1)}(x_i)h - \frac{7f^{(2)}(x_i)h^2}{2!}$$
$$+ \frac{5f^{(3)}(x_i)h^3}{3!} + \frac{17f^{(4)}(x_i)h^4}{4!} \qquad (10.33)$$

This leaves five times the first-order term and five times the third-order term. The series $f(x_i - h)$ will thus need to be multiplied by 5, and added to the other two to cancel out these terms:

$$f(x_i - 3h) - 4f(x_i - 2h) + 5f(x_i - h) = 2f(x_i) - \frac{2f^{(2)}(x_i)h^2}{2!} + \frac{22f^{(4)}(x_i)h^4}{4!}$$
$$(10.34)$$

$$f^{(2)}(x_i) = \frac{-f(x_i - 3h) + 4f(x_i - 2h) - 5f(x_i - h) + 2f(x_i)}{h^2} + O(h^2) \qquad (10.35)$$

The second-order forward divided-difference formula is the same but with a sign difference, and can be derived using the same process:

$$f^{(2)}(x_i) = \frac{-f(x_i + 3h) + 4f(x_i + 2h) - 5f(x_i + h) + 2f(x_i)}{h^2} + O(h^2) \qquad (10.36)$$

Example 10.5
A robot is observed moving in a straight line. It starts off at the 8 m mark, and is measured every second:

Time t (s)	0	1	2	3	4
Position $f(t)$ (m)	8	16	34	62	100

Estimate its current acceleration, at time 4 s.

Solution
Acceleration is the second derivative of position. Since only past measurements and the current-time measurement are available, the second-order

(continued)

Example 10.5 (continued)
backward divided-difference formula can be used. Applying Eq. (10.35) gives:

$$f^{(2)}(4) = \frac{-f(1) + 4f(2) - 5f(3) + 2f(4)}{1^2}$$

$$= \frac{-16 + 4 \times 34 - 5 \times 62 + 2 \times 100}{1}$$

$$= 10 \text{ m/s}^2$$

To verify this result, note that the equation modelling the robot's position is:

$$f(t) = 5t^2 + 3t + 8$$

The second derivative of this equation is trivial to compute, and evaluated at $t=4$ it does give 10 m/s^2.

The speed of the robot at $t=4$ has already been found to be 43 m/s in Example 10.3. Not modelling acceleration, at $t=5$ the position would be assumed to be 143 m and the speed still 43 m/s. But now that the model does include acceleration, the speed and position at $t=5$ will be found to be 53 m/s and 148 m respectively. This shows the importance of including not only the rate of change of parameters, but also the second-derivative change of the rate of change, in engineering models.

10.6   Unevenly Spaced Measurements

All the formulae seen so far have one thing in common: they require measurements taken at equal intervals before or after the target point at which the derivative is required. Unfortunately, such measurements may not always be available. They might have been recorded irregularly because of equipment failure, or lost to a data storage failure, bad record-keeping, or simple human negligence. Another approach will be needed to deal with such cases.

Given measurements at irregular intervals, one simple option is to interpolate a polynomial that fits these measurements using any of the techniques learned in Chap. 6, and then simply compute the derivative of that polynomial. In fact, there's an even better option, namely to include the derivative in the interpolation formula and thus to interpolate the derived equation directly. This can be done easily starting from the Lagrange polynomial formula:

$$f(x) = \sum_{i=0}^{n-1} f(x_i) \frac{(x - x_0) \dots (x - x_{i-1})(x - x_{i+1}) \dots (x - x_{n-1})}{(x_i - x_0) \dots (x_i - x_{i-1})(x_i - x_{i+1}) \dots (x_i - x_{n-1})} \qquad (10.37)$$

To interpolate the derivative instead, take the derivative of the Lagrange formula with respect to x. This is in fact easier than it looks, since x only appears in the numerator:

$$f^{(1)}(x) = \sum_{i=0}^{n-1} f(x_i) \frac{\frac{d}{dx}(x - x_0) \dots (x - x_{i-1})(x - x_{i+1}) \dots (x - x_{n-1})}{(x_i - x_0) \dots (x_i - x_{i-1})(x_i - x_{i+1}) \dots (x_i - x_{n-1})} \qquad (10.38)$$

This is for the first derivative, but higher derivatives can be obtained in the same way. The interpolation method then works as it did back in Chap. 6: for each of the n measurements available, compute the polynomial that results from the multiplications in the numerator, derive it, and sum it with the other polynomials from the other measurements to get the derivative equation. That equation can then be evaluated at any point of interest within the interpolation interval.

Example 10.6
A robot is observed moving in a straight line. It has been measured at the following positions:

Time t (s)	1	3	4
Position $f(t)$ (m)	16	62	100

Estimate its current speed, at time 4 s.

Solution
Speed is the first derivative of position. However, the measurements available do not make it possible to use any of the derived-difference formulae. Instead, Eq. (10.38) can be used to interpolate the speed equation.

$$f^{(1)}(t) = 16 \frac{\frac{d}{dt}(t-3)(t-4)}{(1-3)(1-4)} + 62 \frac{\frac{d}{dt}(t-1)(t-4)}{(3-1)(3-4)} + 100 \frac{\frac{d}{dt}(t-1)(t-3)}{(4-1)(4-3)}$$

$$f^{(1)}(t) = 16 \frac{\frac{d}{dt}(t^2 - 3t - 4t + 12)}{(-2)(-3)} + 62 \frac{\frac{d}{dt}(t^2 - t - 4t + 4)}{(2)(-1)} + 100 \frac{\frac{d}{dt}(t^2 - t - 3t + 3)}{(3)(1)}$$

$$f^{(1)}(t) = 16 \frac{2t - 7}{6} + 62 \frac{2t - 5}{-2} + 100 \frac{2t - 4}{3}$$

$$f^{(1)}(t) = 10t + 3$$

(continued)

Example 10.6 (continued)

This formula can then be used to compute the speed of the robot at any time between the interpolation bounds, from $t = 1$ s to $t = 4$ s. At the requested time of $t = 4$ s, the speed is 43 m/s. From Example 10.3, this is known to be the correct result.

10.7   Inaccurate Measurements

The divided-difference formulae studied in this chapter all estimate the derivative from measurements of a system. So far, these measurements have been assumed to be accurate, and have been used as such. But empirical measurements taken in practice will often have measurement errors, due to inaccurate instrumentation and handling errors. Worse, differentiation can be very unstable in the presence of this noise: the errors get added together and amplified.

Consider the robot tracking data of Example 10.3. Given exact data, the derivative can be computed at any of the five times using the backward, centered, or forward divided-difference formulae, as in Table 10.1:

However, small errors in measurements can have a drastic impact. Table 10.2 runs through the example again, this time introducing 1–4 m of errors on the position measurements. Notice how this error is amplified dramatically in the derivatives:

To further illustrate, Fig. 10.3 compares the real and noisy position measurements, and the real and noisy derivative estimations. A visual inspection of that figure confirms how even a small error in measurements can cause errors in the derivative estimation that are not only much larger in amplitude, but also fluctuate wildly.

Clearly, the divided-difference formulae should be avoided in cases such as this one. An alternative solution is to compute a linear regression on the data, as was learned in Chap. 6, to obtain the best-fitting polynomial that goes through the data.

Table 10.1 Robot speed given exact position measurements

Time t (s)	0	1	2	3	4
Position $f(t)$ (m)	8	16	34	62	100
Speed $f^{(1)}(t)$ (m/s)	3	13	23	33	43

Table 10.2 Robot speed given noisy position measurements

Time t (s)	0	1	2	3	4
Position $f(t)$ (m)	7	17	36	60	104
Position error (%)	12.5	6.3	5.9	3.2	4.0
Speed $f^{(1)}(t)$ (m/s)	5.5	14.5	20.6	34	54
Speed error (%)	83.3	11.5	10.4	3.0	25.6

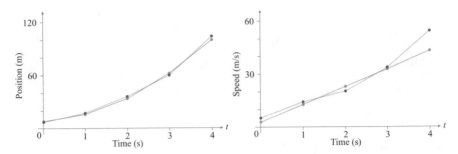

Fig. 10.3 Position (*left*) and speed (*right*) using exact values (*blue*) and noisy values (*red*)

That polynomial can then be derived and used to estimate the derivative value at any point within its interval.

Example 10.7
A robot is observed moving in a straight line. It has been measured, with noise, at the following positions:

Time t (s)	0	1	2	3	4
Position $f(t)$ (m)	7	17	36	60	104

Estimate its current speed, at time 4 s.

Solution
Speed is the first derivative of position. However, since the measurements are noisy, a divided-difference formula cannot be used. Instead, use the Vandermonde method to compute a linear regression of the data. Plotting the data points, as was done in Fig. 10.2 show that they appear to draw an exponential curve, so to compute a regression for a degree-2 polynomial. The Vandermonde matrix and solution vector will be

$$
V = \begin{bmatrix} 1 & 0 & 0 \\ 1 & 1 & 1 \\ 1 & 2 & 4 \\ 1 & 3 & 9 \\ 1 & 4 & 16 \end{bmatrix} \text{ and } y = \begin{bmatrix} 7 \\ 17 \\ 36 \\ 60 \\ 104 \end{bmatrix} ;
$$

and the Vandermonde matrix-vector to solve will thus be:

(continued)

Example 10.7 (continued)

$$\mathbf{V}^T\mathbf{V}\mathbf{c} = \mathbf{V}^T\mathbf{y}$$

$$\begin{bmatrix} 5 & 10 & 30 \\ 10 & 30 & 100 \\ 30 & 100 & 354 \end{bmatrix} \begin{bmatrix} c_0 \\ c_1 \\ c_2 \end{bmatrix} = \begin{bmatrix} 224 \\ 685 \\ 2365 \end{bmatrix}$$

$$\therefore \begin{bmatrix} c_0 \\ c_1 \\ c_2 \end{bmatrix} = \begin{bmatrix} 7.83 \\ 2.84 \\ 5.21 \end{bmatrix}$$

This gives the regressed polynomial for the position of the robot:

$$f(t) = 5.21t^2 + 2.84t + 7.83$$

which is trivial to derive to obtain the equation for the speed. Note that this regressed polynomial is very close to the real polynomial that generated the correct values of the example, which was:

$$f(t) = 5t^2 + 3t + 8$$

Using the derivative of the regressed equation makes it possible to compute the speed at all five measurement times:

Time t (s)	0	1	2	3	4
Speed $f^{(1)}(t)$ (m/s)	2.8	13.3	23.7	34.1	44.5
Speed error (%)	6.7	2.3	3.0	3.3	3.5

These results are clearly much more accurate than those obtained using the divided-difference formula in Table 10.2. To further illustrate the difference, Fig. 10.2 is taken again, this time to include the regressed estimate of the speed (purple dashed line) in addition to the actual value (blue line) and divided-difference estimate (red dashed line). It can be seen that, while the divided-difference estimate fluctuates wildly, the regressed estimate remains linear like the actual derivative, and very close to it in value, even overlapping with it for half a second.

(continued)

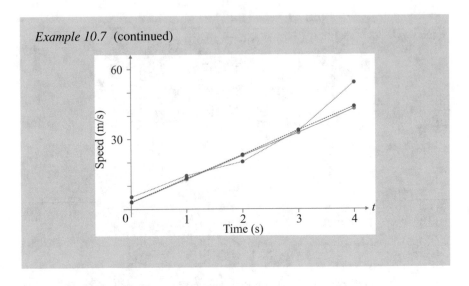

Example 10.7 (continued)

10.8   Engineering Applications

Differentiation is a mathematical tool that allows engineers to measure the rate of change of a parameter of their systems, such as the rate of change in position (speed) of an object, the rate of change of its speed (acceleration), or the rate of change of its acceleration (jerk). This relationship was discussed in Sect. 10.1. But there are countless other engineering applications where differentiation can be useful, and in fact it comes up in many common engineering equations. These include:

- Fourier's law of heat conduction, which models the heat transfer rate through a material as:

$$q_x = -k\frac{dT}{dx} \tag{10.39}$$

where q_x is the heat flux in orientation x, k is the material's conductivity, and dT/dx is the first derivative of the temperature over orientation x.

- Fick's laws of diffusion model the movement of particles of a substance from a region of higher concentration to a region of lower concentration. Fick's first law models the diffusion flux in orientation x, J_x, as:

$$J_x = -D\frac{d\phi}{dx} \tag{10.40}$$

where D is the diffusion coefficient of the medium, and $d\phi/dx$ is the first derivative of the concentration over orientation x. Fick's second law models the rate of change of the concentration over time, $d\phi/dt$, in relationship to the second derivative of the concentration over orientation x:

$$\frac{d\phi}{dt} = D\frac{d^2\phi}{dx^2} \tag{10.41}$$

- The electromotive force ε of an electrical source can be measured by the rate of change of its magnetic flux over time $d\Phi_B/dt$, according to Faraday's law of induction:

$$\varepsilon = -\frac{d\Phi_B}{dt} \tag{10.42}$$

- The current-voltage relationships of electrical components are also the derivative of their performance over time. For a capacitor with capacitance C, that relationship is:

$$I = C\frac{dV}{dt} \tag{10.43}$$

while an inductor of inductance L has the relationship:

$$V = L\frac{dI}{dt} \tag{10.44}$$

This means that the current going through a capacitor is proportional to the rate of change of its voltage over time, while the voltage across an inductor is proportional to the rate of change of the current going through it over time.

In all these examples, as in many others, a value of the system is defined and modelled in relationship to the rate of change of another related parameter. If this parameter can be observed and measured, then the methods seen in this chapter can be used to approximate its rate of change.

10.9   Summary

Engineering models of systems that are not in a steady-state are incomplete if they only include a current snapshot of the values of system parameters. To be complete and accurate, it is necessary to include information about the rate of change of these parameters. With this addition, models are not static pictures but they change, move, or grow, in ways that reflect the changes of the real systems they represent. If a mathematical model of the system is already available, then it is straightforward to compute its derivative and include it in the model. This chapter has focused on the case where such a mathematical model is not available, and presented methods to estimate the derivative using only observed measurements of the system.

Table 10.3 Summary of derivative methods

Method	Requires	Error
Second-order centered divided-difference formula	1 Point before and 1 point after, equally spaced	$O(h^2)$
Second-order backward divided-difference formula	Current point and 2 points before, equally spaced	$O(h^2)$
Second-order forward divided-difference formula	Current point and 2 points after, equally spaced	$O(h^2)$
Fourth-order centered divided-difference formula	2 Points before and 2 points after, equally spaced	$O(h^4)$
Richardson extrapolation	n Times the number of points of the divided-difference formula used with it	$O(h^{2n})$
Interpolation method	n Points, unequally spaced	See Chap. 6
Regression method	n Points, noisy	See Chap. 6

If a set of error-free and equally spaced measurements are available, then one of the divided-difference formulae can be used. The backward, forward, or centered formulae can be used in the case that measurements are available before, after, or around the target point at which the derivative is needed, and more measurements can be used in the formulae to improve the error rate. This chapter presented in detail how new divided-difference formulae can be developed from Taylor series approximations, so whichever set of points are available, it will always be possible to create a custom divided-difference formula to fit them and to know its error rate. And in addition to the divided-difference formulae, Richardson extrapolation was presented as a means to improve the error rate of the derivative estimate.

If measurements are available but they are noisy or unevenly spaced, then the divided-difference formulae cannot be used. Two alternatives were presented to deal with these cases. If the measurements are error-free but unevenly spaced, then an interpolation method can be used to model the derivative of the system. And if the measurements are noisy, whether they are evenly or unevenly spaced, then a regression method should be used to find the best-fitting mathematical model of the system, and the derivative of that model can then be computed. Table 10.3 summarizes all the methods learned in this chapter.

10.10   Exercises

1. The charge of a capacitor is measured every 0.1 s. At the following five measurement times: {7.2, 7.3, 7.4, 7.5, 7.6 s}, the charge is measured at {0.00242759F, 0.00241500F, 0.00240247F, 0.00239001F, 0.00237761F} respectively. Find the rate of change of the charge at 7.4 s using the second-order centered divided-difference formula.

2. The rotation of a satellite is measured at times {3.2, 3.3, 3.4, 3.5, 3.6 s}, and the measured angles are {1.05837, 1.15775, 1.25554, 1.35078, 1.44252 rad} respectively. Approximate the rate of change of the angle at time 3.4 using both the second-order and fourth-order centered divided-difference formulae.

3. Repeat exercise 2 using the second-order backward divided-difference formula.

4. Use $h = 0.5$ approximate the derivative of $f(x) = \tan(x)$ at $x = 1$ to a relative error of 0.00001 using the centered divided-difference formula.

5. Repeat Question 4 but for the function $f(x) = \sin(x)/x$.

6. Perform three iterations of Richardson extrapolation to estimate the derivative of $f(x) = e^x$ at $x = 0$ starting with a step of $h = 1$, using the centered divided-difference formula.

7. Perform three iterations of Richardson extrapolation to estimate the derivative of $f(x) = \sin^2(x)/x$ at $x = 5$ rad starting with a step of $h = 2$, using (a) the second-order centered divided-difference formula; (b) the forward divided-difference formula; (c) the fourth-order centered divided-difference formula.

8. Repeat exercise 7 using the function $f(x) = \cos^{-1}(x)$ at $x = 2$ rad starting with a step of $h = 0.5$. Perform 4 iterations.

9. A runner starts a 40-m sprint at time 0. He passes the 10-m mark after 2 s, the 20-m mark after 3 s, the 30-m mark after 4 s and reaches the finish line after 4.5 s. Estimate his speed at the middle of his run, after 2.25 s.

10. A 3 L container is getting filled. It is one-third filled after an hour, two-thirds filled after 3 h, and full after 6 h. Determine the initial filling rate.

Chapter 11
Integration

11.1 Introduction

Chapter 10 has already introduced the need for differentiation and integration to quantify change in engineering systems. Differentiation measures the rate of change of a parameter, and integration conversely measures the changing value of a given parameter. Chapter 10 demonstrated how important modelling change was to insure that engineering models accurately reflected reality.

Integration does have uses beyond measuring change in parameters. Integration is mathematically the measure of an area under a curve. It can thus be used to model and approximate forces, areas, volumes, and other quantities bounded geometrically. Suppose for example an engineer who needs to model a river; possibly an environmental engineer who needs to model water flow, or a civil engineer who is doing preliminary work to design a dam or a bridge. In all cases, a complete model will need to include the area of a cross-section of the river. So a boat is sent out with a sonar, and it takes depth measurements at regular intervals, to generate a depth map such as the one shown in Fig. 11.1. From these discrete measurements, it is then possible to compute the cross-sectional area of the river. The process to do this computation is an integral: the depth measurements can be seen as points on the curve of a function, the straight horizontal surface of the water is the axis of the graph, and the cross-sectional area is the area under the curve.

As with derivation, computing an exact integral would be a simple calculus problem if the equation of the system were known. This chapter, like Chap. 10, will deal with the case where the equation is not known, and the only information available is discrete measurements of the system. The formulae presented in this chapter are all part of the set of *Newton-Cotes rules* for integration, the general name for the family of formulae that approximate an integral value from a set of equally spaced points, by interpolating a polynomial through these points and computing its area. Most of this chapter will focus on closed Newton-Cotes rules,

© Springer International Publishing Switzerland 2016
R. Khoury, D.W. Harder, *Numerical Methods and Modelling for Engineering*,
DOI 10.1007/978-3-319-21176-3_11

Fig. 11.1 Depth map of a cross-section of a river

which are closed in the sense that the first and last of the equally spaced points are also the integration boundaries. However, the last method presented will be an open Newton-Cotes rule, where the integration boundaries lie beyond the first and last of the equally spaced point.

11.2   Trapezoid Rule

11.2.1   Single Segment

The *trapezoid rule* is the simplest way to approximate an integral, when only two measurements of the system are available. It interpolates a polynomial between the two points—in other words, a single straight diagonal line—and then computes the area of the resulting trapezoid shape.

Suppose two measurements at samples x_0 and x_1, which have the measured values $f(x_0)$ and $f(x_1)$ respectively. An interpolated polynomial would be a straight diagonal line connecting the points $(x_0, f(x_0))$ and $(x_1, f(x_1))$. Then, the integral will be the area under that straight line down to the x-axis, which is the same as the area of a fictional trapezoid that connects the two aforementioned points and the two points on the x-axis, $(x_0, 0)$ and $(x_1, 0)$. This situation is illustrated in Fig. 11.2. The area of that trapezoid can be computed easily, using the simple geometric formula of Eq. (11.1): it is the base width multiplied by the average height.

$$I = \int_{x_0}^{x_1} f(x)dx \approx (x_1 - x_0)\frac{f(x_0) + f(x_1)}{2} \tag{11.1}$$

While the trapezoid rule has an undeniable advantage in simplicity, its downside is potentially a very high error. Indeed, it works by approximating the function being modelled $f(x)$ as a straight line between the measurements x_0 and x_1, and can therefore be very wrong when that is not the case. The value of the integral will be

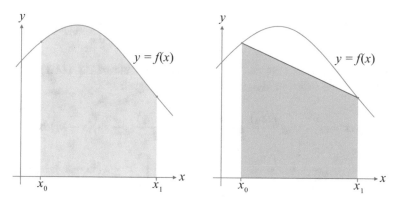

Fig. 11.2 Two measurements at x_0 and x_1 (*left*). A fictional trapezoid, the area of which approximates the integral of the function from x_0 to x_1 (*right*)

Fig. 11.3 Integration error for the example of Fig. 11.2

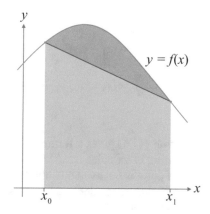

wrong by the area between the straight line and the real curve of the function, as in Fig. 11.3. This graphical representation is a good way to visualize the error, but unfortunately it does not help to compute it.

An alternative way to understand the error of this formula is to recall that the function $f(x)$ is being modelled as a polynomial $p(x)$ interpolated from two points, and then the trapezoid method takes the integral of that interpolation. Consequently, the integration error will be the integral of the interpolation error; and the interpolation error $E(x)$ is one that was already learnt, back in Chap. 6. For an interpolation from two points, the error is:

$$E(x) = \frac{f^{(2)}(x_\varepsilon)}{2}(x - x_0)(x - x_1) \tag{11.2}$$

for a point x_ε in the interval $[x_0, x_1]$. Then the integration error will be obtained by taking the integral of the formula:

$$\int_{x_0}^{x_1} f(x)dx = \int_{x_0}^{x_1} p(x) + E(x)dx = \int_{x_0}^{x_1} p(x)dx + \int_{x_0}^{x_1} E(x)dx \qquad (11.3)$$

The integral of $p(x)$ is obtained from the trapezoid rule of Eq. (11.1). For $E(x)$, Eq. (11.2) can be substituted in to compute the integral:

$$\int_{x_0}^{x_1} f(x)dx = (x_1 - x_0)\frac{f(x_0) + f(x_1)}{2} + \int_{x_0}^{x_1} \frac{f^{(2)}(x_e)}{2}(x - x_0)(x - x_1)dx$$

$$= (x_1 - x_0)\frac{f(x_0) + f(x_1)}{2} + \frac{f^{(2)}(x_e)}{2}\int_{x_0}^{x_1} x^2 - x(x_0 + x_1) + x_0 x_1 dx$$

$$= (x_1 - x_0)\frac{f(x_0) + f(x_1)}{2} + \frac{f^{(2)}(x_e)}{2}\left[\frac{x^3}{3} - \frac{x^2}{2}(x_0 + x_1) + xx_0 x_1\right]_{x_0}^{x_1}$$

$$= (x_1 - x_0)\frac{f(x_0) + f(x_1)}{2} + \frac{f^{(2)}(x_e)}{2}\left[-\frac{(x_1 - x_0)^3}{6}\right]$$

$$= (x_1 - x_0)\frac{f(x_0) + f(x_1)}{2} - \frac{f^{(2)}(x_e)(x_1 - x_0)^3}{12}$$

$$= (x_1 - x_0)\frac{f(x_0) + f(x_1)}{2} - \frac{f^{(2)}(x_e)(x_1 - x_0)^3}{12}$$

$$(11.4)$$

Equation (11.4) gives a formula for the error of the trapezoid method, but it does require an extra point x_e within the interpolation interval in order to compute it. If such a point is not available, the formula can still be used by substituting the exact value of the second derivative at x_e with the average value of the second derivative within the integration interval:

$$\int_{x_0}^{x_1} f(x)dx = (x_1 - x_0)\frac{f(x_0) + f(x_1)}{2} - \frac{\bar{f}^{(2)}(x_1 - x_0)^3}{12} \qquad (11.5)$$

And if this average second derivative is also not available, it can be estimated from the first derivative of the function:

$$\bar{f}^{(2)} = \frac{\int_{x_0}^{x_1} f^{(2)}(x)dx}{x_1 - x_0} = \frac{f^{(1)}(x_1) - f^{(1)}(x_0)}{x_1 - x_0} \qquad (11.6)$$

11.2.2 Composite Trapezoid Rule

The two-point trapezoid rule has some important advantages. It is a simple formula to compute, and the fact that it only requires two measurements of the system is beneficial when measurements are sparse and getting more is difficult. Its major problem is that it also incurs a large error in cases where the straight line between x_0 and x_1 is not a good representation of the function in that interval. To make things worse, the method is not iterative and only depends on the integration bounds x_0 and x_1, which means that the error could not be reduced even if more measurements were available.

The solution to both these problems, if more measurements are available, is to subdivide the integration interval into a set of smaller, nonoverlapping subintervals. Then, compute the integral of each subinterval using the trapezoid rule, and sum them all together to get the integral of the entire interval. Each individual subinterval will have a smaller error, as Fig. 11.4 illustrates. Intuitively, the reason this works is because, at a smaller interval, a function can be more accurately approximated by a straight line, as was mentioned multiple times since Chap. 5. Consequently, the more intermediate points are available and the more subintervals are computed, the more accurate the integration approximation will be.

Provided a set of n measurements, where x_0 and x_{n-1} are the integration bounds, applying the trapezoid rule repeatedly on each of the $m = n - 1$ subsegments will give:

$$\int_{x_0}^{x_{n-1}} f(x)dx \approx (x_1 - x_0) \frac{f(x_0) + f(x_1)}{2} + (x_2 - x_1) \frac{f(x_1) + f(x_2)}{2} + \cdots$$

$$+ (x_{n-1} - x_{n-2}) \frac{f(x_{n-2}) + f(x_{n-1})}{2} \approx \sum_{\check{s}=0}^{n-2} (x_{i+1} - x_i) \frac{f(x_i) + f(x_{i+1})}{2} \tag{11.7}$$

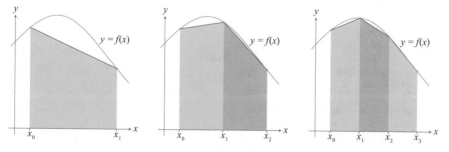

Fig. 11.4 Trapezoid approximation of the integral of Fig. 11.2 with two points and one segment (*left*), with three points and two segments (*center*), and with four points and three segments (*right*)

Moreover, if the measurements are equally spaced, then the length of all the subsegments will be the same fraction of the length of the entire integration interval:

$$(x_{i+1} - x_i) = \frac{(x_{n-1} - x_0)}{m} = h \tag{11.8}$$

And looking closely at Eq. (11.7), it can be seen that all measurements $f(x_i)$ will be summed twice, except for the measurements at the two bounds, $f(x_0)$ and $f(x_{n-1})$. Putting this observation and Eq. (11.8) into the formula of Eq. (11.7) gives the *composite trapezoid rule*:

$$\int_{x_0}^{x_{n-1}} f(x)dx \approx \frac{h}{2}\left(f(x_0) + 2\sum_{i=1}^{n-2} f(x_i) + f(x_{n-1})\right) \tag{11.9}$$

Comparing to the equation for the trapezoid rule with one segment in the previous section, it can be seen that Eq. (11.1) is only a simplification of Eq. (11.9) for the special case where only two measurements at the integration bounds are available. The pseudocode for the composite trapezoid rule is presented in Fig. 11.5. Like Eq. (11.9), this code can also simplify for the one-segment rule, by setting the value of the appropriate input variable.

The error on the composite rule is the sum of the error on each two-point subsegment, and the error of each subsegment can be computed using Eq. (11.7). This means the error of the entire formula will be:

```
xL ← Input lower integration bound
xU ← Input upper integration bound
Segments ← Input number of segments

h ← ( xU - xL ) / Segments
Integral ← F( xL )

x ← xL + h
WHILE (x < xU)
    Integral ← Integral + 2 × F(x)
    x ← x + h
END WHILE

Integral ← Integral + F(xU)
Integral ← Integral × h / 2
RETURN Integral

FUNCTION F(x)
    RETURN evaluation of the target function at point x
END FUNCTION
```

Fig. 11.5 Pseudocode of the composite trapezoid rule

$$\int_{x_0}^{x_{n-1}} f(x)dx = \frac{h}{2}\left(f(x_0) + 2\sum_{i=1}^{n-2}f(x_i) + f(x_{n-1})\right) - \sum_{i=0}^{n-2}\frac{f^{(2)}(x_{\varepsilon i})(x_{i+1} - x_i)^3}{12}$$

$$(11.10)$$

where $x_{\varepsilon i}$ is a point in the interval $[x_i, x_{i+1}]$. Substituting in Eq. (11.8) further simplifies the equation to:

$$\int_{x_0}^{x_{n-1}} f(x)dx = \frac{h}{2}\left(f(x_0) + 2\sum_{i=1}^{n-2}f(x_i) + f(x_{n-1})\right) - \frac{(x_{n-1} - x_0)^3}{12m^3}\sum_{i=0}^{n-2}f^{(2)}(x_{\varepsilon i})$$

$$(11.11)$$

This leaves $n - 2$ instances of the second derivative $f^{(2)}(x_{\varepsilon i})$ to evaluate, one for each of the m segments. But recall that, when computing the error for the two-point trapezoid, one approximation used was that the second derivative at any point in the integration interval could be substituted for the average value of the second derivative in the integration interval. Using the same assumption here makes it possible to replace every instance of $f^{(2)}(x_{\varepsilon i})$ with the average:

$$\sum_{i=0}^{n-2}f^{(2)}(x_{\varepsilon i}) = m\overline{f^{(2)}}$$

$$(11.12)$$

The final equation and error is thus:

$$\int_{x_0}^{x_{n-1}} f(x)dx = \frac{h}{2}\left(f(x_0) + 2\sum_{i=1}^{n-2}f(x_i) + f(x_{n-1})\right) - \frac{\overline{f^{(2)}}(x_{n-1} - x_0)^3}{12m^2}$$

$$(11.13)$$

Notice that the error term is almost the same as it was for the two-point trapezoid rule in Eq. (11.5), since x_{n-1} is the upper integration bound as x_1 was back in Eq. (11.5). The difference is that the error term is divided by m^2, the number of subsegments within the integration interval; that value was $m = 1$ in the case of Eq. (11.5) when the entire integration interval was only one segment. It is however also important to keep in mind that Eqs. (11.5) and (11.13) give approximations of the absolute error, not exact values; if the value by which the integral approximation was wrong could be computed exactly, then it'd be added to the approximation to get the exact integral value! An error approximation is useful rather to design and build safety margins into engineering systems. Equation (11.13) also demonstrates that the error is quadratic in terms of the number of segments. And since the number of segments is directly related to the interval width h in Eq. (11.8), this means the trapezoid formulae have a big O error rate of $O(h^2)$.

Example 11.1
The relationship between the voltage $V(t)$ and current $I(t)$ that goes through a capacitor over a period of time from t_0 to t_{n-1} can described by the following integral, where C is the capacitance value and $V(t_0)$ is the initial voltage across the capacitor:

$$V(t) = \frac{1}{C} \int_{t_0}^{t_{n-1}} I(t)dt + V(t_0)$$

For a supercapacitor of 1 F in a computer system, current measurements were taken at computer boot-up, after 0.5 s, and after 1 s. The measurements are given in the following table:

Time t (s)	0	0.5	1
Current $I(t)$ (A)	0	16.2	1

Compute the voltage going across this supercapacitor using one-segment and two-segment trapezoid rules, and determine the error of each approximation.

Solution
To begin, note that, since the first measurement is at the moment the computer boots up, the initial voltage $V(t_0)$ will be null. With the capacitance value of 1 F, the voltage will be only the result of the integral of the current.

The single-segment, two-point trapezoid rule can be computed using Eq. (11.1):

$$V_1(t) = (1 - 0)\frac{0 + 1}{2} = 0.5\,\text{V}$$

While the two-segment, three-point trapezoid rule can be computed using Eq. (11.9):

$$V_2(t) = \frac{(1 - 0)}{2}\frac{0 + 2 \times 16.2 + 1}{2} = 8.4\,\text{V}$$

To compute the error, it is necessary to know the average value of the second derivative over this one-second interval. That information is not given directly; however, the first derivative can be computed from the measurements using the methods learned in Chap. 10, and then Eq. (11.6) can be used to compute the average second derivative. With three measurements available, the forward and backward divided-difference formulae can be applied:

(continued)

Example 11.1 (continued)

$$I^{(1)}(0) = \frac{-I(1) + 4I(0.5) - 3I(0)}{2 \times 0.5} = \frac{-1 + 4 \times 16.2 - 3 \times 0}{2 \times 0.5} = 63.8\,\text{A/s}$$

$$I^{(1)}(1) = \frac{I(0) + 4I(0.5) - 3I(1)}{2 \times 0.5} = \frac{0 - 4 \times 16.2 + 3 \times 1}{2 \times 0.5} = -61.8\,\text{A/s}$$

Then Eq. (11.6) can be used to compute the average second derivative:

$$\overline{I^{(2)}} \approx \frac{I^{(1)}(1) - I^{(1)}(0)}{1 - 0} = 125.6\,\text{A/s}^2$$

Finally, the error can be computed using Eq. (11.13) as:

$$E_1 = -\frac{\overline{I^{(2)}}(1 - 0)^3}{12} = 10.6\,\text{V}$$

for the single-segment trapezoid rule, and:

$$E_2 = -\frac{\overline{I^{(2)}}(1 - 0)^3}{12 \times 2^2} = 2.6\,\text{V}$$

for the two-segment trapezoid. This is consistent with a quadratic error rate; doubling the number of segments had roughly quartered the error on the approximation.

There is a very large difference between the approximated integral value with one and two segments. The reason for this difference is that the initial and final measurements in the interval give a very poor picture of the current going through the supercapacitor over that time. The current over that period is illustrated in the figure below: it can be seen that, starting from zero, it rises to a peak of almost 40 A before dropping again by the time the final measurement is taken. The single-segment trapezoid, by using only the initial and final measurements, ignores everything that happened in-between those bounds. This corresponds to the straight-line interpolation and the purple area in the figure below; it is clearly a poor representation of the current. The two-segment trapezoid uses an additional measurement in the middle of the time interval, and the resulting two interpolations, in red in the figure below (and including the area in purple), while still inaccurate, nonetheless give a much better approximation of the current over that time.

(continued)

Example 11.1 (continued)

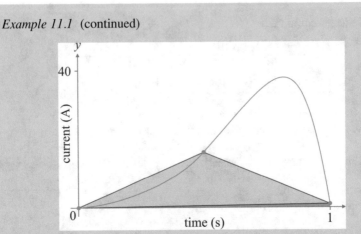

For reference, the actual integral value is 16.5 V. This means that the single-segment trapezoid gave an approximation with an absolute error of 16 V; the error estimate of 10.6 V was in the correct range. Meanwhile, the two-segment approximation had an absolute error of 8.1 V, three times higher than the error estimate of 2.6 V, but still in the correct order of magnitude.

11.3 Romberg Integration Rule

Back in Chap. 10, the Richardson extrapolation method was introduced as a means to iteratively reduce the error rate of the derivative approximation without the risk of subtractive cancellation that would come from taking the difference of two points that are nearer and nearer together. To be sure, the trapezoid rule to approximate the integral does not perform such a difference, and is therefore not susceptible to subtractive cancellation. Nonetheless, an iterative method to improve its approximation accuracy could be very beneficial. The *Romberg integration rule* provides such a method.

Suppose two approximations of an integral I, both obtained using the composite trapezoid rule as written out in Eq. (11.13) but with different numbers of segments. The trapezoid approximation obtained using m_0 segments will be noted $I_{0,0}$, and the other obtained using m_1 segments will be noted $I_{1,0}$.

$$I = I_{0,0} + E_0$$
$$I = I_{1,0} + E_1 \tag{11.14}$$

Moreover, from the discussion in the previous section, it has been noted that doubling the number of segments quarters the error. So if $m_1 = 2m_0$, then

$E_1 = E_0/4$. In that case, it becomes possible to combine the two equations of Eq. (11.14) to express E_0 in terms of the two trapezoid approximations:

$$I_{1,0} - I_{0,0} = \frac{3E_0}{4} \tag{11.15}$$

Next, substitute the value of E_1 of Eq. (11.15) back into the $I_{1,0}$ line of Eq. (11.14) gives:

$$I = I_{0,0} + \frac{4(I_{1,0} - I_{0,0})}{3} = \frac{4I_{1,0} - I_{0,0}}{3} = I_{1,1} \tag{11.16}$$

This integral approximation is labelled $I_{1,1}$; the first subscript 1 is because the best approximation it used from the previous iteration was computed from m_1 segments, and the second subscript 1 is because it is the first iteration (iteration 0 being the trapezoid rule iteration). Moreover, while the approximations of iteration 0 had an error rate of $O(h^2)$, the approximation at iteration 1 has an error rate of $O(h^4)$. This can be shown from the Taylor series, in a proof similar to that of the Richardson extrapolation.

This first iteration can be generalized as follows: given two trapezoid approximations $I_{j,0}$ and $I_{j-1,0}$ computed from 2^j and 2^{j-1} segments respectively using Eq. (11.13) with $O(h^2)$ error rate, then:

$$I_{j,1} = \frac{4I_{j,0} - I_{j-1,0}}{3} + O\left(h^{2+2}\right) \tag{11.17}$$

This process can then be repeated iteratively. For iteration k, the general version of the Romberg integration rule is:

$$I_{j,k} = \frac{4^k I_{j,k-1} - I_{j-1,k-1}}{4^k - 1} + O\left(m^{2+2k}\right) \quad k > 0 \tag{11.18}$$

As with the Richardson extrapolation, this process can be applied iteratively; at each iteration k all new values $I_{j,k}$ are computed by combining successive pairs of approximations from the previous iteration $I_{j,k-1}$ and $I_{j-1,k-1}$. Each iteration will count one less approximation value, until at the last iteration $j = \text{floor}(\log_2(m+1)) + 1$ there will be only one final $I_{j,k}$ with the highest values of j and k, which will be the best possible approximation with $O(h^{2+2k})$. Alternatively, if new values of the function and the trapezoid rule can be computed, the iterative process can go on until a threshold relative error between the approximation of the integral of two iterations is achieved (success condition), or a preset maximum number of iterations is reached (failure condition).

The pseudocode implementing this method, given in Fig. 11.6, also bears a lot of similarity to the implementation of the Richardson extrapolation given in the previous chapter. The Romberg integration rule will also build a table of values

```
xL ← Input lower integration bound
xU ← Input upper integration bound
IterationMaximum ← Input maximum number of iterations
ErrorMinimum ← Input minimum relative error

RombergTable ← empty IterationMaximum × IterationMaximum table
BestValue ← 0

RowIndex ← 0
WHILE (RowIndex < IterationMaximum)
    element at column 0, row RowIndex of RombergTable ← composite
                trapezoid of the target function from xL to xU using (2 to
                the power RowIndex) segments

    ColumnIndex ← 1
    WHILE (ColumnIndex <= RowIndex)
        element at column ColumnIndex, row RowIndex of RombergTable ←
            [ (4 to the power ColumnIndex) × (element at column
            ColumnIndex-1, row RowIndex of RombergTable) - (element at
            column ColumnIndex-1, row RowIndex-1 of RombergTable) ] /
            [(4 to the power ColumnIndex) - 1]
        ColumnIndex ← ColumnIndex + 1
    END WHILE

    PreviousValue ← BestValue
    BestValue ← element at column RowIndex, row RowIndex of RombergTable
    CurrentError ← absolute value of [ (BestValue - PreviousValue) /
                BestValue ]
    IF (CurrentError <= ErrorMinimum)
        RETURN Success, BestValue
    END IF

    RowIndex ← RowIndex + 1
END WHILE
RETURN Failure
```

Fig. 11.6 Pseudocode of the Romberg integration rule

and fill it left to right and top to bottom, where each new column added represents an increment of k in Eq. (11.18) and will contain one less element than the previous column, and each additional row represents a power of 2 of the number of segments, or a value of j in Eq. (11.18).

Example 11.2
The relationship between the voltage $V(t)$ and current $I(t)$ that goes through a capacitor over a period of time from t_0 to t_{n-1} can described by the following integral, where C is the capacitance value and $V(t_0)$ is the initial voltage across the capacitor:

$$V(t) = \frac{1}{C} \int_{t_0}^{t_{n-1}} I(t)dt + V(t_0)$$

(continued)

Example 11.2 (continued)

For a supercapacitor of 1 F in a computer system, current measurements were taken at computer boot-up, after 0.5 s, and after 1 s. The measurements are given in the following table:

Time t (s)	0	0.5	1
Current $I(t)$ (A)	0	16.2	1

Compute the voltage going across this supercapacitor using the best application of the Romberg integration rule possible.

Solution

To begin, note that, since the first measurement is at the moment the computer boots up, the initial voltage $V(t_0)$ will be null. With the capacitance value of 1 F, the voltage will be only the result of the integral of the current.

At iteration 0, two approximations are possible. $I_{0,0}$ is the 2^0 segment trapezoid rule, which has been computed in Example 11.1 as:

$$I_{0,0} = (1 - 0)\frac{0 + 1}{2} = 0.5\,\text{V}$$

The second iteration 0 approximation is $I_{1,0}$, the 2^1 segment trapezoid rule. This has also been computed in Example 11.1:

$$I_{1,0} = \frac{(1 - 0)}{2}\frac{0 + 2 \times 16.2 + 1}{2} = 8.4\,\text{V}$$

At iteration 1, these two approximations can be combined using Eq. (11.18) to get:

$$I_{1,1} = \frac{4^1 I_{1,0} - I_{0,0}}{4^1 - 1} = \frac{4 \times 8.4 - 0.5}{3} = 11.0\,\text{V}$$

For reference, the actual integral value is 16.5 V. As expected, the higher-iteration Romberg rule result generates a better approximation than either of the trapezoid rule approximations it is computed from. In fact, $I_{0,0}$ has a relative error of 97 % and $I_{1,0}$ has a relative error of 49 %, but $I_{1,1}$ has a relative error of only 34 %.

11.4 Simpson's Rules

11.4.1 Simpson's 1/3 Rules

The basic idea of the Newton-Cotes integration rule, as explained back in Sect. 11.1, is to interpolate a simple polynomial to approximate the function being integrated, and then compute its area to approximate the function's integral. Given two measurements of the function, it makes sense to interpolate a straight line, and from that comes the trapezoid rule. Given three measurements, the trapezoid rule uses them as two pairs of measurements and interpolates two straight lines. But is that really the best option in that case?

As was demonstrated in Sect. 11.2, the major source of errors with the trapezoid rule is that a straight line can be a very poor approximation of a complex function over a large interval. The solution proposed by the composite trapezoid rule is to get more measurements and use them to break up the interval, so that the straight-line interpolation is more accurate over each smaller interval. But as was learned back in Chap. 6, with more measurements available it is also possible to interpolate a polynomial of a higher degree than a straight line, and that polynomial will be a more accurate approximation of the real function. Of course, the downside of this approach is that computing the area of the shape created by this more complex polynomial will not be as easy as computing the area of a trapezoid. Consequently, there must be a balance between using a more accurate higher-degree interpolation and having a more complex computation of the integral.

With three measurements, instead of interpolating two straight lines, it is possible to interpolate a second-degree polynomial, a parabola. This will provide a more accurate approximation of the function, while still being a simple enough shape to compute the area. This is the idea that underlies *Simpson's 1/3 rule*. The equation is equally simple to obtain; it stems from approximating the integral of the function between x_0 and x_2 as the integral of its fourth-order Taylor series approximation at x_1:

$$\int_{x_0}^{x_2} f(x)dx = \int_{x_0}^{x_2} f(x_1)dx + \int_{x_0}^{x_2} \left(f^{(1)}(x_1)(x-x_1) + \frac{f^{(2)}(x_1)}{2!}(x-x_1)^2 + \frac{f^{(3)}(x_1)}{3!}(x-x_1)^3 \right) dx$$

$$+ \int_{x_0}^{x_2} \frac{f^{(4)}(x_1)}{4!}(x-x_1)^4 dx$$

$$(11.19)$$

$$= f(x_1)(x_2-x_0) + \left[\frac{f^{(1)}(x_1)}{2}(x-x_1)^2 + \frac{f^{(2)}(x_1)}{3!}(x-x_1)^3 + \frac{f^{(3)}(x_1)}{4!}(x-x_1)^4 \right]\Bigg|_{x_0}^{x_2}$$

$$+ \frac{f^{(4)}(x_1)}{5!}(x-x_1)^5 \Bigg|_{x_0}^{x_2}$$

$$(11.20)$$

Next, recall that the three points x_0, x_1 and x_2 are equally spaced, and the distance between two successive steps is defined as h in Eq. (11.8). As a result, all the even-exponent subtractions in Eq. (11.20) will cancel out, and all the odd-exponent ones will be added together:

$$
\begin{aligned}
(x_2 - x_1)^2 - (x_0 - x_1)^2 &= 0 \\
(x_2 - x_1)^3 - (x_0 - x_1)^3 &= 2h^3 \\
(x_2 - x_1)^4 - (x_0 - x_1)^4 &= 0 \\
(x_2 - x_1)^5 - (x_0 - x_1)^5 &= 2h^5
\end{aligned}
\tag{11.21}
$$

This result simplifies Eq. (11.20) considerably:

$$
\int_{x_0}^{x_2} f(x)dx = f(x_1)2h + \frac{f^{(2)}(x_1)}{3}h^3 + \frac{f^{(4)}(x_1)}{60}h^5
\tag{11.22}
$$

The fourth-order term of the series, which has been kept somewhat separate so far in the equations, will become the error term of the method. This however leaves the second derivative to deal with in the second-order term; after all, the derivative of the function is not known, and only three measurements are available. Fortunately, Chap. 10 has explained how to approximate the derivative of a function from measurements. Specifically, the centered divided-difference formula for the second derivative can be substituted into Eq. (11.22), and subsequently simplified to get the formula for Simpson's 1/3 rule:

$$
\begin{aligned}
\int_{x_0}^{x_2} f(x)dx &= f(x_1)2h + \frac{h^3}{3}\left[\frac{f(x_2) + f(x_0) - 2f(x_1)}{h^2} + O(h^2)\right] + \frac{f^{(4)}(x_1)}{60}h^5 \\
&= f(x_1)\frac{6h}{3} + \frac{h}{3}[f(x_2) + f(x_0) - 2f(x_1)] + O(h^5) + \frac{f^{(4)}(x_1)}{60}h^5 \\
&= \frac{h}{3}[f(x_0) + 4f(x_1) + f(x_2)] + O(h^5) \\
&= (x_2 - x_0)\frac{f(x_0) + 4f(x_1) + f(x_2)}{6} + O(h^5)
\end{aligned}
\tag{11.23}
$$

If more than three measurements are available, then the same idea as for the composite trapezoid applies: group them into triplets of successive points and interpolate multiple smaller and more accurate nonoverlapping parabola, and sum the resulting areas to get a higher-accuracy approximation of the integral. A general form of the equation can be obtained to do this:

$$\int\limits_{x_0}^{x_{n-1}} f(x)dx = (x_{n-1} - x_0) \frac{f(x_0) + 4 \sum\limits_{\substack{i=1,3,5,\dots}}^{n-2} f(x_i) + 2 \sum\limits_{\substack{i=2,4,6,\dots}}^{n-1} f(x_j) + f(x_{n-1})}{3(n-1)}$$

$$+ O\left(\frac{h^5}{n}\right)$$

$$(11.24)$$

Do be careful with the two separate summations that must be computed in the composite equation: each adds every other measurement, but they are multiplied by different constants. Note also that, just like with the composite trapezoid equation, the measurements at the two bounds of the integration interval are only summed once.

Example 11.3
The relationship between the voltage $V(t)$ and current $I(t)$ that goes through a capacitor over a period of time from t_0 to t_{n-1} can described by the following integral, where C is the capacitance value and $V(t_0)$ is the initial voltage across the capacitor:

$$V(t) = \frac{1}{C} \int_{t_0}^{t_{n-1}} I(t)dt + V(t_0)$$

For a supercapacitor of 1 F in a computer system, current measurements were taken at computer boot-up, after 0.5 s, and after 1 s. The measurements are given in the following table:

Time t (s)	0	0.5	1
Current $I(t)$ (A)	0	16.2	1

Compute the voltage going across this supercapacitor using Simpson's 1/3 rule.

Solution
To begin, note that, since the first measurement is at the moment the computer boots up, the initial voltage $V(t_0)$ will be null. With the capacitance value of 1 F, the voltage will be only the result of the integral of the current.
    Since there are three points to compute Simpson's rule from, either Eq. (11.23) or (11.24) could be used; the latter will simplify into the former.

(continued)

Example 11.3 (continued)
  The resulting formula is:

$$\int_0^1 I(t)dt = (1-0)\frac{0+4\times 16.2+1}{6} = 11.0\,\text{V}$$

And this approximation has a relative error of 34% compared to the real integral value of 16.5 V.

Comparing this result to those obtained with the same points in Examples 11.1 and 11.2 shows the improved accuracy of this method. The two-segment trapezoid rule gave a result of 8.3 V with a relative error of 49%, a clear indication that two straight-line interpolations of two halves of the function are not at all as good an approximation as a single parabola over the entire function. The result obtained using the Romberg integration rule was 11.0 V, the same as is obtained here, but with Romberg it was obtained in the second iteration after three steps of computations: two trapezoid rules and one iterative Romberg equation. Here, Simpson's 1/3 rule provides the same result in a single step with much less computation (and thus much less chances for error).

Example 11.4
Two sets of current measurements with different intervals have been taken for the capacitor of Example 11.3. They are:

Time t (s)	0	0.25	0.5	0.75	1
Current $I(t)$ (A)	0	4.4	16.2	35.5	1

and:

Time t (s)	0	0.2	0.4	0.6	0.8	1
Current $I(t)$ (A)	0	3.1	10.1	24.1	37.1	1

Compute the voltage going across this supercapacitor using Simpson's 1/3 rule on each set of points, and compare the results.

Solution
Using Eq. (11.24) on the first set of five measurements gives:

(continued)

Example 11.4 (continued)

$$I_5 = (1 - 0)\frac{0 + 4(4.4 + 35.5) + 2 \times 16.2 + 1}{3 \times 4} + O\left(\frac{h^5}{5}\right) = 16.1\,\text{V}$$

Then using the set of six measurements gives:

$$I_6 = (1 - 0)\frac{0 + 4(3.1 + 24.1) + 2(10.1 + 37.1) + 1}{3 \times 5} + O\left(\frac{h^5}{6}\right) = 13.6\,\text{V}$$

Compared to the real value of this integral of 16.5 V, the result with five points has a relative error of 3 % while the result with six points has a relative error of 18 %. This goes against the intuition built throughout his chapter that more points should make it possible to generate a better approximation of the function and thus a more accurate integral, and it goes against the error rate computed in the equations, which predicts that the approximation with six points should be more accurate.

This problem is the subject of the next section.

11.4.2   Simpson's 3/8 Rule

The previous example demonstrated a problem with Simpson's 1/3 rule: using six measurements yielded an approximation that was considerably worse than when using five measurements, when intuitively the opposite should be true. The problem does not stem from the measurements themselves; both sets of samples are equally accurate. Rather, the problem is the choice of methods. Simpson's 1/3 rule is designed to work with an integration interval broken up in an even number of segments, or an odd number of measurements, and the example that failed used six measurements, an even number that divided the integration interval into an odd number of segments. In practice though, one is limited by the measurements available and the fact they must be equally spaced: if an even number of equally spaced measurements are available, then discarding one to use Simpson's 1/3 rule is not an option. So what can be done in such cases?

The best option is to combine Simpson's 1/3 rule with another rule, called *Simpson's 3/8 rule*. That rule is designed to work with exactly four measurements, or three segments. Given four measurements, it can be used by itself, and given an even number of points greater than 4, it can be used to handle the first or last four points and leave behind an odd number of points for Simpson's 1/3 rule to handle. For example, in Fig. 11.7, an integration interval is divided into five segments using six measurements, much like with Example 11.4. In such a case, the first four measurements can be used in Simpson's 3/8 rule and the last three in Simpson's 1/3 rule, with the fourth measurement thus being used in both formulae. The pseudocode of an algorithm combining both versions of Simpson's rule will be presented in Fig. 11.8.

Fig. 11.7 An integration interval divided into five segments

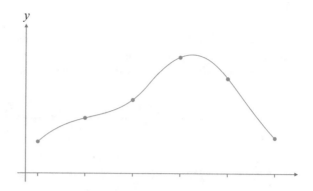

```
xL ← Input lower integration bound
xU ← Input upper integration bound
Segments ← Input number of segments

h ← ( xU - xL ) / Segments

IF (Segments = 1)
    RETURN Failure

ELSE IF (Segments = 2)
    Integral ← (xU - xL) × (F(xL) + F(xL + h) + F(xU)) / 6

ELSE IF (Segments = 3)
    Integral ← (xU - xL) × (F(xL) + 3×F(xL + h)+3×F(xL + 2xh) + F(xU)) / 8

ELSE IF (Segments is odd)
    Integral ← F(xL)
    x ← xL + h
    WHILE (x < xU)
        Integral ← Integral + 4×F(x)+ 2×F(x + h)
        x ← x + 2xh
    END WHILE
    Integral ← Integral + F(xU)
    Integral ← (xU - xL) × Integral / [3 × Segments]

ELSE IF (Segments is even)
    x3 ← xL + 3xh
    Part1 ← (x3 - xL) × (F(xL) + 3×F(xL + h)+3×F(xL + 2xh) + F(x3)) / 8

    Part2 ← F(x3)
    x ← x3 + h
    WHILE (x < xU)
        Part2 ← Part2 + 4×F(x)+ 2×F(x + h)
        x ← x + 2xh
    END WHILE
    Part2 ← Part2 + F(xU)
    Part2 ← (xU - x3) × Part2 / [3 × (Segments-3)]

    Integral ← Part1 + Part2

END IF
RETURN Integral

FUNCTION F(x)
    RETURN evaluation of the target function at point x
END FUNCTION
```

Fig. 11.8 Pseudocode of the Simpson's rules algorithm

The formula for Simpson's 3/8 rule is given below. It can be seen that it has an error rate of $O(h^5)$, just like Simpson's 1/3 rule with three points. Thus, both formulae can be used together without loss of accuracy.

$$\int_{x_0}^{x_3} f(x)dx = (x_3 - x_0)\frac{f(x_0) + 3f(x_1) + 3f(x_2) + f(x_3)}{8} + O(h^5) \qquad (11.25)$$

Example 11.5
The relationship between the voltage $V(t)$ and current $I(t)$ that goes through a capacitor over a period of time from t_0 to t_{n-1} can described by the following integral, where C is the capacitance value and $V(t_0)$ is the initial voltage across the capacitor:

$$V(t) = \frac{1}{C}\int_{t_0}^{t_{n-1}} I(t)dt + V(t_0)$$

For a supercapacitor of 1 F in a computer system, current measurements were taken at computer boot-up and every 0.2 s. The measurements are given in the following table:

Time t (s)	0	0.2	0.4	0.6	0.8	1
Current $I(t)$ (A)	0	3.1	10.1	24.1	37.1	1

Compute the voltage going across this supercapacitor using Simpson's rules.

Solution
To begin, note that, since the first measurement is at the moment the computer boots up, the initial voltage $V(t_0)$ will be null. With the capacitance value of 1 F, the voltage will be only the result of the integral of the current.

With six measurements, two options are available to use Simpson's rules: either apply Simpson's 3/8 rule on the first four and Simpson's 1/3 rule on the last three, or the other way around, apply Simpson's 1/3 rule on the first three measurements and Simpson's 3/8 rule on the last four. There is, a priori, no way to prefer one option over the other. So using the first one, Eq. (11.25) over the first four measurements gives:

$$I_{3/8} = (0.6 - 0) \frac{0 + 3 \times 3.1 + 3 \times 10.1 + 24.1}{8} = 4.8\,V$$

And Eq. (11.24) over the last three measurements gives:

(continued)

Example 11.5 (continued)
$$I_{1/3} = (1 - 0.6)\frac{24.1 + 4 \times 37.1 + 1}{6} = 11.6\,\text{V}$$

The entire integral is the sum of $I_{3/8}$ and $I_{1/3}$, which is 16.4 V, and has a relative error of 0.8 % compared to the real integral value of 16.5 V. This is also an improvement compared to the five-measurement integral approximation of Example 11.4, and especially compared to the erroneous application of Simpson's 1/3 rule on six measurements that was done in Example 11.4.

11.5 Gaussian Quadrature

So far this chapter has introduced several formulae to approximate the integral of a function, using different number of measurements and different number of iterations, and with different error rates. By far, the simplest one was the single-segment trapezoid rule of Eq. (11.1). Unfortunately it was also the method with the worst error rate. As was demonstrated in Example 11.1, the error stems from the selection of the two points; they must be the two integration bounds, and these two bounds may not be representative of the entire function. Since the trapezoid rule is a closed method, those are the two points that must be used.

Wouldn't it be better if there was an open-method equivalent of the trapezoid method, which had the simplicity of using only two points but made it possible to choose points within the integration interval that are more representative of the function, points that make it possible to interpolate a straight line such that the missing area above the interpolated line and the extra area below the interpolated line cancel each other out? Such a method could pick the two points that yield a trapezoid with an area as close as possible to the real function. For example, instead of the two bounds x_0 and x_1 of Fig. 11.2, the two points a and b inside the interval could be selected as in Fig. 11.9 to get a better approximation of the integral value.

Such a method does exist. It is called the *Gaussian quadrature* method, or alternatively the *Gauss-Legendre rule*. To understand where it comes from, it is best to go back to the two-point trapezoid and learn a different way to discover that method.

The single-segment trapezoid of Eq. (11.1) estimates the integral of the function as a combination of the measurement value of that function at two points, x_0 and x_1, the two known integration bounds. The formula can thus be written as:

$$I = \int_{x_0}^{x_1} f(x)dx \approx w_0 f(x_0) + w_1 f(x_1) \tag{11.26}$$

The challenge is then discovering what the weights w_0 and w_1 are. To solve for two unknown values, two equations are needed; that means two equations where the

Fig. 11.9 An open single-segment trapezoid approximation

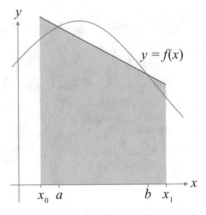

function evaluations $f(x_0)$ and $f(x_1)$ and the total integral value I are known. But these equations don't need to be complicated. They could be, for example, the integral of the straight-line polynomial $p_0(x) = 1$ and the integral of the diagonal line $p_1(x) = x$. Moreover, to further simplify, the polynomials can be centered on the origin, meaning that $x_0 = -x_1 \neq 0$. In that case, the two integrals become:

$$\int_{x_0}^{x_1} p_0(x)dx = w_0 p_0(x_0) + w_1 p_0(x_1)$$

$$\int_{x_0}^{x_1} 1dx = 1w_0 + 1w_1$$

(11.27)

$$x_1 - x_0 = w_0 + w_1$$

$$\int_{x_0}^{x_1} p_1(x)dx = w_0 p_1(x_0) + w_1 p_1(x_1)$$

$$\int_{x_0}^{x_1} x\,dx = w_0 x_0 + w_1 x_1$$

(11.28)

$$0 = w_0 x_0 + w_1 x_1$$

From these two equations, it is simple to solve for the unknown weight values and find that $w_0 = w_1 = (x_1 - x_0)/2$, the same values as in Eq. (11.1).

The Gaussian quadrature formula starts from the same single-segment formula, except the two measurements are taken not at the known bounds but at two unknown points a and b inside the integration interval. The equation thus becomes:

$$I = \int_{x_0}^{x_1} f(x)dx \approx w_0 f(a) + w_1 f(b) \tag{11.29}$$

And there are now four unknown values to discover, namely the two points in addition to the two weights. Four equations will be needed in that case; so in addition to the integrals of the straight line and diagonal line polynomials from earlier, add the polynomials $p_2(x) = x^2$ and $p_3(x) = x^3$. But to simplify, though, assume that the original function $f(x)$ has been transformed to an equivalent function $g(y)$ of the same degree and area but over the integration interval $y_0 = -1$ to $y_1 = 1$. In that case, the four equations become:

$$\int_{-1}^{1} p_0(y)dy = w_0 p_0(a) + w_1 p_0(b)$$

$$\int_{-1}^{1} 1\,dy = 1w_0 + 1w_1 \tag{11.30}$$

$$2 = w_0 + w_1$$

$$\int_{-1}^{1} p_1(y)dx = w_0 p_1(a) + w_1 p_1(b)$$

$$\int_{-1}^{1} y\,dy = w_0 a + w_1 b \tag{11.31}$$

$$0 = w_0 a + w_1 b$$

$$\int_{-1}^{1} p_2(y)dx = w_0 p_2(a) + w_1 p_2(b)$$

$$\int_{-1}^{1} y^2\,dy = w_0 a^2 + w_1 b^2 \tag{11.32}$$

$$\frac{2}{3} = w_0 a^2 + w_1 b^2$$

$$\int_{-1}^{1} p_3(y)dx = w_0 p_3(a) + w_1 p_3(b)$$

$$\int_{-1}^{1} y^3\,dy = w_0 a^3 + w_1 b^3 \tag{11.33}$$

$$0 = w_0 a^3 + w_1 b^3$$

With four equations and four unknowns, it is easy to solve to find that $w_0 = w_1 = 1$ and $a = -b = -1/\sqrt{3}$. Equation (11.29) thus becomes:

$$I = \int_{x_0}^{x_1} f(x)dx = \int_{-1}^{1} g(y)dy = g\left(-\frac{1}{\sqrt{3}}\right) + g\left(\frac{1}{\sqrt{3}}\right) \tag{11.34}$$

The only problem that remains is to convert $f(x)$ into its equivalent form $g(y)$, and Eq. (11.34) will make it easy to compute an approximation of its integral. This transformation will be done with a linear mapping of the form:

$$\begin{aligned} x &= c_0 + c_1 y \\ dx &= c_1 dy \end{aligned} \tag{11.35}$$

This transformation introduces two more unknown coefficients, c_0 and c_1, and thus two equations will be needed to discover their values. Fortunately, two equations are already available thanks to the known mappings of the two bounds from x_0 and x_1 to -1 and 1 respectively:

$$\begin{aligned} x_0 &= c_0 + c_1(-1) \\ x_1 &= c_0 + c_1(1) \end{aligned} \tag{11.36}$$

It is trivial to find the values of c_0 and c_1 as functions of x_0 and x_1 by solving these equations. The conversion of Eq. (11.34) is thus done by using the following substitutions:

$$x = \frac{(x_1 + x_0) + (x_1 - x_0)y}{2} \tag{11.37}$$

$$dx = \frac{(x_1 - x_0)}{2}dy \tag{11.38}$$

To summarize, the Gaussian quadrature method is done in two steps. First, convert the integral of $f(x)$ from x_0 to x_1 into the integral of $g(y)$ from -1 to 1 by substituting x and dx with the values of Eqs. (11.37) and (11.38) respectively. Then second, approximate the value of the integral by adding together the evaluation of g (y) at the two internal points $y = \pm 1/\sqrt{3}$, as indicated in Eq. (11.34). These steps hint at the two major limitations of the Gaussian quadrature method: first, it requires a lot more computation than the other methods seen so far, and second, it requires having access to the actual mathematical model of the system, not just samples, in order to make it possible to perform the variable substitution and to evaluate the function at the two specific internal points required. The trade-off, though, is an accuracy comparable to Simpson's rules using only two measurements of the system.

Example 11.6

By taking samples of the current going through a supercapacitor at every second during seven seconds and computing an interpolation, the following mathematical model of the current has been developed:

$$I(t) = 8t + 42t^2 - 45t^3 + 62t^4 + 286t^5 - 352t^6$$

Determine the voltage going through the supercapacitor over the first second after the system's boot-up using the Gaussian quadrature method, knowing that its capacitance value is 1 F.

Solution

The relationship between the voltage $V(t)$ and current $I(t)$ is the following integral:

$$V(t) = \frac{1}{C} \int_{t_0}^{t_{n-1}} I(t)dt + V(t_0)$$

Since the integration interval starts at the moment the computer boots up the initial voltage $V(t_0)$ will be null, and with the capacitance value of 1 F the voltage will be only the result of the integral of the current.

The first step of applying the Gaussian quadrature method is to convert the integral using the two Eqs. (11.37) and (11.38). With the integration bounds $t_0 = 0$ and $t_1 = 1$, the equations become:

$$t = \frac{1+y}{2} \quad dt = \frac{dy}{2}$$

And the integral becomes:

$$V(t) = \int_0^1 \left(8t + 42t^2 - 45t^3 + 62t^4 + 286t^5 - 352t^6\right)dt$$

$$= \int_{-1}^1 \left(8\left(\frac{1+y}{2}\right) + 42\left(\frac{1+y}{2}\right)^2 - 45\left(\frac{1+y}{2}\right)^3 + 62\left(\frac{1+y}{2}\right)^4 \right.$$

$$\left. +286\left(\frac{1+y}{2}\right)^5 - 352\left(\frac{1+y}{2}\right)^6\right)\frac{dy}{2}$$

$$= \int_{-1}^1 \left(8.1 + 17.7y + 11.9y^2 - 5.4y^3 - 17.0y^4 - 12.0y^5 - 2.8y^6\right)dy$$

The next step is to approximate the value of the integral using Eq. (11.34):

(continued)

Example 11.6 (continued)

$$V(t) = \int_{-1}^{1} g(y)dy$$

$$\approx g\left(-\frac{1}{\sqrt{3}}\right) + g\left(\frac{1}{\sqrt{3}}\right)$$

$$\approx 1.7 + 18.5$$

$$= 20.2 \, V$$

Compared to the real value of the integral of 16.5 V, this approximation has a relative error of 22.2 %. This is a massive improvement compared to the trapezoid rule approximation of Example 11.1, which had a relative error of 97.0 % using the same number of measurements. This approximation is also an improvement compared to the three-measurement approximations obtained by the composite trapezoid rule and Simpson's 1/3 rule, which had relative errors of 49.5 % and 33.7 % respectively, despite being computed using one more measurement than this approximation.

To further illustrate how this equation works, apply Eq. (11.37) again to find that the two points at $y = \pm 1/\sqrt{3}$ correspond to times $t = 0.21$ s and $t = 0.79$ s. This means the integral approximation is computed from the colored area of the trapezoid under the red line in the figure below. Compared to the single-segment and two-segment trapezoids of Example 11.1, included as the light and dark purple lines respectively in this figure, it is clear to see how the Gaussian quadrature gives a superior result. Because it is an open method, it can forego the unrepresentative points at the integration bounds that the two trapezoid rules are forced to use in their computations. The straight-line interpolation resulting from the Gaussian quadrature points is clearly a better linear approximation of the function over a large part of the integration interval than either of the trapezoid interpolations. And even the errors, the large section included under the interpolation line beyond $t = 0.83$ s when the function begins decreasing quickly, is cancelled out in part by the negative area under the curve from $t = 0$ to $t = 1.5$.

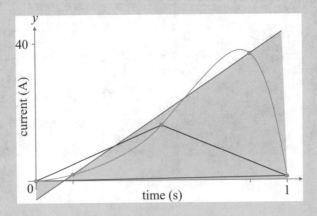

To conclude, it should be noted that the version of Gaussian quadrature presented in this section is actually a special case of the method that uses two points. In fact, the Gaussian quadrature method can be used with any number of points, just like the composite trapezoid rule or Simpson's rules. The general form of the formula is:

$$I = \int_{x_0}^{x_1} f(x)dx = \int_{-1}^{1} g(y)dy = \sum_{k=0}^{n-1} w_k g(y_k) \qquad (11.39)$$

In the case where $n=2$, the weights w_k are always 1 and the evaluated points y_k are $\pm 1/\sqrt{3}$, and Eq. (11.39) reduces to Eq. (11.34). Weights and points for the first four values of n are presented in Table 11.1. Notice from this table that the weights and points are different at every value of n; this means that the complete summation will have to be recomputed from scratch every time the number of points n is increased. The Gaussian quadrature method thus cannot be implemented in an iterative algorithm that increments the number of points to gradually improve the quality of the approximation, in the way the Romberg integration rule did.

Table 11.1 Points and weights for the Gaussian quadrature method with different number of points

Number of points n	Evaluated points y_k	Weights w_k
1	0	2
2	$-\sqrt{\dfrac{1}{3}}$	1
	$\sqrt{\dfrac{1}{3}}$	1
3	$-\sqrt{\dfrac{3}{5}}$	$\dfrac{5}{9}$
	0	$\dfrac{8}{9}$
	$\sqrt{\dfrac{3}{5}}$	$\dfrac{5}{9}$
4	$-\sqrt{\dfrac{3}{7}+\dfrac{2}{7}\sqrt{\dfrac{6}{5}}}$	$\dfrac{18-\sqrt{30}}{36}$
	$-\sqrt{\dfrac{3}{7}-\dfrac{2}{7}\sqrt{\dfrac{6}{5}}}$	$\dfrac{18+\sqrt{30}}{36}$
	$\sqrt{\dfrac{3}{7}-\dfrac{2}{7}\sqrt{\dfrac{6}{5}}}$	$\dfrac{18+\sqrt{30}}{36}$
	$\sqrt{\dfrac{3}{7}+\dfrac{2}{7}\sqrt{\dfrac{6}{5}}}$	$\dfrac{18-\sqrt{30}}{36}$

The development of the error for Eq. (11.39) falls outside the scope of this book, but the final result is:

$$E(x) = \frac{2^{2n+1}(n!)^4}{(2n+1)(2n!)^3} g^{(2n)}(x_\varepsilon) = O\left(h^{2n}\right) \tag{11.40}$$

This shows that the Gaussian quadrature method will compute the exact integral value with no error for a polynomial $f(x)$ of degree $2n - 1$, in which case the $2n$th derivative of $g(x)$ will be 0. In the special case of $n = 2$ that has been the topic of this section, the method will have an error rate of $O(h^4)$, a considerable improvement compared to the $O(h^2)$ error rate of the trapezoid rule with the same number of points.

11.6 Engineering Applications

Like differentiation, integration comes up in countless engineering models of real physical phenomena. Chapter 10 demonstrated this using the relationship between the jerk, the acceleration (the integral of the jerk), the speed (the integral of the acceleration), and the position (the integral of the speed) of a moving object. Section 11.1 demonstrated in turn how integration could also be used to estimate physical areas. Other popular applications include the following:

- The work W performed by a force that varies over position $F(x)$ on an object that is moving in a straight line from position x_0 to position x_1 is given by:

$$W = \int_{x_0}^{x_1} F(x)dx \tag{11.41}$$

- The Fourier transform of a continuous signal over time $s(t)$ into a continuous frequency-domain signal $S(\Omega)$ is done using the equation:

$$S(\Omega) = \int_{-\infty}^{\infty} s(t)e^{-j\Omega t}dt \tag{11.42}$$

  where e and j are Euler's number and the imaginary number, respectively.
- According to Ohm's law, the voltage between two points x_0 and x_1 along a path is given by:

$$V = -\int_{x_0}^{x_1} E\,dx = -\int_{x_0}^{x_1} \rho J\,dx \tag{11.43}$$

  where E is the electric field, J is the current density, and ρ is the resistivity along the path.

- Given a spring of stiffness k that was initially at rest and was gradually stretched or compressed by a length L, the total elastic potential energy transferred into the spring is computed as:

$$U = \int_0^L kx\,dx \tag{11.44}$$

- Finally, integration is also useful to model a nonuniform force being applied on a surface, such as an irregular wind on a sail or the increasing water pressure on the side of a dam.

11.7 Summary

Approximating integrals is a useful tool in any modelling task. In addition to complementing differentiation in modelling change in systems and its literal use in modelling physical areas, numerous physical processes and formulae that may be critical to include in models of the real-world are computed by integration of system parameters.

This chapter has introduced several methods for approximating the integral of a function. They all fall under the Newton-Cotes family of integration formulae, and consequently share the same basic idea: to interpolate a polynomial approximation from measurements of the function over the integration interval, and use the area of the shape formed by this interpolation, the x-axis, and the two integration bounds, as an approximation of the integral. The simplest possible such technique is the trapezoid rule, which only interpolates a straight line between the two integration bounds. Given more measurements, the integration interval can be broken up into smaller intervals and the integral computed using the composite trapezoid rule, and even refined iteratively using the Romberg integration rule. With more measurements, it is also possible to interpolate parabolas, which give a better approximation of the function and therefore of the integral, using Simpson's 1/3 rule and Simpson's 3/8 rule. These techniques are all closed methods, in that they require that the two integration bounds be among the measurements used to approximate the integral. This is a limitation, as the choice of these bounds will usually be constrained by the problem being studied, and not selected for being representative points of the system's behavior. The Gaussian quadrature method is an open method, which uses representative points inside the integration interval and not the bounds. Although that method is more computationally intensive, it does yield results using only two points that rival a three- or four-point application of Simpson's rules. Table 11.2 summarizes the methods covered in this chapter.

Table 11.2 Summary of integration methods

Method	Requires	Error
Trapezoid rule	Measurements at the two integration bounds	$O(h^2)$
Composite trapezoid rule	n measurements of the function	$O(h^2)$
	$n < 2$	
Romberg integration	n measurements of the function and k iterations	$O(h^{2k+2})$
	$n = 2^k$	
Simpson's rules	n measurements of the function	$O(h^5)$
	$n < 3$	
Gaussian quadrature	Two points selected within boundaries	$O(h^4)$

11.8   Exercises

1. Approximate the integral of the function $f(x) = e^{-x}$ over the interval $[0, 10]$ using:

   (a) A single-segment trapezoid rule.
   (b) A 20-segment composite trapezoid rule.
   (c) Romberg's integration rule, starting with one interval and continuing until the absolute error between two approximations is less than 0.000001.
   (d) Simpson's rule with three points.
   (e) Simpson's rule with four points.

2. Using a single-segment trapezoid rule, approximate the integral of the following functions over the specified intervals.

   (a) $f(x) = x^3$ over the interval $[1, 2]$.
   (b) $f(x) = e^{-0.1x}$ over the interval $[2, 5]$.

3. Using a single-segment trapezoid rule, approximate the integral of the following functions over the specified intervals. Then, evaluate their approximate error and their real error.

   (a) $f(x) = x^2$ over the interval $[0, 2]$.
   (b) $f(x) = x^4$ over the interval $[0, 2]$.
   (c) $f(x) = \cos(x)$ over the interval $[0.2, 0.4]$.

4. Approximate the integral of $f(x) = x^3$ over the interval $[1, 2]$ using a four-segment composite trapezoid rule.

5. Approximate the integral of $f(x) = xe^{-x}$ over the interval $[0, 4]$ using a 10-segment composite trapezoid rule.

6. Using four- and eight-segment composite trapezoid rules, approximate the integral of the following functions over the specified intervals. Then, evaluate their approximate error and their real error when using eight segments.

   (a) $f(x) = x^2$ over the interval $[-2, 2]$.
   (b) $f(x) = x^4$ over the interval $[-2, 2]$.

7. Use Romberg integration to approximate the integral of $f(x) = \cos(x)$ over the interval $[0, 3]$, starting with one interval and computing ten iterations.

8. Use Romberg integration to approximate the integral of $f(x) = x^5$ on the interval $[0, 4]$, starting with one interval and until the error on two successive steps is 0.

9. Use Romberg integration to approximate the integral of $f(x) = \sin(x)$ on the interval $[0, \pi]$, starting with one interval and until the error on two successive steps is less than 10^{-5}.

10. Using a three-point Simpson's rule, approximate the integral of the following functions over the specified intervals.

    (a) $f(x) = x^3$ over the interval $[1, 2]$.
    (b) $f(x) = e^{-0.1x}$ over the interval $[2, 5]$.

11. Using a three-point Simpson's rule and a four-point Simpson's rule, approximate the integral of the following functions over the specified intervals.

    (a) $f(x) = x^2$ over the interval $[0, 2]$.
    (b) $f(x) = x^4$ over the interval $[0, 2]$.

Chapter 12
Initial Value Problems

12.1 Introduction

Consider a simple RC circuit such as the one shown in Fig. 12.1. Kirchhoff's law states that this circuit can be modelled by the following equation:

$$\frac{dV}{dt} = -\frac{V(t)}{RC} \tag{12.1}$$

This model would be easy to use if the voltage and the values of the resistor and capacitor are known. But what if the voltage is not known or measurable over time, and only the initial conditions of the system are known? That is to say, only the initial value of the voltage and of its derivative, along with the resistor and capacitor value, are known.

This type of problem is an *initial value problem* (IVP), a situation in which a parameter's change (derivative) equation can be modelled mathematically and initial condition measurements are available, and future values of the parameters need to be estimated. Naturally, if the initial value of a parameter and the equation modelling its change over time are both available, it can be expected that it is possible to predict the value at any time in the system's operation. Different numerical methods to accomplish this, with different levels of complexity and of accuracy, will be presented in this chapter.

To formalize the discussion, an equation such as (12.1), or more generally any equation of the form

$$y^{(1)}(t) = f(t, y(t)) = c_0 + c_1 y(t) + c_2 t \tag{12.2}$$

is called a first-order *ordinary differential equation* (ODE). For an IVP, the initial value $y(t_0) = y_0$ is known, as are the values of the coefficients c_0, c_1, and c_2, and the goal is to determine the value at a future time $y(t_{n-1})$. However, the challenge is that

© Springer International Publishing Switzerland 2016
R. Khoury, D.W. Harder, *Numerical Methods and Modelling for Engineering*,
DOI 10.1007/978-3-319-21176-3_12

Fig. 12.1 A sample RC
circuit

the equation of $y(t)$ is itself unknown. It is thus necessary to approximate its
behavior from knowledge of its derivative only. Later sections of this chapter will
deal with more complex cases, namely with systems of ODEs and with higher-
order ODEs.

It is also pertinent to note in this introduction that time, while a continuous
variable, will be handled as a set of discrete equally spaced time *mesh points*. That
is to say, instead of considering all possible moments in a continuous time interval
going from the initial instant that parameter values of the system are known for to
the target instant the parameter value is needed for:

$$t \in [t_0, t_{n-1}] \tag{12.3}$$

the methods in this chapter will instead consider a set of n discrete time mesh
points:

$$t \in \{t_0, \ldots, t_i, t_{i+1}, \ldots, t_{n-1}\} \tag{12.4}$$

separated by an equal interval h:

$$h = \frac{t_{n-1} - t_0}{n-1} = t_{i+1} - t_i \tag{12.5}$$

From these equations, any mesh point within a problem's time interval can be
written as:

$$t_i = t_0 + ih \tag{12.6}$$

While these definitions may seem simple, and indeed they are, they will also be
fundamental to the numerical methods presented in this chapter. Indeed, they make
the IVP problem simpler: instead of trying to model the behavior of the unknown
function $y(t)$ over the entire time interval of Eq. (12.3), it is only necessary to
approximate it over the finite set of mesh points of Eq. (12.4).

12.2   Euler's Method

It was established, back in Chap. 5, that a function can be approximated as a straight
line over a short interval around a specific point. And moreover, it was learned that
this straight line is the derivative of the function evaluated at that point. This gives
the intuition that underlies the simplest and most intuitive of the IVP methods,

```
y ← Input initial value
tL ← Input lower mesh point
tU ← Input upper mesh point
h ← Input step size

t ← tL
WHILE (t <= tU)
    y ← y + h × F(t,y)
    t ← t + h
END WHILE

RETURN y

FUNCTION F(t,y)
    RETURN evaluation of the derivative of the target function at mesh
           point t and at function point y
END FUNCTION
```

Fig. 12.2 Pseudocode of Euler's method

Euler's method: starting at the initial known mesh point t_0, evaluate the derivative at each mesh point and follow the straight line to approximate the function to the next mesh point, and repeat this process until the target point t_{n-1} is reached. Stated more formally, this method follows the equation:

$$y(t_{i+1}) = y(t_i) + hy^{(1)}(t_i)$$
$$= y(t_i) + hf(t_i, y(t_i)) \tag{12.7}$$

Starting from the known initial conditions of $y(t_0) = y_0$ at time t_0, it is possible to evaluate the ODE to obtain the value of the derivative $y^{(1)}(t_0)$ and to use it to approximate the value of $y(t_1)$. This process is then repeated iteratively at each mesh point until the requested value of $y(t_{n-1})$ is obtained. The pseudocode of an algorithm to do this is presented in Fig. 12.2.

Equation (12.7) should be immediately recognizable as a first-order Taylor series approximation of the function $y(t)$ evaluated at t_{i+1} from the point t_i (and indeed it could have been obtained from the Taylor series instead of the reasoning presented above). This means that the error on this method is proportional to the second-order term of the series:

$$\frac{y^{(2)}(t_i)}{2!}h^2 = O(h^2) \tag{12.8}$$

Euler's method thus has a quadratic error rate, and for example halving the step size h will quarter the approximation error. It should be easy to understand why reducing the step size improves the approximation: as was seen in Chap. 5, the underlying assumption that a function can be approximated by its straight-line first derivative is only valid for a short interval around any given point and becomes more erroneous the farther away from that point the approximation goes.

Example 12.1

Using Kirchhoff's law, the voltage in a circuit has been modelled by the following equation:

$$\frac{dV}{dt} = V(t) - t + 1$$

Given that the initial voltage was of 0.5 V, determine the voltage after 1 s using six steps of Euler's method.

Solution

Using $n = 6$ gives a sample every 0.20 s, following Eq. (12.5). Putting the ODE equation of this problem into Euler's method Eq. (12.7) gives the formula to compute these samples:

$$V(t_{i+1}) = V(t_i) + hV^{(1)}(t_i)$$
$$= V(t_i) + h(V(t_i) - t_i + 1)$$

And this formula can then be used to compute the value at each step of the method:

$$V(0) = 0.5\,\text{V}$$
$$V(0.20) = 0.5 + 0.20(0.5 - 0 + 1) = 0.80\,\text{V}$$
$$V(0.40) = 0.80 + 0.20(0.80 - 0.20 + 1) = 1.12\,\text{V}$$
$$V(0.60) = 1.12 + 0.20(1.12 - 0.40 + 1) = 1.46\,\text{V}$$
$$V(0.80) = 1.46 + 0.20(1.46 - 0.60 + 1) = 1.83\,\text{V}$$
$$V(1) = 1.83 + 0.20(1.83 - 0.80 + 1) = 2.24\,\text{V}$$

To compare, note that the equation for the voltage used in this example was:

$$V(t) = t + \frac{e^t}{2}$$

The actual voltage values computed by this equation are presented in the table below, alongside the values computed by Euler's method and their relative error. It can be seen that the error is small, thanks to the small step size used to in this example. It can also be noted that the error increases in each successive step. This is a consequence of the process implemented by Euler's method, as explained in this section: the new point estimated at each step is computed by following an approximation of the function starting from an approximation of the previous point, and thus errors accumulate step after step.

	Euler's method (V)	Real value (V)	Relative error (%)
$V(0.20)$	0.80	0.81	1.23
$V(0.40)$	1.12	1.15	2.61
$V(0.60)$	1.46	1.51	3.31
$V(0.80)$	1.83	1.91	4.19
$V(1.00)$	2.24	2.36	5.08

(continued)

Example 12.1 (continued)

To further study this example, it is interesting to draw the $(t, V(t))$ plot in the range $[0, 4] \times [-1, 3]$ and to draw the orientation of the derivative at every 0.2 interval in that range. The resulting graph is shown below, and four example runs of Euler's method are marked on it with red dots and lines, with the initial conditions of $y(0) = -0.5$ V, $y(0) = 0$ V, $y(0) = 0.5$ V, and $y(0) = 1.0$ V. The actual functions given these four initial conditions are also marked with blue lines on the figure. These examples show how Euler's method follows the direction indicated by the derivative, step by step, until it reaches the target time. In fact, from a figure like the one below, it looks possible for a human to simply draw and connect the arrows to reach the solution at any time given any initial condition!

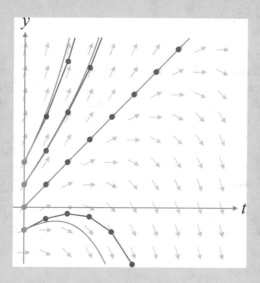

Example 12.2

Given the following IVP, approximate the value of $y(1)$ and $y(0.5)$ using one step of Euler's method for each:

$$y^{(1)}(t) = 1 - ty(t)$$
$$y(0) = 1$$

Solution

Using Eq. (12.7), the result can be computed immediately:

(continued)

Example 12.2 (continued)

$$y(0+h) = y(0) + h(1 - 0 \times y(0))$$
$$y(1) = 1 + 1(1 - 0 \times 1) = 2$$
$$y(0.5) = 1 + 0.5(1 - 0 \times 1) = 1.5$$

It is interesting to note that the real values of the function are: $y(0.5) = 1.34$ and $y(1) = 1.33$. This means the relative error at $t = 0.5$ is 12 % and at $t = 1$ is 50 %. Reducing the step size h by a factor of 2, from 1 to 0.5, has thus reduced the error by approximately a factor of 4. This is exactly what should be expected for a method with a quadratic error rate.

12.3 Heun's Method

It was explained in the previous section that Euler's method approximates the behavior of a function by following the derivative at the current point $y(t_i)$ for one step. But since the function being approximated will normally not be linear, the approximated behavior will diverge from the real function and the estimated next point $y(t_{i+1})$ will be somewhat off. From that point, the function will again be approximated by a straight line, and the following point $y(t_{i+2})$ will be more off compared to the real function's value. These errors will continue to accumulate, step after step. In the case of a convex function such as the one in Fig. 12.3, for example, it will lead to a consistent and increasing underestimation of the values of the function.

The reason for this accumulation of error is that the derivative at $y(t_i)$ is a good approximation of the function's behavior at that point, but not at the next point $y(t_{i+1})$.

Fig. 12.3 Euler's method underestimating a convex function's values

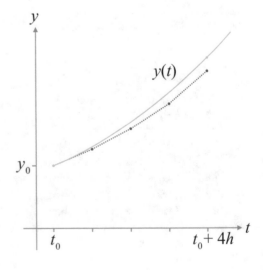

Fig. 12.4 Euler's method overestimating a convex function's values

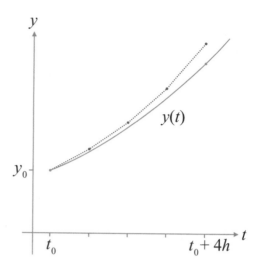

But what if the derivative at $y(t_{i+1})$ was somehow available to be used in Euler's method instead of the derivative at $y(t_i)$? It would give a good approximation of the behavior of the function at $y(t_{i+1})$... but not at $y(t_i)$. The net result would be an accumulation of errors in the opposite direction from before. For the convex function of Fig. 12.3, it would lead to a consistent and increasing overestimation instead of an underestimation, resulting in Fig. 12.4.

Considering the previous discussion, and comparing Figs. 12.3 and 12.4, a solution becomes apparent: to average out the two estimates. Since the derivative at $y(t_i)$ is a good approximation of the behavior of the function at $y(t_i)$ but leads to errors at $y(t_{i+1})$, and vice-versa, an average of the two derivatives should lead to a good representation of the function's behavior on average over the interval from t_i to t_{i+1}. Or, looking at the figures, taking the average of the underestimation of Fig. 12.3 and the overestimation of Fig. 12.4 should give much more accurate estimates, as shown in Fig. 12.5. And a better approximation of the behavior of the function will, in turn, lead to a better approximation of $y(t_{i+1})$.

That is the intuition that underlies Heun's method. Mathematically, it simply consists in rewriting the Euler's method equation of (12.7) to use the average of the two derivatives instead of using only the derivative at the current point:

$$y(t_{i+1}) = y(t_i) + h\frac{y^{(1)}(t_i) + y^{(1)}(t_{i+1})}{2}$$
$$= y(t_i) + h\frac{f(t_i, y(t_i)) + f(t_{i+1}, y(t_{i+1}))}{2} \tag{12.9}$$

There is one problem with Eq. (12.9): it requires the use of the value of the next point $y(t_{i+1})$ in order to compute the derivative at the next point, $f(t_{i+1}, y(t_{i+1}))$, and that next point is exactly what the method is meant to estimate! This circular requirement can be solved easily though, by using Euler's method to get an initial estimate of the value of $y(t_{i+1})$. That initial estimate will be of lesser quality than the

Fig. 12.5 Heun's method
averaging the Euler's
method approximations

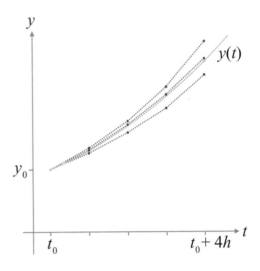

```
y ← Input initial value
tL ← Input lower mesh point
tU ← Input upper mesh point
h ← Input step size

t ← tL
WHILE (t <= tU)
    Euler ← y + h × F(t,y)
    y ← y + h × [ F(t,y) + F(t+h,Euler) ] / 2
    t ← t + h
END WHILE

RETURN y

FUNCTION F(t,y)
    RETURN evaluation of the derivative of the target function at mesh
           point t and at function point y
END FUNCTION
```

Fig. 12.6 Pseudocode of Heun's method

one computed by Heun's method, but it makes the computation of Heun's method possible. Integrating Euler's method into Heun's method alters Eq. (12.9) into:

$$y(t_{i+1}) = y(t_i) + h\frac{f(t_i, y(t_i)) + f(t_{i+1}, y(t_i) + hf(t_i, y(t_i)))}{2} \tag{12.10}$$

The pseudocode for Heun's method is only a small modification of the one presented earlier for Euler's method, as shown in Fig. 12.6.

Just like for Euler's method, the error for Heun's method can be obtained from its Taylor series approximation. Since it was already shown that Euler's method can be obtained from the first-order Taylor series approximation and it was stated that

Heun's method is more accurate, then it can be expected that Heun's method could be obtained from a second-order Taylor series approximation, and thus that its error will be the next term. So begin from a third-order Taylor series approximation:

$$y(t_{i+1}) = y(t_i) + hy^{(1)}(t_i) + \frac{y^{(2)}(t_i)}{2!}h^2 + \frac{y^{(3)}(t_i)}{3!}h^3 \qquad (12.11)$$

The second derivative is a problem, since it has no place in Heun's method equation. However, by isolating the first derivative in a second-order Taylor series approximation (or in other words taking a first-order forward divided difference formula) and taking the derivative of that formula gives an approximation of the second derivative, as follows:

$$y^{(1)}(t_i) = \frac{y(t_{i+1}) - y(t_i)}{h} - \frac{y^{(2)}(t_i)}{2!}h$$
$$y^{(2)}(t_i) = \frac{y^{(1)}(t_{i+1}) - y^{(1)}(t_i)}{h} - \frac{y^{(3)}(t_i)}{2!}h \qquad (12.12)$$

Substituting this second derivative into Eq. (12.11) gives the formula of (12.13), which is only a simplification away from Eq. (12.9) and shows the error term to be $O(h^3)$.

$$y(t_{i+1}) = y^{(1)}(t_i) + hy^{(1)}(t_i) + \frac{y^{(1)}(t_{i+1}) - y^{(1)}(t_i)}{2}h - \frac{y^{(3)}(t_i)}{4}h^3 + \frac{y^{(3)}(t_i)}{3!}h^3 \qquad (12.13)$$

Example 12.3
Using Kirchhoff's law, the voltage in a circuit has been modelled by the following equation:

$$\frac{dV}{dt} = V(t) - t + 1$$

Given that the initial voltage was of 0.5 V, determine the voltage after 1 s using six steps of Heun's method.

Solution
Using $n = 6$ gives a sample every 0.20 s, following Eq. (12.5). Computing the derivative at the initial value, at $t = 0$, gives:

$$V^{(1)}(0) = V(0) - 0 + 1 = 1.5\,V/s$$

(continued)

Example 12.3 (continued)
And the value at the next mesh point is:

$$V(0.2) = V(0) + 0.2V^{(1)}(0) = 0.8\,\text{V}$$

This corresponds to the value computed by Euler's method in the first step of Example 12.1. Next, however, this value is used to compute the derivative at the next mesh point:

$$V^{(1)}(0.2) = V(0.2) - 0.2 + 1 = 1.6\,\text{V/s}$$

And the more accurate approximation of the voltage value at the next mesh point is obtained using Heun's method formula in Eq. (12.9) as the average of these two derivatives:

$$V(0.2) = V(0) + 0.2\frac{V^{(1)}(0) + V^{(1)}(0.2)}{2} = 0.81\,\text{V}$$

The values for all five mesh points to compute, along with the real value computed by the actual voltage equation of $V(t) = t + 0.5e^t$ and the relative error of Heun's approximation, are given in the table below. Note that the relative error in this table was computed using nine decimals of precision instead of the two shown in the table, for added details.

	Heun's method (V)	Real value (V)	Relative error (%)
$V(0.20)$	0.81	0.81	0.09
$V(0.40)$	1.14	1.15	0.15
$V(0.60)$	1.50	1.51	0.21
$V(0.80)$	1.90	1.91	0.27
$V(1.00)$	2.35	2.36	0.33

As with Euler's method, it can be seen that the relative error increases at every step. However, the improved $O(h^3)$ pays off, and even in the final step the relative error is one quarter that of the first step using Euler's method. To further illustrate the improvement, the function is plotted in blue in the figure below, along with Euler's approximation in red and Heun's approximation in green. It can be seen visually that Euler's method diverges from the real function quite quickly, while Heun's method continues to match the real function quite closely over the entire interval.

(continued)

Example 12.3 (continued)

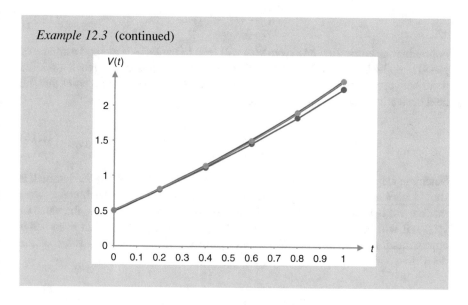

12.4   Fourth-Order Runge–Kutta Method

To summarize the IVP methods learnt so far: following the derivative at $y(t_i)$ for one step generates a poor estimation of the next point $y(t_{i+1})$, while predicting the derivative at $y(t_{i+1})$ and following that for one step generates a poor estimation with the opposite error. Following the first derivative is the idea behind Euler's method, while Heun's method takes the average of both derivatives and cancels out a lot of the errors, thus leading to a much better estimate of $y(t_{i+1})$. It is also known that the error is proportional to h, the step size between two successive mesh points. Taking these ideas together leads to an intuition for a new IVP method: perhaps taking the average derivative at more than two points could lead to a more accurate approximation of the behavior of the function and thus a more accurate computation of $y(t_{i+1})$. And since smaller step sizes help, perhaps this average should include the derivative estimated halfway between t_i and t_i. These are the intuitions that underlie the *fourth-order Runge–Kutta method*.

To lay down foundations for this method, begin by defining a point half a step between two mesh points:

$$t_{i+0.5} = t_i + \frac{h}{2} \qquad (12.14)$$

The Runge–Kutta method begins, like Euler's method and Heun's method, by computing the derivative at the current point. That result will be labelled K_0:

$$K_0 = f(t_i, y(t_i)) \tag{12.15}$$

Following the entire step hK_0 from $y(t_i)$ will lead to the Euler's method estimate of $y(t_{i+1}) = y(t_i) + hf(t_i,y(t_i))$. However, following only half a step will reach a middle point between t_i and t_{i+1}: $y(t_{i+0.5}) = y(t_i) + 0.5hK_0$. It is at this middle point that the next derivative and step are estimated:

$$K_1 = f\left(t_{i+0.5}, y(t_i) + \frac{hK_0}{2}\right) \tag{12.16}$$

Since the step K_1 is measured using the derivative in the middle of the interval, it should normally be a better representation of the behavior of the function over the interval between t_i and t_{i+1}. But instead of using it to approximate the value of $y(t_{i+1})$, it will be used to compute the half-step again and generate an even better approximation of $y(t_{i+0.5}) = y(t_i) + 0.5hK_1$ from which an improved value of the derivative and the step can be computed:

$$K_2 = f\left(t_{i+0.5}, y(t_i) + \frac{hK_1}{2}\right) \tag{12.17}$$

This improved approximation of the derivative is the one that will be used to compute the value of the step along the derivative at $y(t_{i+1}) = y(t_i) + hK_2$:

$$K_3 = f(t_{i+1}, y(t_i) + hK_2) \tag{12.18}$$

Finally, much like with Heun's method, the approximated value of $y(t_{i+1})$ is computed by taking one step along the average value of all the available derivatives. However, in this case it will be a weighted average, with more weight given to K_1 and K_2, the derivatives estimated in the middle of the interval. The reason for this preference is that, as explained before, the derivative at $y(t_i)$ is a poor estimate of the behavior of the function at t_{i+1} and the derivative at $y(t_{i+1})$ is a poor estimate of the behavior of the function at t_i, but the derivatives at $t_{i+0.5}$ offer a good compromise between those two points. The final equation for the fourth-order Runge–Kutta method is thus:

$$y(t_{i+1}) = y(t_i) + h\frac{K_0 + 2K_1 + 2K_2 + K_3}{6} \tag{12.19}$$

A visual representation can help understand this method. Figure 12.7 shows multiple measurements of the derivative $y^{(1)}(t) = 0.5y(t) - t + 1$ in the interval [0, 1.5] × [0, 2], with the black line representing the graph of the function $y(t)$ given the initial condition $y(0) = 0.5$. The top-left figure shows the derivative approximation K_0, the same one that would be used for one full step in Euler's method to generate the next point $y(1)$. But in the top-left figure, only half a step has been followed along that derivative, and the derivative at the middle point is computed and

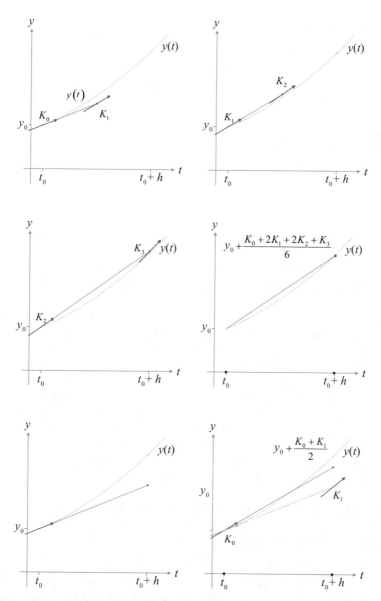

Fig. 12.7 *Top-left*: K_0, aka Euler's method, used to compute K_1. *Top-right*: K_1 used to compute K_2. *Middle-left*: K_2 used to compute K_3. *Middle-right*: K_0 to K_3 used to compute the next point in the fourth-order Runge–Kutta method. *Bottom-left*: Euler's method used to compute the next point. *Bottom-right*: Heun's method used to compute the next point

marked. This gives the value of K_1. Next, in the top-right figure, half a step is taken following K_1 and the derivative at that point gives the value of K_2. It is visually clear in these two graphs that the derivatives measured at the centre of the interval are

```
y ← Input initial value
tL ← Input lower mesh point
tU ← Input upper mesh point
h ← Input step size

t ← tL
WHILE (t <= tU)
    K0 ← F(t,y)
    K1 ← F(t+0.5×h,y+0.5×h×K0)
    K2 ← F(t+0.5×h,y+0.5×h×K1)
    K3 ← F(t+h,y+h×K2)
    y ← y + h × [ K0 + 2×K1 + 2×K2 + K3 ] / 6
    t ← t + h
END WHILE

RETURN y

FUNCTION F(t,y)
    RETURN evaluation of the derivative of the target function at mesh
           point t and at function point y
END FUNCTION
```

Fig. 12.8 Pseudocode of the fourth-order Runge–Kutta method

better approximations of the function than the derivative at the beginning of the interval. In the middle left figure, an entire step along K_2 is taken in order to approximate the derivative at the far end of the interval and measure K_3. The final approximation of $y(1)$ of the fourth-order Runge–Kutta method is obtained by taking the weighted average of these four steps, and is presented in the middle-right graph. By contrast, Euler's method relies only on the initial step K_0 while Heun's method uses only the average of step K_0 and of one step along the derivative approximated the point at the end of K_0, which is actually an approximation of lesser quality than K_3. As a result, both these methods give approximations of $y(1)$ of poorer quality, as shown in the bottom-left and bottom-right graphs. To further help illustrate this process, the pseudocode of the Runge–Kutta method is given in Fig. 12.8.

The computation of the error of the fourth-order Runge–Kutta method is beyond the scope of this book, but note that it is $O(h^5)$, making it considerably more accurate than Euler's method or Heun's method.

Example 12.4
Using Kirchhoff's law, the voltage in a circuit has been modelled by the following equation:

$$\frac{dV}{dt} = V(t) - t + 1$$

Given that the initial voltage was of 0.5 V, determine the voltage after 1 s using six steps of the fourth-order Runge–Kutta method.

(continued)

Example 12.4 (continued)
Solution
Using $n=6$ gives a sample every 0.20 s, following Eq. (12.5). At each step,
Eqs. (12.15), (12.16), (12.17), and (12.18) must be applied to compute the
four steps K_0, K_1, K_2, and K_3. For the first step starting at $t=0$, the steps are:

$$K_0 = (0.5 - 0 + 1) = 1.5$$

$$K_1 = \left(\left(0.5 + \frac{0.2 \times 1.5}{2}\right) - \left(0 + \frac{0.2}{2}\right) + 1\right) = 1.55$$

$$K_2 = \left(\left(0.5 + \frac{0.2 \times 1.55}{2}\right) - \left(0 + \frac{0.2}{2}\right) + 1\right) = 1.555$$

$$K_3 = ((0.5 + 0.2 \times 1.555) - (0 + 0.2) + 1) = 1.611$$

The approximation of $V(0.2)$ is then computed as the weighted average of
these four steps given in Eq. (12.19):

$$V(0.2) = V(0) + 0.2 \times \frac{1.5 + 2 \times 1.55 + 2 \times 1.555 + 1.611}{6} = 0.8107$$

The values for all five mesh points to compute, along with the real value
computed by the actual voltage equation of $V(t) = t + 0.5e^t$ and the relative
error of the Runge–Kutta approximation, are given in the table below.

	Fourth-order Runge–Kutta method (V)	Real value (V)	Relative error (%)
$V(0.20)$	0.810700	0.810701	1.70×10^{-4}
$V(0.40)$	1.145908	1.145912	2.94×10^{-4}
$V(0.60)$	1.511053	1.511059	4.08×10^{-4}
$V(0.80)$	1.912760	1.912770	5.25×10^{-4}
$V(1.00)$	2.359125	2.359141	6.50×10^{-4}

   As with Euler's method and Heun's method, it can be seen that the relative
error increases at every step. However, the error value is three orders of
magnitude smaller than with Heun's method! The improved accuracy
resulting from the inclusion of measurements at half-step intervals in the
weighted average is clear to see. A close-up look at the last approximated
point $V(1.00)$, presented below, gives a visual comparison of the accumulated
error of the three methods seen so far. It can be seen that Runge–Kutta's
approximation (in orange) fits almost perfectly the real function (in blue),
while Heun's approximation (in green) has accumulated a small error, and

(continued)

Example 12.4 (continued)
Euler's approximation (in red) has accumulated enough error to be
visibly off.

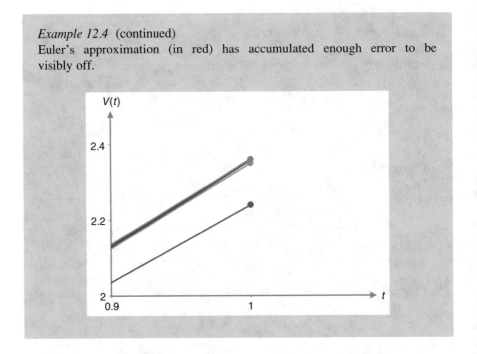

12.5   Backward Euler's Method

There is a class of differential equations on which the IVP methods seen so far will
fail. These are called *stiff ordinary differential equations* and they arise very often
in nature and thus in engineering practice. A stiff differential equation is an
equation of the form of (12.2) where the coefficient c_1 which multiplies the term
$y(t)$ is a lot larger, in absolute value, than either c_0 or c_2. For such equations, the
value of the derivative at points above and below the function, even at points near
the function, will be very large and in opposite directions from each other. This will
cause the steps of the IVP method used to oscillate above and below the actual value
of the function with greater and greater amplitudes.

One solution to this problem is to use one of the three IVP methods seen earlier
and to use very small step values h. This will make it possible for the approximation
at each step to "hug" the actual function value and use its derivative value, and not
diverge. However, this solution is not satisfactory: in addition to not giving any
information on how to compute an appropriate step size, it would also cause the IVP
method to require a massive amount of steps and computations to reach its target
value. Moreover, it is of no help if the step interval cannot be controlled or cannot
be reduced to an appropriately small size. And the risk remains that even the
slightest misstep will still cause the IVP method to diverge.

An alternative solution is a modification of Euler's method, called the *backward
Euler's method*.

The backward Euler's method starts from the same formula as Euler's method in Eq. (12.7), but uses the derivative at the next point $y(t_{i+1})$ instead of the current point $y(t_i)$:

$$
\begin{aligned}
y(t_{i+1}) &= y(t_i) + hy^{(1)}(t_{i+1}) \\
&= y(t_i) + hf(t_{i+1}, y(t_{i+1}))
\end{aligned}
\tag{12.20}
$$

This is reminiscent of Heun's method and the Runge–Kutta method, both of which made use of the next point's derivative. However, these methods actually estimate the value of $y(t_{i+1})$ using Euler's method in order to compute that derivative. When dealing with stiff ODEs, however, using Euler's method to estimate the value of the next point is exactly what must be avoided, because of the risk of divergence. In the backward Euler's method, the value of $y(t_{i+1})$ is left as an unknown in Eq. (12.20). When the problem-specific ODE is used in Eq. (12.2), this unknown value $y(t_{i+1})$ will be present on both sides of the equation, but it will be the only unknown value in the equation. It will thus be possible to solve the equation, using algebra or one of the root-finding methods of Chap. 8, to discover the value of $y(t_{i+1})$. For the algebraic solution, if the derivative formula is of the form of Eq. (12.2), then substituting that equation into Eq. (12.20) and isolating $y(t_{i+1})$ gives:

$$
\begin{aligned}
y(t_{i+1}) &= y(t_i) + h(c_0 + c_1 y(t_{i+1}) + c_2 t_{i+1}) \\
&= \frac{y(t_i) + h(c_0 + c_2 t_{i+1})}{1 - hc_1}
\end{aligned}
\tag{12.21}
$$

The advantage of a formula of the form of (12.21) is that it can directly be implemented into a software system, as is done in the pseudocode of Fig. 12.9.

```
y ← Input initial value
tL ← Input lower mesh point
tU ← Input upper mesh point
h ← Input step size

c1 ← F(0,1) - F(0,0)
Denominator ← 1 - h × c1

t ← tL
WHILE (t <= tU)
    y ← y + h × F(t+h,0) / Denominator
    t ← t + h
END WHILE

RETURN y

FUNCTION F(t,y)
    RETURN evaluation of the derivative of the target function at mesh
           point t and at function point y
END FUNCTION
```

Fig. 12.9 Pseudocode of the backward Euler's method

As explained, the reason Euler's method fails in this case is that it starts from the current point and follows the derivative forward one step to find the next point; however, for a stiff ODE, if the current point is even a little bit wrong then the derivative will be wildly different from the function and the next point will diverge. The backward Euler's method, on the other hand, looks for the next point that, when following the derivative backward one step, will lead back to the current point. This makes it capable of handling these difficult cases.

Notwithstanding the error of the method used to solve Eq. (12.20) for $y(t_{i+1})$, the backward Euler's method is a simple variation of Euler's method, and thus has the same error rate of $O(h^2)$.

Example 12.5

Using Kirchhoff's law, the voltage in a circuit has been modelled by the following equation:

$$\frac{dV}{dt} = -21V(t) + e^{-t}$$

Given that the initial voltage was of 0 V, determine the voltage after 2 s using steps of 0.1 s, using Euler's method and the backward Euler's method.

Solution

To begin, note the large value multiplying $V(t)$; this is a telltale sign that the function is a stiff ODE. Note as well that the correct voltage function that leads to this ODE, and which can be obtained by calculus, is:

$$y(t) = \frac{e^{-t} - e^{-21t}}{20}$$

Using Eq. (12.7), it is easy to compute the first six steps of Euler's method:

$$
\begin{aligned}
V(t_{i+1}) &= V(t_i) + h(-21V(t_i) + e^{-t_i}) \\
V(0.1) &= 0 + 0.1(-21 \times 0 + e^{-0}) = 0.100 \\
V(0.2) &= 0.10 + 0.1(-21 \times 0.10 + e^{-0.1}) = -0.020 \\
V(0.3) &= -0.02 + 0.1(-21 \times -0.02 + e^{-0.2}) = 0.104 \\
V(0.4) &= 0.104 + 0.1(-21 \times 0.104 + e^{-0.3}) = -0.040 \\
V(0.5) &= -0.040 + 0.1(-21 \times -0.040 + e^{-0.4}) = 0.111 \\
V(0.6) &= 0.111 + 0.1(-21 \times 0.111 + e^{-0.5}) = -0.080
\end{aligned}
$$

As predicted, the values are oscillating from positive to negative with greater and greater amplitude. The entire run of 20 steps is presented in the figure below, along with the correct function, to confirm visually that Euler's method is diverging in this case.

(continued)

Example 12.5 (continued)

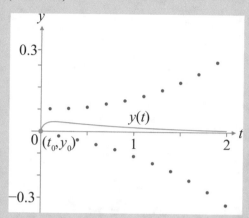

As explained, the problem stems from the fact that the derivative at points above and below the function have large amplitudes in opposite orientations. To visualize, the derivatives at an array of points around the function have been plotted in the graph below. It can be seen from that graph that, saved in the immediate vicinity of the function, the derivatives do not represent the function's behavior at all. As a result, any approximated measurement that is even a little bit inaccurate will generate an erroneous derivative and start the divergence.

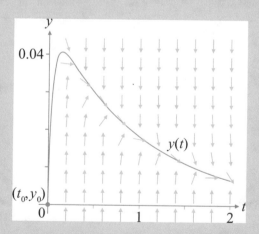

Next, the backward Euler's method formula can be obtained by putting the problem's ODE into Eq. (12.20):

(continued)

Example 12.5 (continued)
$$V(t_{i+1}) = V(t_i) + 0.1(-21V(t_{i+1}) + e^{-t_{i+1}})$$
$$= V(t_i) - 2.1V(t_{i+1}) + 0.1e^{-t_{i+1}}$$
$$= \frac{V(t_i) + 0.1e^{-t_{i+1}}}{3.1}$$

And from that equation, it is easy to compute the approximations:

$$V(0.1) = \frac{0 + 0.1e^{-0.1}}{3.1} = 0.029$$
$$V(0.2) = \frac{0.029 + 0.1e^{-0.2}}{3.1} = 0.036$$
$$V(0.3) = \frac{0.036 + 0.1e^{-0.3}}{3.1} = 0.036$$
$$V(0.4) = \frac{0.036 + 0.1e^{-0.4}}{3.1} = 0.033$$
$$V(0.5) = \frac{0.033 + 0.1e^{-0.5}}{3.1} = 0.030$$
$$V(0.6) = \frac{0.030 + 0.1e^{-0.6}}{3.1} = 0.027$$

The most immediately apparent result is that the approximations no longer oscillate. In fact, plotting the results graphically shows that the approximations are very close to the actual values of the function, as shown below.

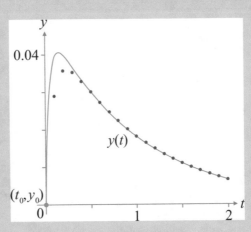

12.6 Systems of IVPs

This chapter has focused so far on handling problems that can be modelled using a single ODE and starting condition. However, many real-life engineering situations are more complex than that, and require the manipulation of multiple ODEs simultaneously, each one with its own initial value. Fortunately, as this section will demonstrate, such problems can be handled using a simple variable substitution and any of the IVP methods seen so far.

A system of IVPs will arise naturally from any multidimensional system, or any system that has multiple interacting variables related in a single model equation. Each of these variables can be represented by a function of a single independent variable (normally time). Assume a system with n such dimensions, $\{y_0(t), y_1(t), \ldots, y_{n-1}(t)\}$. The equation modelling each one is unknown; however, since this is an IVP, an ODE describing the behavior of each one over time in the form of Eq. (12.2) is available:

$$
\begin{aligned}
\frac{dy_0}{dt} &= f_0(t, y_0(t), y_1(t), \ldots, y_{n-1}(t)) \\
\frac{dy_1}{dt} &= f_1(t, y_0(t), y_1(t), \ldots, y_{n-1}(t)) \\
&\vdots \\
\frac{dy_{n-1}}{dt} &= f_{n-1}(t, y_0(t), y_1(t), \ldots, y_{n-1}(t))
\end{aligned}
\tag{12.22}
$$

Likewise, as always for an IVP, initial values of each variable are available: $\{y_0(t_0), y_1(t_0), \ldots, y_{n-1}(t_0)\}$. Note that, since the parameters are interacting in the system, the derivatives of Eq. (12.22) can include terms with any, or all, of the variables of the system. The presence of these multiple variables in the ODEs is what makes this problem difficult to handle; otherwise, it could be dealt with simply by applying an IVP method on each ODE independently of the others.

This situation can be simplified greatly by changing the representation of the system. Instead of a multidimensional problem composed of a set of independent variables, define a single multidimensional variable whose values will be the variables of the system. In other words, a vector of the variables of the system:

$$
\mathbf{u}(t) = \begin{bmatrix} y_0(t) \\ y_1(t) \\ \vdots \\ y_{n-1}(t) \end{bmatrix}
\tag{12.23}
$$

The derivative of this vector is the vector of ODEs of its dimensions:

$$\frac{d\mathbf{u}}{dt} = \begin{bmatrix} \dfrac{dy_0}{dt} \\ \dfrac{dy_1}{dt} \\ \vdots \\ \dfrac{dy_{n-1}}{dt} \end{bmatrix} = \begin{bmatrix} f_0(t, y_0(t), y_1(t), \ldots, y_{n-1}(t)) \\ f_1(t, y_0(t), y_1(t), \ldots, y_{n-1}(t)) \\ \vdots \\ f_{n-1}(t, y_0(t), y_1(t), \ldots, y_{n-1}(t)) \end{bmatrix} = \mathbf{f}(t, \mathbf{u}(t)) \qquad (12.24)$$

And the initial conditions of $\mathbf{u}(t)$ is the vector of initial values of its dimensions, which are all known:

$$\mathbf{u}(t_0) = \begin{bmatrix} y_0(t_0) \\ y_1(t_0) \\ \vdots \\ y_{n-1}(t_0) \end{bmatrix} \qquad (12.25)$$

This simple variable substitution has transformed the multidimensional IVP into a single-variable IVP, albeit a variable in multiple dimensions. But this nonetheless remains an IVP problem that can be solved using any of the IVP methods seen in this chapter, provided the formulae are modified to use a vector variable instead of a scalar. Namely, Euler's method becomes:

$$\mathbf{u}(t_{i+1}) = \mathbf{u}(t_i) + h\mathbf{f}(t_i, \mathbf{u}(t_i)) \qquad (12.26)$$

Heun's method becomes:

$$\mathbf{u}(t_{i+1}) = \mathbf{u}(t_i) + h\frac{\mathbf{f}(t_i, \mathbf{u}(t_i)) + \mathbf{f}(t_{i+1}, \mathbf{u}(t_{i+1}))}{2} \qquad (12.27)$$

The fourth-order Runge–Kutta method becomes:

$$\mathbf{u}(t_{i+1}) = \mathbf{u}(t_i) + \dfrac{\begin{aligned}&h\mathbf{f}(t_i, \mathbf{u}(t_i)) + 2h\mathbf{f}\left(t_{i+0.5}, \mathbf{u}(t_i) + \frac{K_0}{2}\right) + 2h\mathbf{f}\left(t_{i+0.5}, \mathbf{u}(t_i) + \frac{K_1}{2}\right)\\ &\qquad\qquad +h\mathbf{f}(t_{i+1}, \mathbf{u}(t_i) + K_2)\end{aligned}}{6}$$

$$(12.28)$$

And the backward Euler's method becomes:

$$\mathbf{u}(t_{i+1}) = \mathbf{u}(t_i) + h\mathbf{f}(t_{i+1}, \mathbf{u}(t_{i+1})) \qquad (12.29)$$

The error of each method remains unchanged for its scalar version. Likewise, the pseudocode to implement these methods follows the same logical steps as the

```
y ← Input vector of initial values of length N
tL ← Input lower mesh point
tU ← Input upper mesh point
h ← Input step size

t ← tL
WHILE (t <= tU)
    y ← y + h × F(t,y)
    t ← t + h
END WHILE

RETURN y

FUNCTION F(t,y)
    RETURN vector of evaluations of the derivatives of the N target
            function at mesh point t and at the vector of function points y
END FUNCTION
```

Fig. 12.10 Pseudocode of Euler's method for a system of IVP

pseudocode for the scalar versions, except with vectors. For example, the code for Euler's method presented in Fig. 12.2 becomes the vectorial version of Fig. 12.10.

Example 12.6

The *Lorenz equations* are a set of three ODE that describe the chaotic behavior of certain natural systems. These equations arise in many engineering models of real-world systems. The ODE are:

$$\frac{dx}{dt} = \sigma(y - x)$$

$$\frac{dy}{dt} = x(\rho - z) - y$$

$$\frac{dz}{dt} = xy - \beta z$$

Using the system parameters $\sigma = 10$, $\rho = 28$, and $\beta = 2.7$, and the initial values at $t = 0$ of $x = 1$, $y = 1$, and $z = 1$, use Euler's method with a step size of 0.01 to draw the solution of the Lorenz equations.

Solution

This three-dimensional problem can be written using a single three-dimensional vector by applying Eqs. (12.23), (12.24), and (12.25):

(continued)

Example 12.6 (continued)

$$\mathbf{u}(t) = \begin{bmatrix} x(t) \\ y(t) \\ z(t) \end{bmatrix}$$

$$\frac{d\mathbf{u}}{dt} = \begin{bmatrix} \dfrac{dx}{dt} \\ \dfrac{dy}{dt} \\ \dfrac{dz}{dt} \end{bmatrix} = \begin{bmatrix} \sigma(y - x) \\ x(\rho - z) - y \\ xy - \beta z \end{bmatrix} = \mathbf{f}(t, \mathbf{u}(t))$$

$$\mathbf{u}(0) = \begin{bmatrix} 1 \\ 1 \\ 1 \end{bmatrix}$$

It is now possible to apply Euler's method from Eq. (12.26). The results for the first two steps are:

$$\mathbf{u}(0.01) = \begin{bmatrix} 1 \\ 1 \\ 1 \end{bmatrix} + 0.01 \begin{bmatrix} 10(1 - 1) \\ 1(28 - 1) - 1 \\ 1 \times 1 - 2.7 \times 1 \end{bmatrix} = \begin{bmatrix} 1.00 \\ 1.26 \\ 0.98 \end{bmatrix}$$

$$\mathbf{u}(0.02) = \begin{bmatrix} 1.00 \\ 1.26 \\ 0.98 \end{bmatrix} + 0.01 \begin{bmatrix} 10(1.26 - 1.00) \\ 1.00(28 - 0.98) - 1.26 \\ 1.00 \times 1.26 - 2.7 \times 0.98 \end{bmatrix} = \begin{bmatrix} 1.03 \\ 1.52 \\ 0.97 \end{bmatrix}$$

If Euler's method is applied for 1000 steps, the resulting scatter of points will draw the figure below:

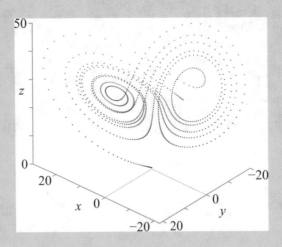

12.7   Higher-Order ODEs

So far, this chapter has only dealt with IVPs that had first-order ODE. It is of course possible, however, for a model to include a higher-order derivative of a parameter. Most commonly, engineering models will include the second derivative (acceleration given position) or third derivative (jerk given position), but it is possible to model an IVP with any degree of derivative.

An IVP with a first-order derivative starts with the ODE of Eq. (12.2) that is a function of time (or some other independent variable the formula is derived with respect to) and of the function (a.k.a. the zeroth-order derivative). Likewise, an IVP with an nth-order ODE starts with an equation that is a function of time and of all lower-order derivatives:

$$y^n(t) = f\left(t, y(t), y^{(1)}(t), \ldots, y^{(n-1)}(t)\right) \tag{12.30}$$

Initial values are available for all n lower-order derivatives in that equation: $\{y(t_0), y^{(1)}(t_0), \ldots, y^{(n-1)}(t_0)\}$.

The method to solve this system is very similar to the method used to solve the system of IVPs in the previous section: by using a variable substitution to reduce the problem to a first-order IVP. Instead of dealing with a problem with multiple derivatives of a variable, define a single multidimensional variable whose values will be the n lower-order derivatives:

$$\mathbf{u}(t) = \begin{bmatrix} y(t) \\ y^{(1)}(t) \\ \vdots \\ y^{(n-1)}(t) \end{bmatrix} \tag{12.31}$$

A single derivative of this variable will introduce the nth-order ODE of the problem into this vector, while all other dimensions are simply shifted up by one position compared to (12.31):

$$\frac{d\mathbf{u}}{dt} = \begin{bmatrix} y^{(1)}(t) \\ y^{(2)}(t) \\ \vdots \\ y^{(n)}(t) = f\left(t, y(t), y^{(1)}(t), \ldots, y^{(n-1)}(t)\right) \end{bmatrix} = \mathbf{f}(t, \mathbf{u}(t)) \tag{12.32}$$

And the initial conditions of $\mathbf{u}(t)$ are the vector of initial values of its dimensions, which are all known:

$$\mathbf{u}(t_0) = \begin{bmatrix} y(t_0) \\ y^{(1)}(t_0) \\ \dots \\ y^{(n-1)}(t_0) \end{bmatrix} \tag{12.33}$$

This simple variable substitution has transformed the nth-order IVP into a first-order IVP, albeit one with a vectorial variable instead of a scalar. But this none-theless remains an IVP problem that can be solved using the vectorial versions of the IVP methods presented in the previous section in Eqs. (12.26), (12.27), (12.28), and (12.29). The error of each method remains unchanged for its scalar version.

Example 12.7
Consider a circuit with a single loop with an inductor of 1 H, a resistor of 10 Ω and a capacitor of 0.25 F. If the system is initially at rest, and at time $t = 0$, a voltage force of $V(t) = \sin(t)$ is applied. Approximate the current $I(t)$ moving through the loop for $t > 0$. This circuit is shown below.

It can be modelled by the following ODE:

$$I^{(2)}(t) + 10I^{(1)}(t) + 4I(t) = \cos(t)$$

Solution
First, rewrite the ODE under the form of Eq. (12.30):

$$I^{(2)}(t) = \cos(t) - 4I(t) - 10I^{(1)}(t)$$

Next, define the vectorial system of Eqs. (12.31), (12.32), and (12.33). Note that the initial conditions are all null, since the system was initially at rest.

$$\mathbf{u}(t) = \begin{bmatrix} I(t) \\ I^{(1)}(t) \end{bmatrix}$$

$$\frac{d\mathbf{u}}{dt} = \begin{bmatrix} I^{(1)}(t) \\ \cos(t) - 4I(t) - 10I^{(1)}(t) \end{bmatrix} = \mathbf{f}(t, \mathbf{u}(t))$$

$$\mathbf{u}(0) = \begin{bmatrix} 0 \\ 0 \end{bmatrix}$$

(continued)

Example 12.7 (continued)

Since no information is given on how to solve the IVP, Heun's method can be used as a trade-off of simplicity and accuracy. The step could be set to 0.1 s. The first iteration gives:

$$\mathbf{f}(0, \mathbf{u}(0)) = \begin{bmatrix} 0 \\ \cos(0) - 4 \times 0 - 10 \times 0 \end{bmatrix} = \begin{bmatrix} 0 \\ 1 \end{bmatrix}$$

$$\mathbf{u}(0.1) = \begin{bmatrix} 0 \\ 0 \end{bmatrix} + 0.1\mathbf{f}(0, \mathbf{u}(0)) = \begin{bmatrix} 0 \\ 0.1 \end{bmatrix}$$

$$\mathbf{f}(0.1, \mathbf{u}(0.1)) = \begin{bmatrix} 0.1 \\ \cos(0.1) - 4 \times 0 - 10 \times 0.1 \end{bmatrix} = \begin{bmatrix} 0.1 \\ -0.005 \end{bmatrix}$$

$$\mathbf{u}(0.1) = \begin{bmatrix} 0 \\ 0 \end{bmatrix} + \frac{0.1}{2}(\mathbf{f}(0, \mathbf{u}(0)) + \mathbf{f}(0.1, \mathbf{u}(0.1))) = \begin{bmatrix} 0.005 \\ 0.050 \end{bmatrix}$$

So the current after the first 0.1 s is $I(0.1) = 0.005$ A. A second iteration gives:

$$\mathbf{f}(0.1, \mathbf{u}(0.1)) = \begin{bmatrix} 0.050 \\ \cos(0.1) - 4 \times 0.005 - 10 \times 0.050 \end{bmatrix} = \begin{bmatrix} 0.050 \\ 0.478 \end{bmatrix}$$

$$\mathbf{u}(0.2) = \begin{bmatrix} 0.005 \\ 0.050 \end{bmatrix} + 0.1\mathbf{f}(0.1, \mathbf{u}(0.1)) = \begin{bmatrix} 0.010 \\ 0.098 \end{bmatrix}$$

$$\mathbf{f}(0.2, \mathbf{u}(0.2)) = \begin{bmatrix} 0.098 \\ \cos(0.2) - 4 \times 0.010 - 10 \times 0.098 \end{bmatrix} = \begin{bmatrix} 0.098 \\ -0.035 \end{bmatrix}$$

$$\mathbf{u}(0.2) = \begin{bmatrix} 0.005 \\ 0.050 \end{bmatrix} + \frac{0.1}{2}(\mathbf{f}(0.1, \mathbf{u}(0.1)) + \mathbf{f}(0.2, \mathbf{u}(0.2))) = \begin{bmatrix} 0.012 \\ 0.072 \end{bmatrix}$$

And so the current has increased to $I(0.2) = 0.012$ A. This method can be applied over and over until an appropriate model of the current is obtained. If 200 steps are done to approximate the current until $t = 20$ s, the results give the following graph:

(continued)

Example 12.7 (continued)

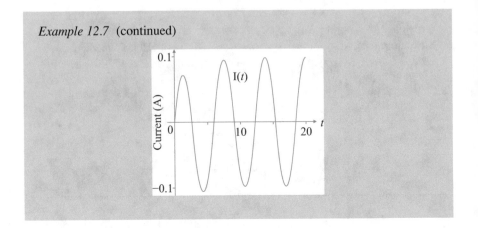

12.8   Engineering Applications

IVP problems arise frequently in engineering modelling; whenever a system is changing (see Chap. 10 on differentiation) but only its initial conditions can be measured. A proper model will need to keep track of the state of the system at all times, and when this information cannot be measured directly, it must be approximated using one of the methods learned in this chapter. Section 12.1 illustrated this situation using a simple RC circuit, but many other scenarios are possible in practice:

- The angle of a swinging pendulum θ is modelled as:

$$\frac{d^2\theta}{dt^2} + \frac{g}{L} \sin(\theta) = 0 \tag{12.34}$$

where L is the length of the pendulum and g is the Earth's gravitational acceleration. Using this equation, it is possible to compute the angle of a pendulum at any time provided the initial angle θ_0 is known.

- Newton's law of cooling models the decreasing temperature T over time of a hot object in a cool environment as:

$$\frac{dT}{dt} = \frac{hA}{C}(T - T_{env}) \tag{12.35}$$

where h is the heat transfer coefficient between the object and the environment, A is the heat transfer surface area between the object and its environment, C is the total heat capacity of the system, and T_{env} is the temperature of the environment.

- The Lotka-Volterra equations describe the dynamic relationship between the populations of two species living in the same environment. One is the prey species, with population size x, and the other is the predator species, with population size y, that hunts the prey. The rate of change of each population is modelled as:

$$\frac{dx}{dt} = \alpha x - \beta xy$$
$$\frac{dy}{dt} = \delta xy - \gamma y \tag{12.36}$$

where the four constants describe the interaction of the two species; α is the growth rate of the prey, β is the rate at which predators kill off preys, δ is the growth rate of predators given their consumption of prey, and γ is the death rate of predators.

- The displacement over time of damped spring in orientation x is described by:

$$\frac{d^2x}{dt^2} + 2\zeta\omega\frac{dx}{dt} + \omega^2 x = 0 \tag{12.37}$$

where ζ is the damping ratio and ω is the angular frequency of the oscillations.

In all these examples, the value of an important parameter of the system can be approximated at any instant from the model, provided only that its initial value is known.

12.9 Summary

Sometimes, in an engineering problem, the equation modelling a system is unknown, but its initial conditions are known, and rate of change is available in the form of an ordinary differential equation (ODE). Such a situation sets up an initial value problem (IVP), where it is necessary to approximate the system's behavior by following its changing parameters step by step, from the initial setup until a target time in the future.

This chapter has introduced three methods to solve IVP. Euler's method does exactly what is described above: it evaluates the ODE at the current point to model the function then follows that model to the next point, continues on step by step until the target time. Heun's method evaluates the ODE at the next point computed by Euler's method, and averages the ODE at the current and next points to get a better model of the function, which it can then use to generate a better approximation of the next point. The fourth-order Runge–Kutta method takes this process even further and computes two approximations of the point halfway between the current and next points to evaluate the ODE within the interval, and computes a weighted average of those four ODEs to model the function and compute the next point. Finally, the backward Euler's method was introduced. A simple variation on Euler's method, it lacks the improved accuracy of Heun's

Table 12.1 Summary of IVP methods

Method	Requires	Error
Euler's method	Initial point and ODE	$O(h^2)$
Heun's method	Initial point and ODE, computes one additional point and derivative	$O(h^3)$
Fourth-order Runge–Kutta method	Initial point and ODE, computes three additional points and derivatives	$O(h^5)$
Backward Euler's method	Initial point and ODE, solves for the next point	$O(h^2)$

method and the fourth-order Runge–Kutta method. However, it is useful for dealing with a special class of problems called stiff ODEs, where the other three methods fail and diverge. Table 12.1 summarizes these techniques.

These methods were developed to evaluate a two-dimensional scalar point using a first-order ODE. Engineering practice, however, is not always so simple, and it is possible one has to work on multidimensional problems where multiple parameters of a system interact in the ODE, or with two-dimensional problems where the ODE uses a higher-order derivative. This chapter has presented a simple technique to deal with either cases using a variable substitution. By writing the multiple dimensions or derivatives of the system as entries in a vectorial variable and changing the equations to use a vectorial variable instead of a scalar, any of the IVP methods seen in this chapter can be used.

12.10   Exercises

1. Given the following IVP, approximate the requested values using one step of Euler's method:

   a. $y(1.5)$.
   b. $y(1)$.
   c. $y(0.75)$

$$y^{(1)}(t) = 1 - ty(t)$$
$$y(0.5) = 2.5$$

2. Using the following IVP and a step $h = 0.5$, approximate $y(0.5)$, $y(1)$, and $y(1.5)$ using Euler's method.

$$y^{(1)}(t) = 1 - 0.25y(t) + 0.2t$$
$$y(0) = 1$$

3. Given the following IVP and a step $h = 0.5$, approximate $y(1.5)$ and $y(2)$ using Euler's method.

$$y^{(1)}(t) = 1 - 0.25y(t) + 0.2t$$
$$y(1) = 2$$

4. Given an IVP with an initial condition $y(0) = y_0$, if the second derivative is bounded by $-8 < y^{(2)}(t) < 8$, on how large an interval can we estimate $y(t)$ using Euler's method if we want to ensure that the error is less than 0.0001?

5. Solve Example 12.2 using Heun's method. Compare your results to the error from Euler's method given in that example.

6. Given the following IVP, approximate $y(1)$ and $y(1.5)$ using one step of Heun's method.

$$y^{(1)}(t) = 1 - ty(t)$$
$$y(0.5) = 2.5$$

7. Using the following IVP and a step of $h = 0.5$, approximate $y(0.5)$, $y(1)$, and $y(1.5)$ using Heun's method.

$$y^{(1)}(t) = 1 - 0.25y(t) + 0.2t$$
$$y(0) = 1$$

8. Approximate $y(1.5)$ and $y(2)$ with Heun's method, using the IVP below and a step of $h = 0.5$.

$$y^{(1)}(t) = 1 - 0.25y(t) + 0.2t$$
$$y(1) = 2$$

9. Given an IVP with an initial condition $y(0) = y_0$, if the second derivative is bounded by $-8 < y^{(2)}(t) < 8$, on how large an interval can we estimate $y(t)$ using Heun's method if we want to ensure that the error is less than 0.0001? Compare this with the range for Question 4.

10. Solve Example 12.2 using the fourth-order Runge–Kutta method. Compare your results to the error from Euler's method given in that example.

11. Given the following IVP, approximate $y(1)$ and $y(1.5)$ using one step of the fourth-order Runge–Kutta method.

$$y^{(1)}(t) = 1 - ty(t)$$
$$y(0.5) = 2.5$$

12. Given the following IVP, approximate $y(0.5)$, $y(1)$, and $y(1.5)$ using the fourth-order Runge–Kutta method and a step of $h = 0.5$.

$$y^{(1)}(t) = 1 - 0.25y(t) + 0.2t$$
$$y(0) = 1$$

13. Given the following IVP, approximate $y(1.5)$ and $y(2)$ using the fourth-order Runge–Kutta method and a step of $h = 0.5$.

$$y^{(1)}(t) = 1 - 0.25y(t) + 0.2t$$
$$y(1) = 2$$

14. Approximate $y(1)$ for the following IVP using four steps of Euler's method, Heun's method, and the fourth-order Runge–Kutta method, given the initial condition that $y(0) = 1$.

$$y^{(1)}(t) = 1 - 0.25y(t) + 0.2t$$
$$y(0) = 1$$

15. Approximate $y(1)$ for the following IVP using four steps of Euler's method, Heun's method, and the fourth-order Runge–Kutta method.

$$y^{(1)}(t) = ty(t) + t - 1$$
$$y(0) = 1$$

16. Solve Example 12.2 using the backward Euler's method. Compare your results to the error from Euler's method given in that example.

17. Given the following IVP, approximate $y(1)$ and $y(1.5)$ using the backward Euler's method.

$$y^{(1)}(t) = 1 - ty(t)$$
$$y(0.5) = 2.5$$

18. Approximate $y(0.5)$, $y(1)$, and $y(1.5)$ using the backward Euler's method for the following IVP.

$$y^{(1)}(t) = 1 - 0.25y(t) + 0.2t$$
$$y(0) = 1$$

19. Given the following IVP, approximate $y(1.5)$ and $y(2)$ using the backward Euler's method.

$$y^{(1)}(t) = 1 - 0.25y(t) + 0.2t$$
$$y(1) = 2$$

20. Given the following system of IVPs, compute the first two steps using $h=0.1$.

$$x^{(1)}(t) = -0.6x(t) + 1.8y(t)$$
$$y^{(1)}(t) = -2.6x(t) - 0.3y(t)$$
$$x(0) = 1$$
$$y(0) = 1$$

21. Find the current moving through the circuit of Example 12.7 after 0.1 s, 0.2 s, and 0.3 s:

    a. When the initial current is -1 A.
    b. When the voltage function is $v(t) = e^{-t}$ for $t \geq 0$ s.

22. Consider the second-order IVP below. Perform three steps of Euler's method using $h=0.1$.

$$y^{(2)}(t) = 4 - \sin(t) + y(t) - 2y^{(1)}(t)$$
$$y(0) = 1$$
$$y^{(1)}(0) = 2$$

23. Consider the third-order IVP below. Perform two steps of Heun's method using $h=0.1$.

$$y^{(3)}(t) = y(t) - ty^{(1)}(t) + 4y^{(2)}(t)$$
$$y(2) = 1$$
$$y^{(1)}(2) = 2$$
$$y^{(2)}(2) = 3$$

24. Van der Pol's second-order ODE, given below, is an example of a higher-order stiff ODE. Using $\mu=0.3$ and $h=0.1$, and with the initial conditions $y(0)=0.7$ and $y^{(1)}(0)=1.2$, compute four steps of this IVP.

$$y^{(2)}(t) - \mu\left(1 - y(t)^2\right)y^{(1)}(t) + y(t) = 0$$

25. Given the following IVP, approximate $y(1)$ by using

    a. One step of Euler's method.
    b. Two steps of Euler's method.
    c. Four steps of Euler's method.

$$y^{(3)}(t) + y^{(2)}(t) - y^{(1)}(t) + y(t) + 3 = 0$$
$$y(0) = 3$$
$$y^{(1)}(0) = 2$$
$$y^{(2)}(0) = 1$$

26. Given the following IVP, approximate $y(1)$ by using

    a. One step of Euler's method.
    b. Two steps of Euler's method.
    c. Four steps of Euler's method.

$$y^{(3)}(t) + ty^{(2)}(t) - y^{(1)}(t) + y(t) + 3t = 0$$
$$y(0) = 3$$
$$y^{(1)}(0) = 2$$
$$y^{(2)}(0) = 1$$

Chapter 13
Boundary Value Problems

13.1 Introduction

Consider a simple circuit loop with an inductor of 1 H, a resistor of 10 Ω and a capacitor of 0.25 F, as shown in Fig. 13.1. This circuit can be modelled by the following ODE:

$$I^{(2)}(t) + 10I^{(1)}(t) + 4I(t) = \cos(t) \tag{13.1}$$

This model would be easy to use if the initial value of the current and its derivative was known. But what if the rate of change of the current is not known, and the only information available is the current in the circuit the moment the system was turned on and when it was shut down? That is to say, only the initial and final values of the current are known, while its derivatives are unknown.

This type of problem is a *boundary value problem* (BVP), a situation in which a parameter's value is known at the two boundaries of an interval, called the *boundary conditions* of the system, and its value within that interval must be estimated. The boundary conditions typically represent initial and terminal conditions of the system. And the interval might be over time, as in the previous example using the current at the moment the system was turned on and off, or over space, for example if one was measuring the voltage at the starting and terminal positions along a long transmission line.

One can immediately see the similarities and differences with the higher-order initial value problem (IVP) situation presented in the previous chapter. Both setups are very similar: both start with an nth order ODE that is a function of an independent variable (we will use time, for simplicity) and of all lower-order derivatives:

© Springer International Publishing Switzerland 2016
R. Khoury, D.W. Harder, *Numerical Methods and Modelling for Engineering*,
DOI 10.1007/978-3-319-21176-3_13

Fig. 13.1 A sample circuit

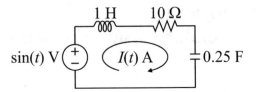

$$y^{(n)}(t) = f\left(t, y(t), y^{(1)}(t), \ldots, y^{(n-1)}(t)\right) \tag{13.2}$$

In both cases, the challenge is to approximate the values of a system's parameters at mesh points over an interval going from t_0 to t_{n-1} with n equal steps h such that:

$$h = \frac{t_{n-1} - t_0}{n-1} = t_{i+1} - t_i \tag{13.3}$$

And in both cases, this approximation must be computed without knowledge of the system's model equation, only of its higher-order derivative equation as a function of the lower-order derivatives. In the IVP case, initial values $\{y(t_0), y^{(1)}(t_0), \ldots, y^{(n-1)}(t_0)\}$ for the parameters all lower-order derivatives are available, and the ultimate aim is to get the final values of these parameters at the end of the interval; values computed within the interval are merely a means to that end. In the BVP case, the initial and final values $\{y(t_0), y(t_{n-1})\}$ are available for the parameters, but the derivative values are not, and the aim is to approximate the parameter values within the interval.

13.2 Shooting Method

Given the similarities between higher-order IVPs and BVPs, it would be nice if it were possible to use the higher-order IVP method learnt in the previous chapter in order to solve BVPs. Recall from Chap. 12 that this would require rewriting the system of Eq. (13.2) as a vector of lower-order derivative values and taking that vector's derivative:

$$\mathbf{u}(t) = \begin{bmatrix} y(t) \\ y^{(1)}(t) \\ \vdots \\ y^{(n-1)}(t) \end{bmatrix} \tag{13.4}$$

$$\frac{d\mathbf{u}}{dt} = \begin{bmatrix} y^{(1)}(t) \\ y^{(2)}(t) \\ \vdots \\ y^{(n)}(t) = f\left(t, y(t), y^{(1)}(t), \ldots, y^{(n-1)}(t)\right) \end{bmatrix} = \mathbf{f}(t, \mathbf{u}(t)) \qquad (13.5)$$

Then, given a vector of initial values $\mathbf{u}(t_0)$, vectorial versions of the IVP methods of Chap. 12 can be used to solve the problem. The only real issue is that these initial values of the derivatives required to complete vector $\mathbf{u}(t_0)$ and to compute the initial step of the IVP methods are unknown in the BVP case. Without them, the IVP methods cannot start.

However, with the initial and final values of that parameter known and a model of the derivative equation available, a simple solution presents itself: take a shot at guessing the initial value of the derivatives. Using some random value for the initial value of the derivatives in $\mathbf{u}(t_0)$, it is possible to apply any of the IVP methods and compute what the approximated final value will be. Then, compare this approximation to the real target final value to see how far off the mark the shot was, and refine the guess to take another shot. Actually getting the correct initial derivative value in this manner is unlikely; but after N guesses, a set of N n-dimensional points will have been generated, each combining initial derivative values and a final parameter value. It now becomes possible to use these points to compute an equation modelling the relationship between the initial derivative values and the final parameter value using any of the interpolation and regression methods of Chap. 6, and to use this equation to compute the correct initial derivative corresponding to the real final value of the parameter. Finally, once the correct initial derivative value is known, the chosen IVP method can be used one last time to compute the correct approximations of the parameter within the interval.

This method is called the *shooting method*. Its biggest advantage is that it uses only interpolation and IVP methods that were learnt in previous chapters. Its error rate will likewise be directly related to the interpolation and IVP methods chosen. Using a greater number N of shots will lead to a better interpolation or regression and to a better approximation of the initial derivative, which can be used in turn in a more accurate IVP method. Moreover, since the shooting method is a bounded method, with the boundary conditions constraining the initial and final value of the IVP method, the absolute error of the IVP approximation will increase the farther a point is from the two boundaries, and will be at its maximum near the middle of the interval. The biggest disadvantage of the shooting method is its runtime: for N random shots, the method will need to apply the IVP method $N+1$ times, or once for every shot plus one final time using the final estimate of the initial derivative. This in turn limits its applicability to problems in more than two dimensions, which will require a large number of shots in order to generate enough points to estimate the multidimensional function and the correct derivative values.

Figure 13.2 presents the pseudocode for a version of the shooting method for a second-order BVP of the form of Eq. (13.1). There are two loops in this version of

```
yL ← Input initial value
yU ← Input final value
tL ← Input lower mesh point
tU ← Input upper mesh point
h ← Input step size

WHILE (yLow >= yU)
    DerivativeLow ← guess an initial derivative value
    yLow ← solve IVA problem for F using yL and LowDerivative from tL to
           tU using step size h
END WHILE

WHILE (yHigh <= yU)
    DerivativeHigh ← guess an initial derivative value
    yHigh ← solve IVA problem for F using yL and LowDerivative from tL to
            tU using step size h
END WHILE

DerivativeFunction ← interpolate a function from point(DerivativeLow,yLOW)
                     to point (DerivativeHigh,yHigh)
Derivative ← DerivativeFunction(yH)

Steps ← solve IVA problem for F using yL and Derivative from tL to tU
        using step size h

RETURN Steps

FUNCTION F(t,y,d)
    RETURN evaluation of the second derivative of the target function at
           mesh point t and at function point y with first derivative d
END FUNCTION
```

Fig. 13.2 Pseudocode of the shooting method

the algorithm; the first generates guesses of the first derivative until one shoots below the final value of the parameter, and the second generates guesses of the first derivative until one shoots above that final value. Next, the two 2D points generated are used to interpolate a function and to get the correct value of the first derivative, which is then used in a final application of the IVP method to get the steps that solve the BVP.

Example 13.1
Consider a circuit with a single loop with an inductor of 1 H, a resistor of 3 Ω and a capacitor of 0.5 F, as shown in Fig. 13.1. Initially, a current of 1.28 A is going through the loop. At time $t = 0$ s, a voltage force of $V(t) = \sin(t)$ is applied. After 1 s, the current is measured to be 0.991 A. Approximate the current $I(t)$ moving through the loop at every 0.2 s using Euler's method. This situation can be modelled by the following ODE:

$$I^{(2)}(t) + 3I^{(1)}(t) + 2I(t) = \cos(t)$$

(continued)

Example 13.1 (continued)
Solution
To apply the higher-order IVP, first, rewrite the ODE under the form of
Eq. (13.2):

$$I^{(2)}(t) = \cos(t) - 2I(t) - 3I^{(1)}(t)$$

Next, define the vectorial system of Eqs. (13.4) and (13.5). Note that the
initial value of the first derivative is unknown, and has been noted by α.

$$\mathbf{u}(t) = \begin{bmatrix} I(t) \\ I^{(1)}(t) \end{bmatrix}$$

$$\frac{d\mathbf{u}}{dt} = \begin{bmatrix} I^{(1)}(t) \\ \cos(t) - 2I(t) - 3I^{(1)}(t) \end{bmatrix} = \mathbf{f}(t, \mathbf{u}(t))$$

$$\mathbf{u}(0) = \begin{bmatrix} 1.28 \\ \alpha \end{bmatrix}$$

A first guess for the value of α could be simply 0. This would give the
following steps using Euler's method:

$$\mathbf{u}(0.2) = \mathbf{u}(0) + h\mathbf{f}(0, \mathbf{u}(0)) = \begin{bmatrix} 1.28 \\ 0 \end{bmatrix} + 0.2\begin{bmatrix} 0 \\ -1.56 \end{bmatrix} = \begin{bmatrix} 1.28 \\ -0.312 \end{bmatrix}$$

$$\mathbf{u}(0.4) = \mathbf{u}(0.2) + h\mathbf{f}(0.2, \mathbf{u}(0.2)) = \begin{bmatrix} 1.28 \\ -0.312 \end{bmatrix} + 0.2\begin{bmatrix} -0.312 \\ -0.644 \end{bmatrix} = \begin{bmatrix} 1.218 \\ -0.441 \end{bmatrix}$$

$$\mathbf{u}(0.6) = \mathbf{u}(0.4) + h\mathbf{f}(0.4, \mathbf{u}(0.4)) = \begin{bmatrix} 1.218 \\ -0.441 \end{bmatrix} + 0.2\begin{bmatrix} -0.441 \\ -0.192 \end{bmatrix} = \begin{bmatrix} 1.129 \\ -0.479 \end{bmatrix}$$

$$\mathbf{u}(0.8) = \mathbf{u}(0.6) + h\mathbf{f}(0.6, \mathbf{u}(0.6)) = \begin{bmatrix} 1.129 \\ -0.479 \end{bmatrix} + 0.2\begin{bmatrix} -0.479 \\ -0.004 \end{bmatrix} = \begin{bmatrix} 1.034 \\ -0.478 \end{bmatrix}$$

$$\mathbf{u}(1.0) = \mathbf{u}(0.8) + h\mathbf{f}(0.8, \mathbf{u}(0.8)) = \begin{bmatrix} 1.034 \\ -0.478 \end{bmatrix} + 0.2\begin{bmatrix} -0.478 \\ -0.065 \end{bmatrix} = \begin{bmatrix} 0.938 \\ -0.465 \end{bmatrix}$$

The estimated current at $t = 1$ s with this shot is 0.938 A, less than the actual
value measured in the system. The derivative can be increased for a second
shot, for example to $\alpha = 0.5$. In that case, the steps will be:

(continued)

Example 13.1 (continued)

$$\mathbf{u}(0.2) = \begin{bmatrix} 1.38 \\ -0.112 \end{bmatrix} \quad \mathbf{u}(0.4) = \begin{bmatrix} 1.358 \\ -0.401 \end{bmatrix}$$

$$\mathbf{u}(0.6) = \begin{bmatrix} 1.278 \\ -0.519 \end{bmatrix} \quad \mathbf{u}(0.8) = \begin{bmatrix} 1.174 \\ -0.554 \end{bmatrix} \quad \mathbf{u}(1.0) = \begin{bmatrix} 1.063 \\ -0.552 \end{bmatrix}$$

In this shot, the estimated current at $t = 1$ s is 1.063 A, too high. However, these two shots give two 2D points relating the derivative value α to the final current value: (0, 0.938) and (0.5, 1.063). A straight-line interpolation of the relationship can be obtained from these two points. The interpolated line is:

$$A = 0.250\alpha + 0.938$$

And using this line, the value of α that gives a current of 0.991 A can easily be found to be 0.212. With that initial value, the steps can once again be computed using Euler's method, and found to be:

$$\mathbf{u}(0.2) = \begin{bmatrix} 1.322 \\ -0.227 \end{bmatrix} \quad \mathbf{u}(0.4) = \begin{bmatrix} 1.277 \\ -0.424 \end{bmatrix}$$

$$\mathbf{u}(0.6) = \begin{bmatrix} 1.192 \\ -0.496 \end{bmatrix} \quad \mathbf{u}(0.8) = \begin{bmatrix} 1.093 \\ -0.510 \end{bmatrix} \quad \mathbf{u}(1.0) = \begin{bmatrix} 0.991 \\ -0.502 \end{bmatrix}$$

In this shot, the estimated current at $t = 1$ s is exactly 0.998 A, as required by the system measurements in the question. The four values at 0.2 s steps, which were the information actually required in the problem statement, have also been computed in this last shot.

The actual value of the first derivative of this system was 0.2, which means the interpolated value had a relative error of 6 %. Using that value, the correct steps and their relative errors are given in the table below.

t (s)	Shooting method (A)	Real value (A)	Relative error (%)
0.0	1.280	1.280	0.00
0.2	1.322	1.291	2.47
0.4	1.277	1.245	2.55
0.6	1.192	1.171	1.780
0.8	1.093	1.084	0.820
1.0	0.991	1.991	0.00

(continued)

Example 13.1 (continued)

The current function over time and the three approximations can be plotted together to compare visually. The undershooting approximation (in purple), which used a shot for the derivative that was too low, clearly underestimates the real function (in blue) in this interval. Likewise, the overshooting approximation (in green), which used a shot that was too high, overestimates the function. The interpolated shot (in red) also overestimates the real function, but remains a much closer approximation than either of the shots.

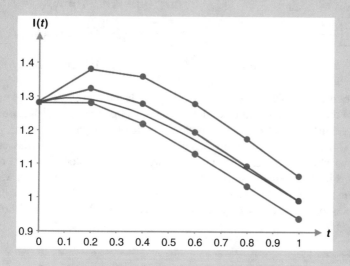

13.3   Finite Difference Method

13.3.1   One-Dimensional Functions

The problem at the core of this chapter is to try to solve a high-order ODE model without any knowledge of the system's equation or its derivatives, and with only the initial and final value of the system. One solution is to guess the value of the derivatives; that yields the shooting method of the previous section. Another option is to approximate the derivatives using only parameter values. This gives the *finite difference method*, which is the topic of this section.

In fact, methods to approximate a derivative using only its function values were the topic of Chap. 10. Recall in particular the second-order centered divided-difference formulae for the first and second derivatives:

$$y^{(1)}(t_i) = \frac{y(t_{i+1}) - y(t_{i-1})}{2h} + O(h^2) \tag{13.6}$$

$$y^{(2)}(t_i) = \frac{y(t_{i+1}) - 2y(t_i) + y(t_{i-1})}{h^2} + O(h^2) \tag{13.7}$$

Using these approximation formulae, it is possible to rewrite a second-order ODE equation of the form of Eqs. (13.1) and (13.2) as:

$$c_2 y^{(2)}(t_i) + c_1 y^{(1)}(t_i) + c_0 y(t_i) = f(t_i)$$

$$c_2 \left(\frac{y(t_{i+1}) - 2y(t_i) + y(t_{i-1})}{h^2} \right) + c_1 \left(\frac{y(t_{i+1}) - y(t_{i-1})}{2h} \right) + c_0 y(t_i) = f(t_i)$$

$$(2c_2 - hc_1)y(t_{i-1}) + (2h^2 c_0 - 4c_2)y(t_i) + (hc_1 + 2c_2)y(t_{i+1}) = 2h^2 f(t_i) \tag{13.8}$$

The second-order ODE system has now been reduced to a function that only uses values of the parameter at three of the mesh points that are needed! It is possible to write $n-2$ instances of Eq. (13.8), with the value of $y(t)$ going from $y(t_{i-1}) = y(0)$ to $y(t_{i+1}) = y(t_{n-1})$ to cover all n mesh points. Moreover, since the boundary conditions $y(0)$ and $y(t_{n-1})$ are known, only the $n-2$ internal mesh points are unknowns in these equations. In other words, there are $n-2$ equations with $n-2$ unknowns, a problem that can easily be solved using the $\mathbf{Mx} = \mathbf{b}$ methods learnt, back in Chap. 4. The $\mathbf{Mx} = \mathbf{b}$ system created by the $n-2$ equations of the form of Eq. (13.8) is:

$$\begin{bmatrix}
2h^2 c_0 - 4c_2 & hc_1 + 2c_2 & 0 & 0 & \cdots & 0 & 0 & 0 & 0 \\
2c_2 - hc_1 & 2h^2 c_0 - 4c_2 & hc_1 + 2c_2 & 0 & \cdots & 0 & 0 & 0 & 0 \\
0 & 2c_2 - hc_1 & 2h^2 c_0 - 4c_2 & hc_1 + 2c_2 & \cdots & 0 & 0 & 0 & 0 \\
\vdots & \vdots & \vdots & \vdots & \ddots & \vdots & \vdots & \vdots & \vdots \\
0 & 0 & 0 & 0 & \cdots & 2c_2 - hc_1 & 2h^2 c_0 - 4c_2 & hc_1 + 2c_2 & 0 \\
0 & 0 & 0 & 0 & \cdots & 0 & 2c_2 - hc_1 & 2h^2 c_0 - 4c_2 & hc_1 + 2c_2 \\
0 & 0 & 0 & 0 & \cdots & 0 & 0 & 2c_2 - hc_1 & 2h^2 c_0 - 4c_2
\end{bmatrix}$$

$$\begin{bmatrix}
y(1) \\
y(2) \\
y(3) \\
\vdots \\
y(t_{n-4}) \\
y(t_{n-3}) \\
y(t_{n-2})
\end{bmatrix} = \begin{bmatrix}
2h^2 f(1) - (2c_2 - hc_1)y(0) \\
2h^2 f(2) \\
2h^2 f(3) \\
\vdots \\
2h^2 f(t_{n-4}) \\
2h^2 f(t_{n-3}) \\
2h^2 f(t_{n-2}) - (hc_1 + 2c_2)y(t_{n-1})
\end{bmatrix}$$

$$\tag{13.9}$$

The pseudocode for an algorithm to solve a second-order ODE by building the matrix-vector system of Eq. (13.9) is presented in Fig. 13.3. Note that, given an equation of the form $c_2 y^{(2)}(t) + c_1 y^{(1)}(t) + c_0 y(t) = f(t)$, the value of each individual coefficient can be obtained by the algorithm (rather than input manually by the user) by setting in turn either $y(t)$ or one of its derivatives to 1 and the other two values to 0.

```
yL ← Input initial value
yU ← Input final value
tL ← Input lower mesh point
tU ← Input upper mesh point
h ← Input step size

c0 ← F(1,0,0)
c1 ← F(0,1,0)
c2 ← F(0,0,1)

Lower ← 2 × c2 - h × c1
Diagonal ← 2 × h × h × c0 - 4 × c2
Upper ← h × c1 + 2 × c2
Result ← 2 × h × h

NumberOfSteps ← [ (tU - tL) / h ] - 1
M ← zero matrix of size NumberOfSteps × NumberOfSteps
B ← zero vector of size NumberOfSteps

RowIndex ← 0
WHILE (RowIndex <= NumberOfSteps)

    ColumnIndex ← 0
    WHILE (ColumnIndex <= NumberOfSteps)

        IF (ColumnIndex = RowIndex - 1)
            element at column ColumnIndex, row RowIndex in M ← Lower
        ELSE IF (ColumnIndex = RowIndex + 1)
            element at column ColumnIndex, row RowIndex in M ← Upper
        ELSE IF (ColumnIndex = RowIndex)
            element at column ColumnIndex, row RowIndex in M ← Diagonal
        END IF

        ColumnIndex ← ColumnIndex + 1
    END WHILE

    IF (RowIndex = 0)
        element at row RowIndex in B ← Result - Lower × yL
    ELSE IF (RowIndex = NumberOfSteps)
        element at row RowIndex in B ← Result - Upper × yU
    ELSE
        element at row RowIndex in B ← Result
    END IF

    RowIndex ← RowIndex + 1
END WHILE

Steps ← solve the matrix-vector system M × Steps = B
RETURN Steps

FUNCTION F(y,d,d2)
    RETURN evaluation of the target function using function point y with
           first derivative d and second derivative d2
END FUNCTION
```

Fig. 13.3 Pseudocode of the finite difference method

As with the shooting method, the main advantage of this method is that it relies exclusively on methods learnt in previous chapters, namely Chaps. 4 and 10, and requires no new mathematical tools. Moreover, it again gives the designer control on the error rate of the method; a higher error rate can be achieved by approximating the ODE using higher-order divided-difference formulae, albeit at the cost of having more terms to handle in the new version of Eq. (13.8) and the new matrix of Eq. (13.9). The main disadvantage is its cost overhead: before Eq. (13.8) can be developed, it is necessary to write out the divided-difference formula at the required error rate for every derivative order in the system's ODE by using the Taylor series technique presented in Chap. 10.

Example 13.2
Consider a circuit with a single loop with an inductor of 1 H, a resistor of 3 Ω and a capacitor of 0.5 F, as shown in Fig. 13.1. Initially, a current of 1.28 A is going through the loop. At time $t=0$, a voltage force of $V(t) = \sin(t)$ is applied. After 1 s, the current is measured to be 0.988 A. Approximate the current $I(t)$ moving through the loop at every 0.2 s using Euler's method. This situation can be modelled by the following ODE:

$$I^{(2)}(t) + 3I^{(1)}(t) + 2I(t) = \cos(t)$$

Solution
Using $h=0.2, c_0=2, c_1=3$, and $c_2=1$, the ODE can be rewritten in the form of Eq. (13.8):

$$1.40I(t_{i-1}) - 3.84I(t_i) + 2.60I(t_{i+1}) = 0.08 \cos(t_i)$$

This new equation can then be duplicated for each mesh point and used to write a matrix-vector system of the form of Eq. (13.9):

$$\begin{bmatrix} -3.84 & 2.60 & 0 & 0 \\ 1.40 & -3.84 & 2.60 & 0 \\ 0 & 1.40 & -3.84 & 2.60 \\ 0 & 0 & 1.40 & -3.84 \end{bmatrix} \begin{bmatrix} y(0.2) \\ y(0.4) \\ y(0.6) \\ y(0.8) \end{bmatrix} = \begin{bmatrix} -1.714 \\ 0.074 \\ 0.066 \\ -2.521 \end{bmatrix}$$

And finally, this system can be solved using any of the methods seen in Chap. 4, or by straightforward backward elimination, to get the values of the four internal mesh points. These values are presented in the table below, with their correct equivalents and the relative errors. It can be seen that the errors are a lot smaller than they were in Example 13.1, despite the fact that both examples use $O(h^2)$ methods, namely Euler's IVP method in the previous example and the second-order divided-difference formula in this one. The

(continued)

Example 13.2 (continued)
difference comes from the fact that this method uses a better approximation of the derivative for its computations, obtained from the divided-difference formula. By contrast, the shooting method uses a lower-quality approximation of the derivative obtained from a straight-line interpolation of the previous two random shots, thus leading to the loss of quality in the results.

t (s)	Finite difference method (A)	Real value (A)	Relative error (%)
0.0	1.280	1.280	0.00
0.2	1.286	1.291	0.35
0.4	1.240	1.245	0.42
0.6	1.169	1.171	0.20
0.8	1.082	1.084	0.19
1.0	0.991	0.991	0.00

The current function over time and the finite difference approximation can be plotted together, along with the best approximation from the shooting method of Example 13.1. This figure confirms visually that the finite difference method (in purple), gives a much closer approximation of the real function (in blue) than the shooting method (in red).

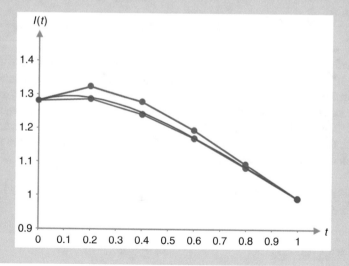

13.3.2 Multidimensional Functions

The previous section expounded the finite difference method in the case where the function derived is a one-dimensional function $y(t)$. This same method can also be expanded to handle multidimensional functions of the form $y(x_0, x_1, \ldots, x_{m-1})$.

In the multidimensional BVP, each independent variable x_i takes values over an interval $[x_{i,0}, x_{i,n-1}]$. This interval is broken up into discrete mesh points as before, at step intervals of length h. The step value h is the same for all independent variables. Boundary conditions are available for each independent variable at all boundary positions; that is to say, at every point where $x_i = x_{i,0}$ or $x_i = x_{i,n-1}$, for all values of all other independent variables. As for the system itself, it is composed of partial derivatives of the function with respect to each independent variable:

$$
\begin{aligned}
&c_9 \frac{\partial^n y(x_0, x_1, \ldots, x_{m-1})}{\partial x_0{}^n} + \cdots + c_8 \frac{\partial^2 y(x_0, x_1, \ldots, x_{m-1})}{\partial x_0{}^2} + \\
&c_7 \frac{\partial y(x_0, x_1, \ldots, x_{m-1})}{\partial x_0} + c_6 \frac{\partial^n y(x_0, x_1, \ldots, x_{m-1})}{\partial x_1{}^n} + \cdots + \\
&c_5 \frac{\partial^2 y(x_0, x_1, \ldots, x_{m-1})}{\partial x_1{}^2} + c_4 \frac{\partial y(x_0, x_1, \ldots, x_{m-1})}{\partial x_1} + \cdots + \\
&c_3 \frac{\partial^n y(x_0, x_1, \ldots, x_{m-1})}{\partial x_{m-1}{}^n} + \cdots + c_2 \frac{\partial^2 y(x_0, x_1, \ldots, x_{m-1})}{\partial x_{m-1}{}^2} + \\
&c_1 \frac{\partial y(x_0, x_1, \ldots, x_{m-1})}{\partial x_{m-1}} + c_0 y(x_0, x_1, \ldots, x_{m-1}) = f(x_0, x_1, \ldots, x_{m-1})
\end{aligned}
\tag{13.10}
$$

As with the case of a single independent variable, each partial derivative in Eq. (13.10) can be approximated using a divided-difference formula such as the ones learnt in Chap. 10. In fact, since each partial derivative is taken with respect to a single variable, the divided-difference formulae are the same as before, save for the fact that the function has independent variables that remain constant. For example, the second-order centered divided-difference formulae for the first and second derivatives with respect to x_0 are:

$$
\frac{\delta y(x_{0,i}, x_{1,j}, \ldots, x_{m-1,k})}{\delta x_0} = \frac{y(x_{0,i+1}, x_{1,j}, \ldots, x_{m-1,k}) - y(x_{0,i-1}, x_{1,j}, \ldots, x_{m-1,k})}{2h} + O(h^2)
\tag{13.11}
$$

$$
\begin{aligned}
&\frac{\delta^2 y(x_{0,i}, x_{1,j}, \ldots, x_{m-1,k})}{\delta x_0{}^2} \\
&= \frac{y(x_{0,i+1}, x_{1,j}, \ldots, x_{m-1,k}) - 2y(x_{0,i}, x_{1,j}, \ldots, x_{m-1,k}) + y(x_{0,i-1}, x_{1,j}, \ldots, x_{m-1,k})}{h^2} + O(h^2)
\end{aligned}
\tag{13.12}
$$

Comparing these to Eqs. (13.6) and (13.7), it can be seen that the variable x_0 that the derivative is taken with respect to changes as before, while the other independent variables x_1 to x_{m-1} remain constant.

Once the derivatives in the system Eq. (13.10) are approximated using divided-difference formulae, the resulting equation can be written out for every internal mesh point that needs to be approximated. As before, this will generate an equal number of equations and unknown mesh point values, which can be written into an $\mathbf{Mx} = \mathbf{b}$ system and solved.

The previous explanation was for the general case with m independent variables and an equation using up to the nth derivative. So it might feel a bit abstract. It is worthwhile to expound the method by considering the simplest but common case of working in three dimensions, with the function $y(x_0, x_1)$ of two independent variables to model. The variable x_0 takes values in the interval $[x_{0,0}, x_{0,m-1}]$ and the variable x_1 takes values in the interval $[x_{1,0}, x_{1,n-1}]$, both at regular steps of h, so that:

$$h = \frac{x_{0,m-1} - x_{0,0}}{m - 1} = \frac{x_{1,n-1} - x_{1,0}}{n - 1} \qquad (13.13)$$

The boundary conditions for this problem are the set of values the function takes at every point where either x_0 or x_1 is at its minimum or maximum value: $y(x_{0,0}, x_1)$ and $y(x_{0,m-1}, x_1)$ for all values of x_1, and $y(x_0, x_{1,0})$ and $y(x_0, x_{1,n-1})$ for all values of x_0. The problem is to determine the values for all mesh points of the function that are not on the boundary. This situation is illustrated in Fig. 13.4. The value of $y(x_0, x_1)$ is known at the blue points on the boundary in that figure, and must be estimated for the red points inside the interval.

Finally, a formula is known to model the system using its partial derivatives; again for simplicity, assume it only uses up to the second-order derivatives:

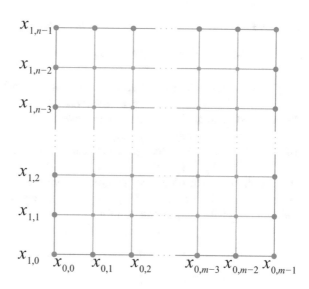

Fig. 13.4 A visualization of a two-dimensional BVP

$$c_4 \frac{\delta^2 y(x_0, x_1)}{\delta x_0^2} + c_3 \frac{\delta^2 y(x_0, x_1)}{\delta x_1^2} + c_2 \frac{\delta y(x_0, x_1)}{\delta x_0} + c_1 \frac{\delta y(x_0, x_1)}{\delta x_1} + c_0 y(x_0, x_1)$$
$$= f(x_0, x_1) \tag{13.14}$$

Since the partial derivative takes one parameter to be constant, each can be taken to be a single-variable derivative. The centered divided-difference formulae of Eqs. (13.6) and (13.7) can then be modified to use a function of two independent variables where one of the variables is constant. The resulting equations are:

$$\frac{\delta y(x_{0,i}, x_{1,j})}{\delta x_0} = \frac{y(x_{0,i+1}, x_{1,j}) - y(x_{0,i-1}, x_{1,j})}{2h} + O(h^2) \tag{13.15}$$

$$\frac{\delta^2 y(x_{0,i}, x_{1,j})}{\delta x_0^2} = \frac{y(x_{0,i+1}, x_{1,j}) - 2y(x_{0,i}, x_{1,j}) + y(x_{0,i-1}, x_{1,j})}{h^2} + O(h^2) \tag{13.16}$$

$$\frac{\delta y(x_{0,i}, x_{1,j})}{\delta x_1} = \frac{y(x_{0,i}, x_{1,j+1}) - y(x_{0,i}, x_{1,j-1})}{2h} + O(h^2) \tag{13.17}$$

$$\frac{\delta^2 y(x_{0,i}, x_{1,j})}{\delta x_1^2} = \frac{y(x_{0,i}, x_{1,j+1}) - 2y(x_{0,i}, x_{1,j}) + y(x_{0,i}, x_{1,j-1})}{h^2} + O(h^2) \tag{13.18}$$

These can be substituted back into the model equation to write it as only a function of data points and eliminate the derivatives:

$$(2c_4 + hc_2)y(x_{0,i+1}, x_{1,j}) + (2c_3 + hc_1)y(x_{0,i}, x_{1,j+1}) +$$
$$(-hc_2 + 2c_4)y(x_{0,i-1}, x_{1,j}) + (-hc_1 + 2c_3)y(x_{0,i}, x_{1,j-1}) + \tag{13.19}$$
$$(-4c_4 - 4c_3 + 2h^2 c_0)y(x_{0,i}, x_{1,j}) = 2h^2 f(x_{0,i}, x_{1,j})$$

This model equation can then be written out for every internal mesh point, to create a set of $(m \times n) - m - n$ equations with as many unknowns. This set can then be arranged in an $\mathbf{Mx} = \mathbf{b}$ system to be solved:

The matrix-vector system is a lot larger than it was in the two-dimensional case of the finite difference method presented earlier, but the methodology used to obtain it is the same. Solving the three-dimensional BVP is no more difficult than solving the two-dimensional BVP, only longer. Likewise, the pseudocode for this version of the finite difference method will not be substantially different from that of the two-dimensional case presented in Fig. 13.3.

$$
\mathbf{M} =
\begin{bmatrix}
-4c_4 - 4c_3 + 2h^2c_0 & 2c_4 + hc_2 & 0 & 0 & 2c_3 + hc_1 & 0 & 0 & \cdots & 0 & 0 & 0 \\
-hc_2 + 2c_4 & -4c_4 - 4c_3 + 2h^2c_0 & 2c_4 + hc_2 & 0 & 0 & 2c_3 + hc_1 & 0 & \cdots & 0 & 0 & 0 \\
0 & -hc_2 + 2c_4 & -4c_4 - 4c_3 + 2h^2c_0 & 2c_4 + hc_2 & 0 & 0 & 2c_3 + hc_1 & \cdots & 0 & 0 & 0 \\
\vdots & \vdots & \vdots & \vdots & \vdots & \vdots & \vdots & \ddots & \vdots & \vdots & \vdots \\
0 & 0 & 0 & 0 & 0 & 0 & 0 & \cdots & -hc_2 + 2c_4 & -hc_1 + 2c_3 & -4c_4 - 4c_3 + 2h^2c_0
\end{bmatrix}
$$

$$
\times \mathbf{x} =
\begin{bmatrix}
y(x_{0,1}, x_{1,1}) \\
y(x_{0,2}, x_{1,1}) \\
y(x_{0,3}, x_{1,1}) \\
y(x_{0,4}, x_{1,1}) \\
y(x_{0,1}, x_{1,2}) \\
y(x_{0,2}, x_{1,2}) \\
y(x_{0,3}, x_{1,2}) \\
\vdots \\
y(x_{0,1}, x_{1,n-2}) \\
y(x_{0,2}, x_{1,n-3}) \\
y(x_{0,m-2}, x_{1,n-2})
\end{bmatrix}
$$

$$
= \mathbf{b} =
\begin{bmatrix}
2h^2 f(x_{0,1}, x_{1,1}) - (-hc_2 + 2c_4)y(x_{0,0}, x_{1,1}) - (-hc_1 + 2c_3)y(x_{0,1}, x_{1,0}) \\
2h^2 f(x_{0,2}, x_{1,1}) - (-hc_1 + 2c_3)y(x_{0,2}, x_{1,0}) \\
2h^2 f(x_{0,3}, x_{1,1}) - (-hc_1 + 2c_3)y(x_{0,3}, x_{1,0}) \\
\vdots \\
2h^2 f(x_{0,m-2}, x_{1,n-2}) - (2c_4 + hc_2)y(x_{0,m-1}, x_{1,n-2}) - (2c_3 + hc_1)y(x_{0,m-2}, x_{1,n-1})
\end{bmatrix}
$$

$$(13.20)$$

Example 13.3

Consider Laplace's equation of a function of two independent variables, $z = f(x, y)$:

$$\frac{\delta^2 f(x, y)}{\delta x^2} + \frac{\delta^2 f(x, y)}{\delta y^2} = 0$$

This can be modelled as a BVP problem. Use the domains $x \in [0, 1]$ and $y \in [0, 1]$ with a step of $h = 0.25$. The boundary conditions will be:

$$f(x, y) = \begin{cases} 0 & x = 0 \\ 0 & y = 0 \\ 1 & x = 1, y \neq 0 \\ 1 & y = 1, x \neq 0 \end{cases}$$

Compute the value for all internal mesh points.

Solution

Applying the centered divided-difference formulae, Laplace's equation becomes:

$$f\left(x_{i+1}, y_j\right) + f\left(x_{i-1}, y_j\right) + f\left(x_i, y_{j+1}\right) + f\left(x_i, y_{j-1}\right) - 4f\left(x_i, y_j\right) = 0$$

Now this equation can be written out for each internal point. In this example, this is a simple matter, since there are only nine internal mesh points, namely (0.25, 0.25), (0.25, 0.50), (0.25, 0.75), (0.50, 0.25), (0.50, 0.50), (0.50, 0.75), (0.75, 0.25), (0.50, 0.75), and (0.75, 0.75). The values of the 16 boundary points are already known. This setup is represented in the figure below:

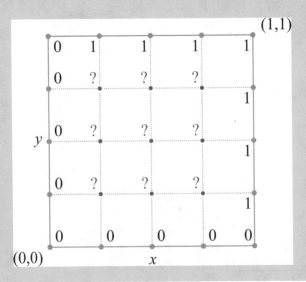

(continued)

Example 13.3 (continued)

The equation for each of the mesh points, in the order listed above, is:

$$f(0.50,0.25)+f(0.25,0.50)-4f(0.25,0.25)=0-f(0.00,0.25)$$
$$-f(0.25,0.00)=0$$
$$f(0.50,0.50)+f(0.25,0.75)+f(0.25,0.25)-4f(0.25,0.50)=0$$
$$-f(0.00,0.50)=0$$
$$f(0.50,0.75)+f(0.25,0.50)-4f(0.25,0.75)=0-f(0.00,0.75)$$
$$-f(0.25,1.00)=-1$$
$$f(0.75,0.25)+f(0.25,0.25)+f(0.50,0.50)-4f(0.50,0.25)=0$$
$$-f(0.50,0.00)=0$$
$$f(0.75,0.50)+f(0.25,0.50)+f(0.50,0.75)+f(0.50,0.25)$$
$$-4f(0.50,0.50)=0f(0.75,0.75)+f(0.25,0.75)$$
$$+f(0.50,0.50)-4f(0.50,0.75)=0-f(0.50,1.00)=-1$$
$$f(0.50,0.25)+f(0.75,0.50)-4f(0.75,0.25)=0-f(1.00,0.25)$$
$$-f(0.75,0.00)=-1$$
$$f(0.50,0.50)+f(0.75,0.75)+f(0.75,0.25)-4f(0.75,0.50)=0$$
$$-f(1.00,0.50)=-1$$
$$f(0.50,0.75)+f(0.75,0.50)-4f(0.75,0.75)=0$$
$$-f(1.00,0.75)-f(0.75,1.00)=-2$$

These equations can in turn be used to construct an $\mathbf{Mx}=\mathbf{b}$ system:

$$
\begin{bmatrix}
-4 & 1 & 0 & 1 & 0 & 0 & 0 & 0 & 0 \\
1 & -4 & 1 & 0 & 1 & 0 & 0 & 0 & 0 \\
0 & 1 & -4 & 0 & 0 & 1 & 0 & 0 & 0 \\
1 & 0 & 0 & -4 & 1 & 0 & 1 & 0 & 0 \\
0 & 1 & 0 & 1 & -4 & 1 & 0 & 1 & 0 \\
0 & 0 & 1 & 0 & 1 & -4 & 0 & 0 & 1 \\
0 & 0 & 0 & 1 & 0 & 0 & -4 & 1 & 0 \\
0 & 0 & 0 & 0 & 1 & 0 & 1 & -4 & 1 \\
0 & 0 & 0 & 0 & 0 & 1 & 0 & 1 & -4
\end{bmatrix}
\begin{bmatrix}
f(0.25,0.25) \\
f(0.25,0.50) \\
f(0.25,0.75) \\
f(0.50,0.25) \\
f(0.50,0.50) \\
f(0.50,0.75) \\
f(0.75,0.25) \\
f(0.75,0.50) \\
f(0.75,0.75)
\end{bmatrix}
=
\begin{bmatrix}
0 \\
0 \\
-1 \\
0 \\
0 \\
-1 \\
-1 \\
-1 \\
-2
\end{bmatrix}
$$

And this system can be solved using any of the methods learnt in Chap. 4 to find the solution vector:

$$
\begin{bmatrix}
f(0.25,0.25) \\
f(0.25,0.50) \\
f(0.25,0.75) \\
f(0.50,0.25) \\
f(0.50,0.50) \\
f(0.50,0.75) \\
f(0.75,0.25) \\
f(0.75,0.50) \\
f(0.75,0.75)
\end{bmatrix}
=
\begin{bmatrix}
0.14 \\
0.29 \\
0.50 \\
0.29 \\
0.50 \\
0.71 \\
0.50 \\
0.71 \\
0.86
\end{bmatrix}
$$

(continued)

Example 13.3 (continued)
Graphically, the solution has the following shape:

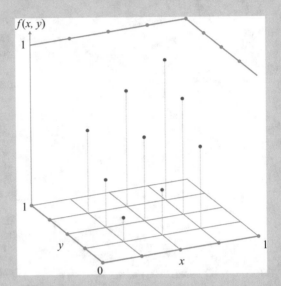

which is indeed a good approximation of the behavior of Laplace's equation,
if a rough one due to the large step value used. Solving the BVP using a
smaller value of *h* will give a better approximation. For example, the figures
below illustrate the solutions computed using the technique of this example
and a step size of $h = 0.05$ (left) and $h = 0.01$ (right).

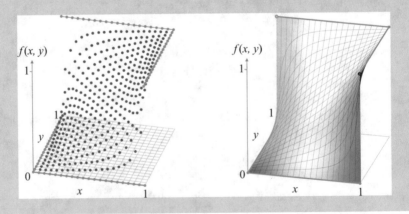

13.4 Engineering Applications

The BVPs of this chapter are closely related to the IVPs of Chap. 12, with the difference that in this case there are measurements at the start and end of the system that the solution must respect. For example, in Sect. 13.1, the current has been measured in the circuit both when it was turned on and when it was turned off, and the solution, the current at intermediate moments, must be approximated in a way that is consistent with both these measurements. Other examples include:

- Newton's law of cooling models the decreasing temperature T over distance of a hot object in a cool environment as:

$$\frac{dT}{dx} = \frac{hA}{C}(T - T_{env}) \qquad (13.21)$$

  where h is the heat transfer coefficient between the object and the environment, A is the heat transfer surface area between the object and its environment, C is the total heat capacity of the system, and T_{env} is the temperature of the environment. Provided the object is heated to a different temperature at each end, this model makes it possible to compute the intermediate temperatures along the length of the object.
- The Lotka-Volterra equations describe the dynamic relationship between the populations of two species living in the same environment. One is the prey species, with population size x, and the other is the predator species, with population size y, that hunts the prey. The rate of change of each population is modelled as:

$$\begin{aligned} \frac{dx}{dt} &= \alpha x - \beta xy \\ \frac{dy}{dt} &= \delta xy - \gamma y \end{aligned} \qquad (13.22)$$

  where the four constants describe the interaction of the two species; α is the growth rate of the prey, β is the rate at which predators kill off preys, δ is the growth rate of predators given their consumption of prey, and γ is the death rate of predators. Provided the predator and prey population sizes have been observed at two different times, it becomes possible to use this model to estimate the population history.
- The displacement over time of damped spring in orientation x is described by:

$$\frac{d^2x}{dt^2} + 2\zeta\omega\frac{dx}{dt} + \omega^2 x = 0 \qquad (13.23)$$

where ζ is the damping ratio and ω is the angular frequency of the oscillations. Knowing the position at two different instants makes it possible to approximate the complete motion of the spring.

13.5   Summary

Sometimes, in an engineering problem, one has to develop a model for a "black box" type of system, where the input and output can be measured and an ODE gives an idea of the behavior of the system, but the internal states are completely inaccessible and the equation modelling the system is unknown. Such situations include, for example, systems where measurements can only be taken at the moment they are turned on or off but not during their operation, or geographically spread-out systems where measurements can only be taken at the physical end points. These measurements, that are only accessible at the boundaries of the system's operations, are called boundary conditions, and the problem of modelling them is called a BVP.

This chapter has introduced two methods to solve BVP, both of which combine pairs of methods learnt in the previous chapters. The shooting method makes guesses at the initial value of the derivatives of the system's parameters, and shoots using an IVP method to see where the final value would be with each guess. After multiple such guesses, an interpolation or regression method can be applied to find a better approximation of the derivatives, which can then be inputted into the IVP method again to compute approximations of the internal points. The finite difference method instead replaces the derivatives in the ODE with derivative approximation equations from Chap. 10, and builds a matrix-vector system that can be solved using linear algebra methods. While both methods have the same error rate and are computationally expensive, this second one has the benefit of being easier to extend to multidimensional problems. Table 13.1 summarizes these techniques.

Table 13.1 Summary of BVP methods

Method	Requires	Error
Shooting method	Boundary conditions, multiple shots IVP method and interpolation or regression method	$O(h^2)$ or better
Finite difference method	Boundary conditions, divided-difference formula and linear algebra method	$O(h^2)$ or better

13.6   Exercises

1. Solve the following BVP using $h = 0.5$ using the method of your choice.

$$y^{(2)}(t) + 2y(t) = 1$$
$$y(0) = 0$$
$$y(2) = 1$$

2. Using the shooting method and the shots $y^{(1)}(0) = 10$ and $y^{(1)}(0) = 20$, solve the following BVP using $h = 0.5$.

$$y^{(2)}(t) + 3y^{(1)}(t) + 8y(t) = 0$$
$$y(0) = 1$$
$$y(1) = 2$$

3. Approximate nine interior mesh points using the finite difference method for the following BVP.

$$y^{(2)}(t) + 3y^{(1)}(t) + 8y(t) = 0$$
$$y(0) = 1$$
$$y(1) = 2$$

4. Solve the following BVP using $h = 1$ and applying the shooting method. For the shots, use $y^{(1)}(0) = 0$ and $y^{(1)}(0) = 1$. For the IVP method, use the Fourth-Order Runge Kutta method.

$$2y^{(2)}(t) + y(t) = 0$$
$$y(0) = 5$$
$$y(5) = -5$$

5. Repeat Exercise 4 using the finite difference method.
6. Solve the following BVP using $h = 0.2$ and applying the shooting method. For the shots, use $y^{(1)}(0) = -1$ and $y^{(1)}(0) = -2$. For the IVP method, use the Fourth-Order Runge Kutta method.

$$y^{(2)}(t) - 2y^{(1)}(t) + 3y(t) = 4$$
$$y(0) = 2$$
$$y(1) = -2$$

7. Repeat Exercise 6 using the finite difference method and using $h = 0.1$.

8. Solve the following BVP using $h = 1$ and applying the shooting method. For the shots, use $y^{(1)}(0) = 0$ and $y^{(1)}(0) = 2$. For the IVP method, use Euler's method.

$$4y^{(2)}(t) - 8y^{(1)}(t) + 2y(t) = -7$$
$$y(0) = 0$$
$$y(5) = 10$$

9. Repeat Exercise 8 using the Fourth-Order Runge Kutta method.
10. Repeat Exercise 8 using the finite difference method.
11. Repeat Example 13.3 with the following boundary conditions:

$$\frac{\delta^2 f(x, y)}{\delta x^2} + \frac{\delta^2 f(x, y)}{\delta y^2} = \cos(\pi(x + y))$$

Appendix A: Code and Pseudocode

A.1 Introduction

The focus of this textbook is not just on teaching how to perform the computations of the mathematical tools and numerical methods that will be presented, but also on demonstrating how to implement these tools and methods as computer software. It is indeed necessary for a modern engineer to be able not just to understand the theory behind numerical methods or to use a calculator in which they are preprogrammed, but to be able to write the software to compute a method when it is not available or to verify this software when it is given. Programming has become a fundamental part of a successful engineering career.

Most people today write software in a language in the C programming language family. This is a very large and diverse family that includes C++, C#, Objective-C, Java, Matlab, Python, and countless other languages. Writing out each algorithm in every language, one is likely to use would be an endless task! Instead, this book presents the algorithms in *pseudocode*. Pseudocode is a middle-ground between English and programming, that makes it possible to plan out a program's structure and logical steps in a way that is understandable to humans and easily translatable to a programming language without being tied to one specific language over all others. In fact, writing out complex algorithms in pseudocode is considered an important step in software development projects and an integral part of a software system's documentation.

For example, consider a software program that takes in the length of the side of a square and computes and displays the area and perimeter of that square. The pseudocode of that program could be the one presented in Fig. A.1.

Note the use of an arrow to assign values to the variables Side, Perimeter, and Area. This is done to avoid confusion with the equal sign, which could be interpreted as either an assignment or an equality test. Note as well the use of human terms for commands, such as Input and Display. These commands are used to abstract away the technical details of specific languages. This pseudocode

© Springer International Publishing Switzerland 2016
R. Khoury, D.W. Harder, *Numerical Methods and Modelling for Engineering*,
DOI 10.1007/978-3-319-21176-3

Fig. A.1 Pseudocode of the
square program

```
Display "Enter the side length"
Side ← Input a number

Perimeter ← Side × 4
Area ← Side × Side

Display "The perimeter is", Perimeter
Display "The area is", Area
```

```
#include <iostream>
using namespace std;
int main() {
    int Side, Perimeter, Area;
    cout << "Enter the side length ";
    cin >> Side;


    Perimeter = Side * 4;
    Area = Side * Side;

    cout << "The perimeter is " << Perimeter << endl;
    cout << "The area is " << area << endl;

    return 0;
}
prompt = 'Enter the side length ';
Side = input(prompt)

Perimeter = Side * 4
Area = Side * Side

disp(['The perimeter is ', num2str(Perimeter)])
disp(['The area is ', num2str(Area)])
Side = int(input("Enter the side length "))

Perimeter = Side * 4
Area = Side * Side

print "The perimeter is", Perimeter
print "The area is", Area
```

Fig. A.2 Code of the square program in C++ (*top*), Matlab (*middle*), and Python (*bottom*)

will never run on a computer, nor is it meant to. But it is simple to translate into a
variety of programming languages:

All the functions in Fig. A.2 will run in their respective programming environ-
ments. Notice how the pseudocode human commands `Input` and `Display` were
replaced by the language-specific commands `cin` and `cout` in C++, `input` and
`disp` in Matlab, and `input` and `print` in Python, and how unique language-
specific actions needed to be added to each program, such as the explicit declaration
of the variable type `int` in C++, the square-bracketed array display in Matlab, or the

int command to convert the user input into an integer in Python. These language-specific technical details are simplified away using pseudocode, in order to keep the reader's focus on the "big picture," the language-independent functionalities of the algorithm.

A.2 Control Statements

Control statements are programming commands that allow developers to control the execution flow of their software. A software program can have different execution paths, with each path having different commands that the computer will execute. When the execution reaches a control statement, a condition is evaluated, and one of the paths is selected to be executed based on the result of the condition. While the specific syntax that must be obeyed to use a control statement will differ a lot from one programming language to the next, there are two basic control statements that are universal to all languages in the C programming language family and always behave in the same way. They are the IF statement and the WHILE statement. Note that, by convention for clarity, these control statements are written in uppercase in the pseudocode.

A.2.1 IF Control Statements

The IF control statement evaluates an expression and, if the expression is true, it executes a set of commands; otherwise the commands are skipped entirely. It is also possible to use subsequent ELSE IF control statements to evaluate alternative conditions in the case that the first condition is false. In fact, it is possible to use an unlimited sequence of ELSE IF statements after an initial IF; each one will be evaluated in turn until one is found to be true, at which point its set of commands will be executed and all subsequent statements will be ignored. Finally, it is possible (and often recommended) to include a final ELSE statement that executes unconditionally if all preceding IF and ELSE IF statements evaluated as false. That ELSE statement defines the default behavior of the program.

Consider the example pseudocode given in Fig. A.3, which will display one of the five different messages depending on the value input by the user. The control block of code begins with the initial IF word and ends at the final END IF line (this line doesn't execute anything in the program, but is simply used to mark explicitly the end of the IF code). There are five possible paths in the program, four of which have explicit conditions that need to be evaluated, while the fifth is an unconditional default behavior. Each condition causes a unique set of commands to be run (in this example, one display command each); each set of commands is marked by being tabulated to the right and ends when the next control statement begins. Given a new user value, each condition is evaluated in turn, and as soon as one is found to be true

```
Display "Enter a value"
Value ← Input a number

IF (Value < 0)
    Display "This is a negative value"
ELSE IF (Value = 0)
    Display "This value is null"
ELSE IF (Value = 1)
    Display "This is the loneliest number"
ELSE IF (Value < 10)
    Display "This is a small positive value"
ELSE
    Display "This is a big value"
END IF
```

Fig. A.3 Pseudocode using an IF control statement

(or correct) the corresponding lines of code are executed, then the program jumps to the END IF line without checking the other conditions. That is why it is not necessary to re-check that previous conditions are false in later conditions; if the later conditions are being checked at all, then all previous conditions must have been false. For example, in the line "ELSE IF (Value < 10)", it is not necessary to check that the value is greater than zero before displaying that the value is positive, since there has already been the line "IF (Value < 0)" which must have been evaluated as false. A negative value would have evaluated to true and led to the execution of the code corresponding to that condition and consequently would have never reached the less-than-10 evaluation. The only way a program will reach the less-than-10 line is if the value is not less than zero and not equal to 0 and not equal to 1.

All C family programming languages will have the IF and ELSE control statements, and the ELSE IF control can be written in two words (as in C++ and Java), one word (ELSEIF, as in Matlab) or an abbreviation (ELIF, as in Python). The condition may be required to be between parentheses (C, C++, C#) or not (Matlab, Python). And the code to be executed might be required to be between curly brackets (C++, unless the code is exactly one line) or tabulated (Python) or require no special markers at all. The END IF termination of the block can be marked by closing the curly brackets (C++, Java) or de-tabulating the lines (Python), or by an explicit END command (Matlab). These variations are illustrated in Fig. A.4, which gives three functional implementations of the pseudocode of Fig. A.3. Finally, some languages offer alternative controls as well, namely the SWITCH-CASE control which is useful when the value of the same variable is evaluated in all conditions, and the ? : operator for cases where only two outcomes are possible. These additional controls provide fundamentally the same functionalities as the IF command, but are made available to ease code writing in some common special cases.

```cpp
#include <iostream>
using namespace std;
int main() {
    int Value;
    cout << "Enter a value";
    cin >> Value;

    if (Value < 0)
    { cout << "This is a negative value\n"; }
    else if (Value == 0)
    { cout << "This value is null\n"; }
    else if (Value == 1)
    { cout << "This is the loneliest number\n"; }
    else if (Value < 10)
    { cout << "This is a small positive value\n"; }
    else
    { cout << "This is a big value\n"; }

    return 0;
}
```

```matlab
prompt = 'Enter a value';
Value = input(prompt)

if (Value < 0)
    disp('This is a negative value')
elseif (Value == 0)
    disp('This value is null')
elseif (Value == 1)
    disp('This is the loneliest number')
elseif (Value < 10)
    disp('This is a small positive value')
else
    disp('This is a big value')
end
```

```python
Value = int(input("Enter a value"))

if Value < 0:
    print "This is a negative value"
elif Value == 0:
    print "This value is null"
elif Value == 1:
    print "This is the loneliest number"
elif Value < 10:
    print "This is a small positive value"
else:
    print "This is a big value"
```

Fig. A.4 IF control statement implemented in C++ (*top*), Matlab (*middle*), and Python (*bottom*)

A.2.2 WHILE Control Statements

The WHILE control statement is used to create loops in the program by executing a block of code over and over again multiple times. Just like the IF command, it will evaluate a condition and execute a block of code if that condition is true. But unlike

Fig. A.5 Pseudocode using
a WHILE control statement

```
Display "Enter a value"
UserValue ← Input a number
CurrentValue ← 1

WHILE (CurrentValue < UserValue)
    Display CurrentValue
    CurrentValue ← CurrentValue + 1
END WHILE

Display "Thank you, and goodbye!"
```

the IF command, once the block of code is completed, the condition will be evaluated again and, if it is still true, the block of code will run again. This will go on until the condition evaluates as false. Note that if the condition is initially false, then the block of code will not be executed even once.

Consider the example pseudocode in Fig. A.5, which is meant to display sequences of numbers. The user inputs a value, and the program will display all numbers from 1 until that value. This is done by using a WHILE control statement that evaluates whether the value to display is less than the user-specified target. If it is, the code inside the WHILE is executed: the value is displayed and incremented by 1. Once the code inside the WHILE has been completely executed, the program returns to the beginning of the loop and evaluates the condition again. This repeats until the condition evaluates to false (meaning that the incremented value has become equal or greater than the user-specified maximum), at which point the loop ends, the code inside the WHILE is skipped, and the program continues on to the goodbye message. As well, if the user inputs a value that is less than 1 (such as 0), the WHILE control statement will initially evaluate to false and will be skipped, and nothing will be displayed.

All C family programming languages will have the WHILE control statements and the FOR control statement. Both of them allow programmers to create loops, and simply offer different syntaxes. As well, many languages will have a DO-WHILE control statement, which works as a WHILE with the difference that the evaluation of the condition comes after the block of code is executed instead of before (meaning that one execution of the block of code is guaranteed unconditionally). The condition may be required to be between parenthesis (C, C++, C#) or not (Matlab, Python). And the code to be executed might be required to be between curly brackets (C++, unless the code is exactly one line) or tabulated (Python) or require no special markers at all. The END WHILE termination of the block can be marked by closing the curly brackets (C++, Java) or de-tabulating the lines (Python), or by an explicit END command (Matlab). These differences are illustrated in Fig. A.6.

```cpp
#include <iostream>
using namespace std;
int main() {
    int UserValue;
    cout << "Enter a value";
    cin >> UserValue;
    int CurrentValue = 1;

    while (CurrentValue < UserValue)
    {
        cout << CurrentValue << " ";
        CurrentValue++;
    }

    cout << "\n Thank you, and goodbye!\n";

    return 0;
}
```

```matlab
prompt = 'Enter a value';
UserValue = input(prompt)
CurrentrValue = 1;

while (CurrentValue < UserValue)
    disp([num2str(CurrentValue) ' '])
    CurrentValue = CurrentValue + 1;
end

disp('Thank you, and goodbye!')
```

```python
UserValue = int(input("Enter a value"))
CurrentValue = 1

while CurrentValue < UserValue:
    print CurrentValue,
    CurrentValue = CurrentValue + 1

print "\nThank you, and goodbye!"
```

Fig. A.6 WHILE control statement implemented in C++ (*top*), Matlab (*middle*), and Python (*bottom*)

A.2.3   CONTINUE and BREAK Control Statements

Two more control statements are worth noting: they are the CONTINUE control and the BREAK control. They are both used inside WHILE blocks and in conjunction with IF controls. The CONTINUE control is used to skip the rest of the WHILE block and jump to the next evaluation of the loop. The BREAK control is used to skip the rest of the WHILE block and exit the WHILE loop regardless of the value of the condition, to continue the program unconditionally. These two control statements are not necessary—the same behaviors could be created using finely crafted IF and WHILE conditions—but they are very useful to create simple and clear control paths in the program. The code in Fig. A.7 uses both control statements in the code for a simple two-player "guess the number" game. The game runs an

```
Display "Player 1: enter a number"
PlayerOneValue ← Input a number


WHILE (TRUE)
    Display "Player 2: guess the number"
    PlayerTwoValue ← Input a number

    IF (PlayerTwoValue < PlayerOneValue)
        Display "Too low!"
        CONTINUE
    END IF

    IF (PlayerTwoValue > PlayerOneValue)
        Display "You think too big!"
        CONTINUE
    END IF

    Display "You got it!"
    BREAK

    Display "How did you get here?"

END WHILE

DISPLAY "Thank you for playing!"
```

Fig. A.7 Pseudocode using the CONTINUE and BREAK control statements

infinite loop (the WHILE (TRUE) control statement, which will always evaluate to true and execute the code), and in each loop Player 2 is asked to input a guess as to the number Player 1 selected. There are two IF blocks; each one evaluates whether Player 2's guess is too low or too high and displays an appropriate message, and then encounters a CONTINUE control statement that immediately stops executing the block of code and jumps back to the WHILE command to evaluate the condition (which is true) and start again. If neither of these IF command statement conditions evaluate to true (meaning Player 2's guess is neither too low nor too high), then the success message is displayed and a BREAK command statement is reached. At that point, the execution of the block of code terminates immediately and the execution leaves the WHILE loop entirely (even though the WHILE condition still evaluates to true), and the program continues from the END WHILE line to display the thank-you message. The hidden display line in the WHILE block of code after the BREAK cannot possibly be reached by the execution, and will never be displayed to the user. In fact, some compilers will even display a warning if such a line is present in the code.

A.3 Functions

Oftentimes, it will be impossible to write a program that does everything that is required in a single sequence, even using IF and WHILE control statements. Such a program would simply be too long and complicated to be usable. The solution is to break up the sequence into individual blocks that are their own programs, and to write one program that *calls*, or executes, the other ones in sequence. These smaller programs are called *functions*. Breaking your program into functions makes it easier to conceptualize, to write, to maintain, and to reuse. The one inconvenient of dividing your program this way is that it becomes a little less efficient at runtime, but that is usually not a problem, except in critical real-time applications.

The internal workings of each function are completely isolated from the rest of the program. Each function thus has its own variables that only it can use and that are invisible to the rest of the program. These are called *local variables*. Local variables are very useful: the function can create and modify them as it wants without affecting the rest of the program in any way, and in turn a developer studying the main program can ignore completely the details internal to the individual functions. Local variables are so called to differentiate them from *global variables*, or variables that are accessible by all functions of a program and can be modified by all functions, and where the modifications done by one function will be visible by all others. Usage of global functions is counter-indicated in good programming practice.

But of course, in order to be useful, a function must perform work on the main program's data and give back its results to the main program. The function cannot access the program's data or the variables of another function. However, when a function is called, it can be given the current values of another function's variables as *parameters*. Copies of these variables become local variables of the function and can be worked on. And since they are only copies, the original variables of the function that called it are not affected in any way. Similarly, the final result of a function can be sent back to the function that called it by giving it a copy of the final values of local variables. These variables are called *return values*, and again the copies passed become local variables of the function receiving them. In both cases, if multiple variables are being copied from one function to another, the order of the variables matters and the names do not. For example, if function A calls function B and gives its variables x and y as parameters in that order, and function B receives as parameters variables y and x in that order, then the value x in function A (passed first) is now the value y in function B (received first) and the value y in function A (passed second) is now value x in function B (received second).

The pseudocode control statement to call a function is simply CALL followed by the name of the function and the list of variables that are sent as parameters. The function itself begins by the control statement FUNCTION and ends with END FUNCTION, and can return values to the program that called it using the RETURN control statement. Note that RETURN works like BREAK: the execution of the function is immediately terminated and the program jumps back to the CALL

```
Display "Enter your name"
User ← Input a string
Display "Enter a Fibonacci sequence length (minimum 2)"
Length ← Input a number
Final ← CALL Fibonacci(Length)
Display "Dear", User, "the last digit in the sequence is", Final


FUNCTION Fibonacci(Value)
    F0 ← 0
    F1 ← 1
    Counter ← 1
    User ← "Leonardo Fibonacci"
    WHILE (Counter < Value)
        F2 ← F0 + F1
        F0 ← F1
        F1 ← F2
        Counter ← Counter + 1
    END WHILE
    RETURN F2
END FUNCTION
```

Fig. A.8 Pseudocode calling a function

line. Figure A.8 gives an example of a function call. The function `Fibonacci` computes a Fibonacci sequence to a value specified in parameter, and returns the final number of that sequence. That is all the information that a developer who uses that function needs to know about it. The fact that it creates four additional local variables (`F0`, `F1`, `User`, and `Counter`) is transparent to the calling function, since these variables are never returned. These variables will be destroyed after the function `Fibonacci` returns. Likewise, the user's name is a local variable of the main program that is not passed to the `Fibonacci` function, and is thus invisible to that function. The fact that there are two variables with the same name `User`, one local to the main program and one local to the `Fibonacci` function, is not a problem at all. These remain two different and unconnected variables, and the value of `User` displayed at the end of the main program will be the one input by the user at the beginning of that program, not the one created in the `Fibonacci` function.

Like the other control statements, having functions is standard in all languages of the C family, but the exact syntax and keywords used to define them vary greatly. In C and C++, there is no keyword to define a function, but the type of the variable returned must be specified ahead of it (so the function in Fig. A.8 would be "int `Fibonacci`" for example). Other languages in the family do use special keywords to declare that a function definition begins, such as `def` in Python and `function` in Matlab. These differences are illustrated in Fig. A.9.

```cpp
#include <iostream>
using namespace std;
int main() {
    string User;
    cout << "Enter your name";
    cin >> User;
    int Length;
    cout << "Enter a Fibonacci sequence length (minimum 2)";
    cin >> Length;
    int Final = Fibonacci(Length);
    cout << "Dear " << User << " the last digit in the sequence
        is " << Final << endl;

    return 0;
}

int Fibonacci(int Value) {
    int F0 = 0, F1 = 1, Counter = 1, F2 = -1;
    string User = "Leonardo Fibonacci";
    while (Counter < Value)
    {
        F2 = F0 + F1;
        F0 = F1;
        F1 = F2;
        Counter++;
    }
    return F2;
}
```

```matlab
prompt = 'Enter your name';
User = input(prompt)
prompt = 'Enter a Fibonacci sequence length (minimum 2)';
Length = input(prompt)
Final = Fibonacci(Length);
disp(['Dear ' User ' the last digit in the sequence is '
        num2str(Final)]);

function F2 = Fibonacci(Value)
    F0 = 0;
    F1 = 1;
    Counter = 1;
    User = 'Leonardo Fibonacci';
    while (Counter < Value)
        F2 = F0 + F1;
        F0 = F1;
        F1 = F2;
        Counter = Counter + 1;
    end
end
```

```python
User = input("Enter your name")
Length = int(input("Enter a Fibonacci sequence length (minimum
        2)"))
Final = Fibonacci(Length);
print "Dear", User, "the last digit in the sequence is",
    str(Final);

def Fibonacci(Value):
    F0 = 0
    F1 = 1
    Counter = 1
    User = "Leonardo Fibonacci"
    while Counter < Value:
        F2 = F0 + F1
        F0 = F1
        F1 = F2
        Counter = Counter + 1
    return F2
```

Fig. A.9 Functions implemented in C++ (*top*), Matlab (*middle*), and Python (*bottom*)

Appendix B: Answers to Exercises

Chapter 1

1. This level of precision is impossible with a ruler marked at only every 0.1 cm. Decimals lesser than this are noise.
2. The second number has higher precision, but the first is more accurate.
3. First number is more precise, second number is more accurate.
4. The lower precision on the distance nullifies the higher precision of the conversion. The decimals should not have been kept.
5. Absolute error ≈ 0.001593. Relative error $\approx 0.05\,\%$.
6. Absolute error ≈ 0.0013. Relative error $\approx 0.04\,\%$. Three significant digits.
7. Absolute error $\approx 2.7 \times 10^{-7}$. Relative error $\approx 8.5 \times 10^{-6}\,\%$. Six significant digits.
8. Absolute error $\approx 3.3\ \Omega$. Relative error $\approx 1.4\,\%$.
9. Absolute error ≈ 0.3 MV. Relative error $\approx 14\,\%$.
10. Absolute error ≈ 8.2 mF. Relative error $\approx 7.6\,\%$. Zero significant digits.
11. 3.1415 has four significant digits. 3.1416 has five significant digits.
12. One significant digit.
13. Two significant digits.

Chapter 2

1. 5.232345×10^2 or 5.232345e2.
2. The value 12300000000 suggests an implied precision of 0.5, which has a maximum relative error of 4.1×10^{-11}, while the original scientific notation suggests an implied precision of 50000000, which has a maximum relative error of 0.0041.
3. 1101101_2
4. 10110010000_2
5. 111.0001_2
6. 11010.111001_2
7. 110000110_2

© Springer International Publishing Switzerland 2016
R. Khoury, D.W. Harder, *Numerical Methods and Modelling for Engineering*,
DOI 10.1007/978-3-319-21176-3

8. 11101011001_2
9. 11.1101111_2
10. 1100.0011_2
11. 11101.011001_2
12. 18.1875
13. -7.984375
14. (a) 0.09323
    (b) -9.323
    (c) 93.23
    (d) -932300

15. (a) [1233.5, 1234.5]
    (b) [2344.5, 2345.5]

16. (a) 0491414
    (b) 0551000
    (c) 1444540

17. 0011111111000000000000000000000000000000000000000000000000000000
18. 1100000000011101011000000000000000000000000000000000000000000000
    and -7.34375
19. (a) 153.484375
    (b) 0.765625

20. 4014000000000000
21. (a) 0.0004999
    (b) 0.05000
    (c) 0.04999

22. Starting at $n = 100000$ and decrementing to avoid the non-associativity problem.

Chapter 3

1. Converges to $x = 7/12 = 0.58333\ldots$ Because of the infinite decimals, it is impossible to reach the equality condition.
2. Computing the relative errors shows that $\cos(x)$ converges at approximately twice the rate as $\sin(x)$.
3. The function converges to positive zero for $0 \leq x < 1$, to negative zero for $-1 < x < 0$, and the function diverges towards positive infinity for $x > 1$ and diverges towards negative infinity for $x < -1$.
4. $x_1 = 0.979$, $E_1 = 42.4\%$. $x_2 = 1.810$, $E_2 = 11.5\%$. $x_3 = 2.781$, $E_3 = 0.2\%$. $x_4 = 3.134$, $E_4 = 2.5 \times 10^{-6}\%$. $x_5 = 3.1415926$, $E_5 = 0\%$.
5. $x_1 = 3.85$, $E_1 = 53\%$. $x_2 = 3.07$, $E_2 = 28\%$. $x_3 = 2.55$, $E_3 = 12\%$. $x_4 = 2.24$, $E_4 = 4\%$. $x_5 = 2.08$, $E_5 = 1\%$.
6. (a) 1.566
   (b) -4
   (c) 2 and 10.

(d) -1 and 0.4

(e) -1.4, 3, and 8.7

Chapter 4

1. (a) $\mathbf{P}^T = \begin{bmatrix} 0 & 0 & 1 \\ 1 & 0 & 0 \\ 0 & 1 & 0 \end{bmatrix}$ $\mathbf{L} = \begin{bmatrix} 1 & 0 & 0 \\ 0.2 & 1 & 0 \\ 0.1 & 0.3 & 1 \end{bmatrix}$ $\mathbf{U} = \begin{bmatrix} 6.0 & -2.0 & 1.0 \\ 0.0 & -8.0 & -3.0 \\ 0 & 0 & -5.0 \end{bmatrix}$ $\mathbf{x} = \begin{bmatrix} 2.0 \\ -1.0 \\ -2.0 \end{bmatrix}$

(b) $\mathbf{P}^T = \begin{bmatrix} 0 & 0 & 1 & 0 \\ 0 & 0 & 0 & 1 \\ 0 & 1 & 0 & 0 \\ 1 & 0 & 0 & 0 \end{bmatrix}$ $\mathbf{L} = \begin{bmatrix} 1 & 0 & 0 & 0 \\ 0.2 & 1 & 0 & 0 \\ 0 & 0.4 & 1 & 0 \\ 0.5 & 0 & -0.3 & 1 \end{bmatrix}$ $\mathbf{U} = \begin{bmatrix} 6.0 & 2.0 & 1.0 & 0 \\ 0 & -5.0 & 0 & 3.0 \\ 0 & 0 & 7.0 & -2.0 \\ 0 & 0 & 0 & 8.0 \end{bmatrix}$

$\mathbf{x} = \begin{bmatrix} -1.3 \\ 0.4 \\ 3.4 \\ 0.9 \end{bmatrix}$

(c) $\mathbf{P}^T = \begin{bmatrix} 1 & 0 & 0 & 0 \\ 0 & 0 & 1 & 0 \\ 0 & 0 & 0 & 1 \\ 0 & 1 & 0 & 0 \end{bmatrix}$ $\mathbf{L} = \begin{bmatrix} 1 & 0 & 0 & 0 \\ 0.2 & 1 & 0 & 0 \\ 0.1 & 0 & 1 & 0 \\ -0.2 & 0.5 & -0.1 & 1 \end{bmatrix}$ $\mathbf{U} = \begin{bmatrix} 10.0 & 3.0 & 2.0 & 3.0 \\ 0 & 8.0 & 1.0 & 2.0 \\ 0 & 0 & -12.0 & 3.0 \\ 0 & 0 & 0 & 8.0 \end{bmatrix}$

$\mathbf{x} = \begin{bmatrix} 1.0 \\ 0.0 \\ -1.0 \\ 2.0 \end{bmatrix}$

2. (a) $\mathbf{L} = \begin{bmatrix} 9 & 0 & 0 \\ 2 & 8 & 0 \\ 1 & -2 & 6 \end{bmatrix}$ $\mathbf{x} = \begin{bmatrix} 10 \\ 5 \\ 1 \end{bmatrix}$

(b) $\mathbf{L} = \begin{bmatrix} 7 & 0 & 0 \\ -2 & 9 & 0 \\ 4 & 6 & 11 \end{bmatrix}$ $\mathbf{x} = \begin{bmatrix} 2 \\ -1 \\ 3 \end{bmatrix}$

(c) $\mathbf{L} = \begin{bmatrix} 3.00 & 0 & 0 & 0 \\ 0.20 & 4.00 & 0 & 0 \\ -0.10 & 0.30 & 2.00 & 0 \\ 0.50 & -0.40 & -0.20 & 5.00 \end{bmatrix}$ $\mathbf{x} = \begin{bmatrix} 0.30 \\ 0.00 \\ 0.20 \\ -0.10 \end{bmatrix}$

(d) $\mathbf{L} = \begin{bmatrix} 7 & 0 & 0 & 0 \\ 2 & 5 & 0 & 0 \\ -1 & -2 & 6 & 0 \\ 1 & 0 & -3 & 5 \end{bmatrix}$ $\mathbf{x} = \begin{bmatrix} 1 \\ -2 \\ 0 \\ -1 \end{bmatrix}$

(e) $\mathbf{L} = \begin{bmatrix} 2.00 & 0 & 0 & 0 \\ 0.20 & 1.00 & 0 & 0 \\ 0.40 & -0.20 & 3.00 & 0 \\ -0.10 & 0.30 & 0.50 & 2.00 \end{bmatrix}$ $\mathbf{x} = \begin{bmatrix} 0.00 \\ -1.00 \\ 1.00 \\ 3.00 \end{bmatrix}$

3. (a) $\mathbf{x} = \begin{bmatrix} 0.42 \\ 0.16 \end{bmatrix}$

(b) $\mathbf{x} = \begin{bmatrix} 0.51 \\ -0.31 \\ -0.13 \end{bmatrix}$

4. See answers to Exercise 3
5. 12.5 %
6. 89.7 %
7. 0.6180, which is φ, the golden ratio

Chapter 5

1. $f(x_0 + h) = f(x_0) + f^{(1)}(x_0)h + \frac{f^{(2)}(x_0)}{2!}h^2 + \frac{f^{(3)}(x_0)}{3!}h^3$
2. 0.8955012154 and 0.0048
3. 0.89129
4. (a) $[0.125, 0.125\ e^{0.5}] \approx [0.125, 0.20609]$
   (b) $[0.125\ e^{0.5}, 0.125\ e] \approx [0.20609, 0.33979]$

5. (a) $f(x+h) = 0$; error approximation $= -1$; absolute error $= 0.75$
      $f(x+h) = -1$; error approximation $= 0.25$; absolute error $= 0.25$
      $f(x+h) = -0.75$; error approximation $= 0$; absolute error $= 0$
   (b) $f(x+h) = 3$; error approximation $= 3.5$; absolute error $= 6.375$
      $f(x+h) = 6.5$; error approximation $= 2.5$; absolute error $= 0.2875$
      $f(x+h) = 9$; error approximation $= 0.378$; absolute error $= 0.375$
   (c) $f(x+h) = 1$; error approximation $= -1.5$; absolute error $= -6.813$
      $f(x+h) = -0.5$; error approximation $= -2.5$; absolute error $= -5.313$
      $f(x+h) = -3$; error approximation $= -2.142$; absolute error $= -2.813$

Chapter 6

1.
   (a) $f(x) = 0.33 + 1.33x$
   (b) $f(x) = 2 + 3x + x^2$
   (c) $f(x) = 1 + 2x - 3x^3$
   (d) $f(x) = 3 + 2x$
   (e) $f(x) = 0.25327 + 0.97516x + 0.64266x^2 - 0.12578x^3 - 0.03181x^4$
   (f) $f(x) = 0.13341 - 0.64121x + 6.48945x^2 - 4.54270x^3 - 0.99000x^4 - 0.82168x_5$

2. No, they only need to be consistent from row to row. They can compute the coefficients of the polynomial in any column (exponent) order.
3. $f(x) = 0.03525\sin(x) + 0.72182\cos(x)$
4. $f(x) = -0.96 - 1.86\sin(x) + 0.24\cos(x)$
5. (a) $f(x) = -1(x - 5) - 2(x - 2)$
   (b) $f(x) = -1(x - 3) + 2(x - 1)$
   (c) $f(x) = 3(x - 3)(x - 5) + 7(x - 2)(x - 5) - 4(x - 2)(x - 3)$
   (d) $f(x) = 0.33(x - 1)(x - 3) + 0 + 0.66x(x - 1)$

(e) $f(x) = -0.66(x - 7.3) + 0.33(x - 5.3)$

(f) $f(x) = 0 - 2x(x - 2) + 18x(x - 1)$

(g) $f(x) = 0 + x(x - 2)(x - 3) - 18x(x - 1)(x - 3) + 42x(x - 1)(x - 2)$

6. (a) $f(x) = 3 + 1.66(x - 2)$

   (b) $f(x) = 2 - (x - 2) + 0.5(x - 2)(x - 3)$

   (c) $f(x) = -39 + 21(x + 2) - 6(x + 2)x + 2(x + 2)x(x - 1)$

   (d) $f(x) = 21 - 10(x + 2) + 3(x + 2)x - 3(x + 2)x(x - 1)$.

   (e) $f(x) = 5 + 2(x - 1)$

   (f) $f(x) = 0.51 - 0.6438(x - 1.3) - 0.2450(x - 1.3)(x - 0.57)$
   $-3.3677(x - 1.3)(x - 0.57)(x + 0.33) + 1.0159(x - 1.3)(x - 0.57)(x + 0.33)(x + 1.2)$
   $+0.8223(x - 1.3)(x - 0.57)(x + 0.33)(x + 1.2)(x - 0.36)$

7. No, the values can be inputted in any order.

8. $f(x) = -2 + 12(x - 3) - 3(x - 3)(x - 4) + (x - 3)(x - 4)(x - 5)$

9. (a) $f(x, y) = 5 - 2x - y + 4xy$

   (b) $f(x, y) = 60 - 22x - 11y + 5xy$

   (c) $f(x, y) = -12 + 5x + 6y - xy$

   (d) $f(x, y) = 22.75 - 2.625x - 1.5833y + 0.5417xy$

   (e) $f(x, y) = 18.500 - 14.700y + 3.400y^2 - 21.550x + 19.825xy - 3.925xy^2$
   $+5.950x^2 - 5.275x^2y + 0.975x^2y^2$

   (f) $f(x, y, z) = 1.078 + 0.311x - 0.478y + 1.489z$

10. (a) $f(x) = -0.5 + 0.6x$

    (b) $f(x) = 0.75027 - 0.28714x$

    (c) $f(x) = 1.889 + 0.539x$

    (d) $f(x) = 2.440 + 1.278x$

11. The measurement at $x = 8$ is very different from the others, and likely wrong. Reasonable solutions include taking a new measurement at $x = 8$ or discarding the measurement.

12. (a) $f(x) = 0 + 0.4x + x^2$

    (b) $f(x) = -0.20 + 0.44x + 0.26x^2$

    (c) $f(x) = 2.5 + 2.4x + 1.9x^2$

13. (a) $f(x) = 2.3492e^{-0.53033x}$

    (b) $f(x) = 0.52118 e^{-3.27260x}$

    (c) $f(x) = 0.71798 e^{-0.51986x}$

14. (a) $f(x) = 2.321\cos(0.4x) - 0.6921 \sin(0.4x)$

    (b) $f(x) = -0.006986 + 2.318\cos(0.4x) - 0.6860\sin(0.4x)$

    (c) Being of much smaller magnitude than the other coefficients, it is most likely unnecessary.

15. (a) $y = -6.5$

    (b) $y = 3.36$

    (c) $y = 2.1951$

16. $y = 10.323$

17. Time $= 2.2785$ s

Chapter 7

No exercises.

Chapter 8

1. (a) 1.4375
   (b) 3.15625
   (c) 3.2812

2. The interval is [40.84070158, 40.84070742] after 24 iterations. Note however that $\sin(x)$ has 31 roots on the interval [1, 99], however the bisection method neither suggests that more roots exist nor gives any suggestion as to where they may be.

3. (a) 1.4267
   (b) 3.16
   (c) 3.3010

4. $x = 1.57079632679490$ after five iterations.

5. $x = 0.4585$ after two iterations.

6. $x_1 = 3/2$, $x_2 = 17/12$, $x_3 = 577/408$

7. (a) $\mathbf{x}_1 = [0.6666667, 1.833333]^T, \mathbf{x}_2 = [0.5833333, 1.643939]^T, \mathbf{x}_3 = [0.5773810, 1.633030]^T$
   (b) $\mathbf{x}_1 = [-1.375, 0.575]^T$, $\mathbf{x}_2 = [-1.36921, 0.577912]^T$, $\mathbf{x}_3 = [-1.36921, 0.577918]^T$

8. $x_1 = 2$, $x_2 = 4/3$, $x_3 = 7/5$

9. $x_1 = -0.5136$, $x_2 = -0.6100$, $x_3 = -0.6514$, $x_4 = -0.6582$

10. $x = 0.4585$ after three iterations.

11. $x_1 = -1.14864$, $\quad$ $x_2 = -0.56812$, $\quad$ $x_3 = -0.66963$, $\quad$ $x_4 = -0.70285$, $x_5 = -0.70686$, $x_6 = -0.70683$.

Chapter 9

1. (a) The final bounds are [−0.0213, 0.0426] after eight iterations.
   (b) The final bounds are [0.0344, 0.0902] after five iterations.
   (c) The final bounds are [0.5411, 0.6099] after seven iterations.
   (d) The final bounds are [4.6737, 4.7639] after five iterations.
   (e) The final bounds are [1.2918, 1.3820] after five iterations.
   (f) The final bounds are [3.8754, 3.9442] after seven iterations.

2. $\lceil \log_{\varphi - 1}(\varepsilon/h) \rceil$

3. (a) $x_1 = 0$
   (b) $x_1 = 0$
   (c) $x_2 = 0.5744$
   (d) $x_3 = 4.7124$
   (e) $x_3 = 1.3333$
   (f) $x_3 = 3.926990816$

4. (a) $x_1 = 0$
   (b) $x_1 = 0.5969$, $x_2 = 0.5019$, $x_3 = 0.4117$, $x_4 = 0.3310$

(c) $x_2 = 0.5735$
(d) $x_5 = 4.712388984477041$
(e) $x_2 = 1.3316$
(f) $x_7 = 3.9269$

5. $\mathbf{x}_1 = [0.2857, 1.3571]^T$, $\mathbf{x}_2 = [0.10714, 1]^T$

Chapter 10

1. -0.00012493 F/s
2. 0.96519 rad/s and 0.96679 rad/s
3. 0.970285 rad/s
4. 3.425518831
5. -0.3011686798
6. $D_3(0.25) = 1.000003$
7. (a) $D_3(0.5) = -0.1441$
   (b) $D_3(0.5) = -0.1911$
   (c) $D_3(0.5) = -0.1585$

8. (d) $D_3(0.03) = 5.272$
   (e) $D_3(0.03) = 4.775$
   (f) $D_3(0.03) = 5.680$

9. 5.38 m/s (using a regression for a degree-2 polynomial)
10. 0.8485 L per hour (using a regression for a degree-2 polynomial)

Chapter 11

1. (a) 5.00022699964881
   (b) 1.02070069942442
   (c) 0.999954724240937 after 13 iterations
   (d) 1.711661979876841
   (e) 1.388606601719423

2. (a) 4.5
   (b) 2.137892120

3. (a) Integral $= 4$; estimated error $= -4/3$; real error $= -4/3$
   (b) Integral $= 16$; estimated error $= -32/3$; real error $= -48/5$
   (c) Integral $= 0.1901127572$; estimated error $= -0.0006358300384$;
      real error $= -0.0006362543$

4. 3.76171875
5. 0.8944624935
6. (a) Four segments $= 6$; 8 segments $= 5.5$; estimated error $= -1/6$; real error $= -1/6$
   (b) Four segments $= 18$; 8 segments $= 14.125$; estimated error $= -4/3$; real error $= -1.325$

7. 0.141120007827708
8. 682.666666666667

9. 1.9999999945872902 after four iterations
10. (a) 3.75
    (b) 2.122006886

11. (a) Four segments $= 8/3$; 8 segments $= 8/3$
    (b) Four segments $= 20/3$; 8 segments $= 176/27$

Chapter 12

1. (a) $y(1.5) = 2.25$
   (b) $y(1) = 2.375$
   (c) $y(0.75) = 2.4375$

2. $y(0.5) = 1.375, y(1) = 1.750, y(1.5) = 2.125$
3. $y(1) = 2.35, y(1.5) = 2.7$
4. $[-0.005, 0.005]$
5. $y(1) = 1$, Relative error $= 25\%$; $y(0.5) = 1.31$, Relative error $= 2\%$
6. $y(1) = 2.09$; $y(1.5) = 1.19$
7. $y(0.5) = 1.3765625, y(1) = 1.75625, y(1.5) = 2.1390625$
8. $y(1.5) = 2.353125, y(2) = 2.7125$
9. $[-0.0531, 0.0531]$
10. $y(1) = 1.32$, Relative error $= 0.8\%$; $y(0.5) = 1.34$, Relative error $= 0\%$
11. $y(1) = 2.13$; $y(1.5) = 1.47$
12. $y(0.5) = 1.376499430, y(1) = 1.755761719, y(1.5) = 2.137469483$
13. $y(1.5) = 2.352998861, y(2) = 2.711523438$
14. Euler: 1.754495239, Heun's: 1.755786729, Runge-Kutta: 1.755760161
15. Euler: $y(0.25) = 0.75$; $y(0.5) = 0.609375$; $y(0.75) = 0.560546875$; $y(1) = 0.603149414$
    Heun's: $y(0.25) = 0.8046875000$; $y(0.5) = 0.7153015137$; $y(0.75) = 0.7299810648$; $y(1) = 0.8677108540$
    Runge-Kutta: $y(0.25) = 0.8082167307$; $y(0.5) = 0.7224772847$; $y(0.75) = 0.7417768882$; $y(1) = 0.8867587481$
16. $y(1) = 1$, Relative error $= 24.8\%$; $y(0.5) = 1.2$, Relative error $= 10.4\%$
17. $y(1) = 2$; $y(1.5) = 1.4$
18. $y(0.5) = 1.377777778, y(1) = 1.76, y(1.5) = 2.145454545$
19. $y(1.5) = 2.355555556, y(2) = 2.72$
20. $\mathbf{u}(0.1) = [1.12, 0.71]^T$, $\mathbf{u}_2 = [1.1806, 0.3975]^T$
21. (a) $I(0.1) = -1.0$ A, $I(0.2) = -0.95$ A, $I(0.3) = -0.90005$ A
    (b) $I(0.1) = 0$ A, $I(0.2) = -0.01$ A, $I(0.3) = -0.021052$ A

22. $\mathbf{u}(0.1) = [1, 2.1]^T$; $\mathbf{u}(0.2) = [1.2, 2.1900166583]^T$; $\mathbf{u}(0.3) = [1.62900166583, 2.2731463936]^T$
23. $\mathbf{u}(0.1) = [1.215, 2.365, 4.5585]^T$; $\mathbf{u}(0.2) = [1.4742925, 2.9169125, 6.8594415]^T$
24. $\mathbf{u}(0.1) = [0.82, 1.14836]^T$; $\mathbf{u}(0.2) = [0.93484, 1.07765]^T$; $\mathbf{u}(0.3) = [1.04260, 0.98824]^T$; $\mathbf{u}(0.4) = [1.14142, 0.88120]^T$
25. (a) $\mathbf{u}(1) = [5, 3, 2]^T$
    (b) $\mathbf{u}(0.5) = [4, 2.5, 1.5]^T$; $\mathbf{u}(1) = (5.25, 3.25, 1.5)^T$

(c) $\mathbf{u}(0.25) = [3.5, 2.25, 1.25]^T$; $\mathbf{u}(0.5) = [4.0625, 2.5625, 1.375]^T$; $\mathbf{u}(0.75) = [4.703125, \quad 2.90625, \quad 1.40625]^T$; $\quad \mathbf{u}(1) = [5.4296875, \quad 3.2578125, 1.35546875]^T$

26. (a) $\mathbf{u}(1) = [5, 3, 0]^T$

(b) $\mathbf{u}(0.5) = [4, 2.5, 0.5]^T$; $\mathbf{u}(1) = [5.25, 2.75, 0.375]^T$

(c) $\mathbf{u}(0.25) = [3.5, \quad 2.25, \quad 0.75]^T$; $\quad \mathbf{u}(0.5) = [4.0625, \quad 2.4375, \quad 0.578125]^T$; $\mathbf{u}(0.75) = [4.671875, 2.58203125, 0.474609375]^T$; $\mathbf{u}(1) = [5.3173828125, 2.70068359375, 0.42565917975]^T$

Chapter 13

1. $(0.5, 0.16667)$, $(1.0, 0.5)$, $(1.5, 0.83333)$.
2. $y^{(1)}(0) = 12$; $y(0.5) = 7$

3.
$$\begin{bmatrix} y(0.1) \\ y(0.2) \\ y(0.3) \\ y(0.4) \\ y(0.5) \\ y(0.6) \\ y(0.7) \\ y(0.8) \\ y(0.9) \end{bmatrix} = \begin{bmatrix} 3.87 \\ 5.72 \\ 6.70 \\ 6.95 \\ 6.65 \\ 5.97 \\ 5.05 \\ 4.02 \\ 2.98 \end{bmatrix}$$

4. $y^{(1)}(0) = 0.73$,
$$\begin{bmatrix} y(1) \\ y(2) \\ y(3) \\ y(4) \end{bmatrix} = \begin{bmatrix} 4.47 \\ 1.81 \\ -1.71 \\ -4.41 \end{bmatrix}$$

5.
$$\begin{bmatrix} y(1) \\ y(2) \\ y(3) \\ y(4) \end{bmatrix} = \begin{bmatrix} 4.55 \\ 1.82 \\ -1.82 \\ -4.55 \end{bmatrix}$$

6. $y^{(1)}(0) = -1.24$,
$$\begin{bmatrix} y(0.2) \\ y(0.4) \\ y(0.6) \\ y(0.8) \end{bmatrix} = \begin{bmatrix} 1.66 \\ 1.11 \\ 0.31 \\ -0.74 \end{bmatrix}$$

7.
$$\begin{bmatrix} y(0.1) \\ y(0.2) \\ y(0.3) \\ y(0.4) \\ y(0.5) \\ y(0.6) \\ y(0.7) \\ y(0.8) \\ y(0.9) \end{bmatrix} = \begin{bmatrix} 1.85 \\ 1.66 \\ 1.41 \\ 1.10 \\ 0.73 \\ 0.30 \\ -0.20 \\ -0.75 \\ -1.35 \end{bmatrix}$$

8. $y^{(1)}(0) = 1.03$, $\begin{bmatrix} y(1) \\ y(2) \\ y(3) \\ y(4) \end{bmatrix} = \begin{bmatrix} 1.03 \\ 2.38 \\ 4.17 \\ 6.59 \end{bmatrix}$

9. $y^{(1)}(0) = 1.02$, $\begin{bmatrix} y(1) \\ y(2) \\ y(3) \\ y(4) \end{bmatrix} = \begin{bmatrix} 1.19 \\ 2.78 \\ 4.87 \\ 7.49 \end{bmatrix}$

10. $\begin{bmatrix} y(1) \\ y(2) \\ y(3) \\ y(4) \end{bmatrix} = \begin{bmatrix} 1.17 \\ 2.72 \\ 4.80 \\ 7.56 \end{bmatrix}$

11. $\begin{bmatrix} f(0.25, 0.25) \\ f(0.25, 0.50) \\ f(0.25, 0.75) \\ f(0.50, 0.25) \\ f(0.50, 0.50) \\ f(0.50, 0.75) \\ f(0.75, 0.25) \\ f(0.75, 0.50) \\ f(0.75, 0.75) \end{bmatrix} = \begin{bmatrix} 0.35 \\ 0 \\ -0.35 \\ 0 \\ 0 \\ 0 \\ -0.35 \\ 0 \\ 0.35 \end{bmatrix}$

References

Beeler, M., Gosper, R.W., Schroeppel, R.: HAKMEM. MIT AI Memo 239, 1972. Item 140

Bradie, B.: A Friendly Introduction to Numerical Analysis. Pearson Prentice Hall, Upper Saddle River (2006)

Chapra, S.C.: Numerical Methods for Engineers, 4th edn. McGraw Hill, New York (2002)

Ferziger, J.H.: Numerical Methods for Engineering Applications, 2nd edn. Wiley, New York (1998)

Goldstine, H.H.: A History of Numerical Analysis. Springer, New York (1977)

Griffits, D.V., Smith, I.M.: Numerical Methods for Engineers, 2nd edn. Chapman & Hall/CRC, New York (2006)

Hämmerlin, G., Hoffmann, K.-H.: Numerical Mathematics. Springer, New York (1991)

James, G.: Modern Engineering Mathematics, 3rd edn. Pearson Prentice Hall, Englewood Cliffs (2004)

Mathews, J.H., Fink, K.D.: Numerical Methods Using Matlab, 4th edn. Pearson Prentice Hall, Upper South River (2004)

Stoer, J., Bulirsch, R.: Introduction to Numerical Analysis. Springer, New York (1993)

Weisstein, E.W.: MathWorld. Wolfram Web Resource. http://mathworld.wolfram.com/

© Springer International Publishing Switzerland 2016
R. Khoury, D.W. Harder, *Numerical Methods and Modelling for Engineering*,
DOI 10.1007/978-3-319-21176-3

Index

© Springer International Publishing Switzerland 2016
R. Khoury, D.W. Harder, *Numerical Methods and Modelling for Engineering*,
DOI 10.1007/978-3-319-21176-3

Printed in the United States
By Bookmasters